1 HYDROGEN **H** 1·008 1	2 HELIUM **He** 4·00 2
			5 BORON **B** 10·81 2.3	6 CARBON **C** 12·01 2.4	7 NITROGEN **N** 14·01 2.5	8 OXYGEN **O** 16·00 2.6	9 FLUORINE **F** 19·00 2.7	10 NEON **Ne** 20·18 2.8
			13 ALUMINIUM **Al** 26·98 2.8.3	14 SILICON **Si** 28·09 2.8.4	15 PHOSPHORUS **P** 30·97 2.8.5	16 SULPHUR **S** 32·06 2.8.6	17 CHLORINE **Cl** 35·45 2.8.7	18 ARGON **Ar** 39·95 2.8.8
EL 1 2	29 COPPER **Cu** 63·54 2.8.18.1	30 ZINC **Zn** 65·37 2.8.18.2	31 GALLIUM **Ga** 69·72 2.8.18.3	32 GERMANIUM **Ge** 72·59 2.8.18.4	33 ARSENIC **As** 74·92 2.8.18.5	34 SELENIUM **Se** 78·96 2.8.18.6	35 BROMINE **Br** 79·91 2.8.18.7	36 KRYPTON **Kr** 83·80 2.8.18.8
IUM 4 .18	47 SILVER **Ag** 107·87 2.8.18.18.1	48 CADMIUM **Cd** 112·40 2.8.18.18.2	49 INDIUM **In** 114·82 2.8.18.18.3	50 TIN **Sn** 118·69 2.8.18.18.4	51 ANTIMONY **Sb** 121·75 2.8.18.18.5	52 TELLURIUM **Te** 127·60 2.8.18.18.6	53 IODINE **I** 126·90 2.8.18.18.7	54 XENON **Xe** 131·30 2.8.18.18.8
UM 09 17.1	79 GOLD **Au** 196·97 2.8.18.32.18.1	80 MERCURY **Hg** 200·59 2.8.18.32.18.2	81 THALLIUM **Tl** 204·37 2.8.18.32.18.3	82 LEAD **Pb** 207·19 2.8.18.32.18.4	83 BISMUTH **Bi** 208·98 2.8.18.32.18.5	84 POLONIUM **Po** [209] 2.8.18.32.18.6	85 ASTATINE **At** [210] 2.8.18.32.18.7	86 RADON **Rn** [222] 2.8.18.32.18.8

IUM 96 5.8.2	64 GADOLINIUM **Gd** 157·25 2.8.18.25.9.2	65 TERBIUM **Tb** 158·92 2.8.18.27.8.2	66 DYSPROSIUM **Dy** 162·50 2.8.18.28.8.2	67 HOLMIUM **Ho** 164·93 2.8.18.29.8.2	68 ERBIUM **Er** 167·26 2.8.18.30.8.2	69 THULIUM **Tm** 168·93 2.8.18.31.8.2	70 YTTERBIUM **Yb** 173·04 2.8.18.32.8.2	71 LUTETIUM **Lu** 174·97 2.8.18.32.9.2
CIUM n 3] 24.9.2	96 CURIUM **Cm** [247] 2.8.18.32.25.9.2	97 BERKELIUM **Bk** [247] 2.8.18.32.26.9.2	98 CALIFORNIUM **Cf** [251] 2.8.18.32.27.9.2	99 EINSTEINIUM **Es** [254] 2.8.18.32.28.9.2	100 FERMIUM **Fm** [257] 2.8.18.32.29.9.2	101 MENDELEVIUM **Md** [256] 2.8.18.32.30.9.2	102 NOBELIUM **No** [255] 2.8.18.32.31.9.2	103 LAWRENCIUM **Lw** [257] 2.8.18.32.32.9.2

mass numbers of the isotopes with longest half-lives.

INTRODUCTION TO MATERIALS SCIENCE

Introduction to Materials Science

B. R. SCHLENKER

JOHN WILEY & SONS AUSTRALASIA PTY LTD
Sydney New York London Toronto

Seventh Printing 1972

SBN 471 76170 2

Library of Congress catalog number: 73-75656
National Library of Australia registry number: AUS 68-2757

Composition by Craftsmen Type-Setters Pty. Ltd., Sydney
Printed in Singapore by Tien Wah Press Limited

Preface

THE purpose of this book is to provide a general introduction to the science of materials for students who have had no previous training in the field. Although the book is self-contained, a basic knowledge of the elements of chemistry, physics, and geology will assist the student to reach a full understanding of the subject matter covered.

The need for a book at this level became apparent after the introduction of Industrial Arts as a subject in the Higher School Certificate examination in New South Wales, materials science forming one of the four strands of this subject. The main problem, however, in writing a materials science textbook at this level is to provide a basic understanding of the scientific principles involved, particularly in connection with the structure of materials, as a basis for subsequent discussion of the behaviour of materials. Chapter 3 attempts to make this relationship clear by presenting at a mainly descriptive level—and I hope with reasonable success—a lengthy discussion of the structure of the solid state.

The concept of the book grew from a series of in-service training lectures on materials science given by the author in collaboration with several other teachers after the introduction of the subject at the Higher School Certificate level. In writing this book I have considerably developed and modified the material originally presented in the in-service lectures.

The book has three main objectives: to introduce the student to the wide variety of materials available today; to demonstrate some of the important relationships existing between the structure of a material and its properties; and to consider some of the ways in which materials are shaped and formed into articles designed to increase the material wealth of our society. Of these three objectives the second, *viz.* the relationships between structure and properties, is the central theme of the book.

Recognising the fact that materials science does present conceptual difficulties to beginning students, the text is supplemented by over 200 illustrations selected to highlight in visual form the main points made in the text. Understanding will also be aided by the 51 laboratory experiments and numerous review questions placed at the end of each chapter. Glossaries of terms met for the first time in the text are also placed at the end of each chapter.

It will be obvious that a book which attempts a detailed coverage at the elementary level of a field as large as materials science will inevitably be faced with the problem of selection. In this case fuels, explosives, and water are three important materials which have suffered from the process of selection. While the main criterion of selection has been the existing materials science syllabus of the New South Wales Department of Education, additional material has been included wherever necessary to provide continuity in the text or to encourage the student to extend his knowledge beyond the scope of that syllabus.

ACKNOWLEDGEMENTS

I wish to record my thanks to Associate Professor W. A. Rachinger of the Department of Physics at Monash University who read the draft and made many helpful suggestions.

The following organisations have willingly assisted in providing specialised photographs and micrographs to illustrate the book:

Australian Consolidated Industries Ltd.
Bohler Steels Pty. Ltd.
Broken Hill Proprietary Co. Ltd.
Copper and Brass Information Centre
Duly and Hansford Ltd.
N.S.W. Department of Railways
United States Steel Corporation

Thanks are also due to those publishers who have kindly given their permission to reproduce various copyrighted illustrations. Acknowledgement is given in each case in the appropriate figure legends.

In conclusion I wish to express my appreciation to the teachers attending the in-service training courses whose comments and criticisms have been of great value to me in writing this book.

Sydney
September 1968

B. R. Schlenker

Foreword

A SIMPLE hand tool like a cold chisel must exhibit certain properties. It must be strong, so that it resists deformation when it is being used, and hard, so that it resists wear. At the same time it must possess toughness, or resistance to fracture, so that it does not shatter when in use. In order to achieve these properties steel cold chisels are usually heat treated. They are heated to about 850°C, quenched in water and then tempered at temperatures around 400°C.

If, after heating to 850°C, a particular cold chisel was slowly cooled it would be much too soft; if quenched, but not tempered, it would be glass hard but would have insufficient toughness; and if tempered after quenching it could exhibit properties ranging from hard-and-brittle to soft-and-ductile depending on the tempering temperature employed.

This example serves to illustrate the fact that a wide range of properties may be obtained with a single material—in this case an alloy of iron and carbon with about one half per cent carbon, purely by changing the structure of the material, by heat treatment.

It is the relationship between structure and properties which forms the basis of materials science.

All solid materials are composed, ultimately, of atoms bonded together in one of a large number of possible ways. The properties of materials reflect, to some extent, the chemical and physical characteristics of the main atomic species of which they are composed, but they also show a very direct dependence on the way in which the atoms, as fundamental building blocks, are assembled, physically, to form the bulk material. They may be assembled as atoms, or as ions, into crystals, or combined chemically with other atoms into molecules, chains, plates or cells. The bulk material will then consist of an assembly of crystals or other building units with some form of secondary binding holding these together.

The notion that the structure of a material determines its properties and that properties may therefore be controlled by controlling structure originated in metallurgy. However, it need not be restricted to metals and alloys. The same principle can be applied universally to ceramic, polymer and mineral materials, whether natural or artificial. Recognising this fact, it was natural for various specialists working with these materials to come together to pool their knowledge and it was in this way that materials science was born. The materials scientist is one who works to discover the principles governing the relationship between structure and properties, with a view to using these in the design or synthesis of new materials with particular properties.

Materials science has developed rapidly during the last twenty years to meet the needs for better materials to provide higher standards of performance and reliability in service in machines, structures and equipment for modern civilisation. But its main support has come from the newer technologies associated with missiles and space research, high-speed flight, nuclear engineering, computers, electronics and control engineering. From these fields have come demands for materials to withstand conditions not previously experienced by man-made components and to perform functions not previously envisaged. Some of these requirements were able to be met by improvements in existing methods of manufacture and treatment of well-tried materials, but for others, rule-of-thumb methods were quite unsatisfactory and it was recognised that a new fundamental approach to materials was essential. The new approach of the materials scientist has paid off handsomely in many ways. It has solved a multitude of problems in rocketry, ranging from the design of materials for ultra-high-temperature service in heat shields, nose cones and rocket nozzles, to materials for minor mechanical components which must operate reliably, without liquid film lubrication, under the high vacuum conditions of space.

It has produced a multitude of "solid-state" electronic devices, beginning with the transistor diode, which have opened up a new world of miniaturisation and efficiency in equipment for communications and control, and has brought the possibility of super-conducting electrical circuits for heavy current applications almost to the point of commercial feasibility. Materials science has also given us a number of types of ultra-high-strength materials based on our knowledge of the mechanisms involved in deformation and fracture and of methods of controlling these.

While the increasing sophistication of materials has important implications for advanced engineering and technology, its implications for the man in the street are also significant. The fitter, machinist, motor mechanic and builder, and even the housewife, are increasingly faced with the necessity to employ new materials in place of well-tried conventional materials.

Intelligent use of the new materials, and, for that matter, of new drugs, implements, machines and labour-saving devices demands today a reasonably high standard of scientific literacy. This is reflected in the changing patterns of primary and secondary education, and in respect of materials in particular it is very significant that introductory studies in materials science are being introduced into high school curricula in a number of countries.

In New South Wales materials science is one of the strands in the subject Industrial Arts of the Higher School Certificate curriculum. This book is designed primarily for students taking Industrial Arts. It provides an introduction to the study of *materials* rather than, as the title suggests *materials science*, since, like the materials science syllabus in Industrial Arts, it is more concerned with the development of background knowledge of the *phenomenology* of materials than with the complex *principles* and *concepts* on which an understanding of the phenomena might be based. The latter is more appropriately a task for a university course, and is in fact being undertaken in a number of Australian universities.

Quite apart from its utilitarian purpose the book will be a success if it manages to transmit to its readers some of the fascination which the study of the structure of solids can provide, and if it induces some of them to undertake further studies in metallurgy, ceramics, polymer science or materials science in tertiary education. In my opinion there are few areas of science which offer the same excitement and challenge for future development as those related to materials.

Sydney
January 1969

Professor Hugh Muir
Head of the School of Metallurgy
University of New South Wales

Table of Contents

1

Materials Through the Ages

TECHNOLOGY, that body of knowledge comprising all of man's attempts to master his environment, has a complex and often ill-defined history. However, since the progressive utilisation and exploitation of all types of materials is the very framework upon which technological advancement has been built, a brief look back into history may provide a convenient reference point in the study of engineering materials.

The beginnings of technology lie in the attempts of primitive man to survive in what was an exceedingly hostile environment. Having no means of transport and no tools he had to seek food and shelter from that tiny part of the earth's surface on which he lived. Thus the men of the Palaeolithic (Old Stone) Age lived in caves, ate fruit, berries, roots and parts of small animals that they were able to kill, and were mainly hunters and food collectors. Man's first technological achievement occurred in this period with the discovery and taming of fire, which was used to cook food, to add to his material comfort by providing him with warmth and to harden wood for weapons and tools.

The first tools used by man were sticks and stones picked up from the ground, but it was not long before tool manufacture began. For instance, sticks and bones were ground into spears and other sharp-pointed implements by continued abrasion on hard rock, while flint and other hard stones were carefully chipped down to fine edges for axes, knives and arrowheads.

The Neolithic period saw man slowly emerging from savagery; population increased rapidly, and man the hunter became man the herder of cattle and sheep and the sower of crops. Technology, although lacking the advantages bestowed on later civilisations by the discovery and utilisation of metals, had advanced to the stage where the implements then in use could be shaped by grinding, polishing and even drilling, thus allowing effective utilisation of materials such as flint, obsidian (volcanic glass), bone, horn and wood.

Gold, silver and copper, the less active metals, were the first metals to be used by man, since they were fairly plentiful in their "free" states. It is probable that meteoric iron was also used during this early period, although in very small quantities due to the nature of the source of the supply. The advantages offered by some metals were quite obvious to Neolithic man in that they were harder and stronger than stone and retained a sharp edge for longer periods. Gold, probably first discovered about 1800 B.C., was being used extensively in Egypt by 1200 B.C., and was probably the first metal to be used by man. It was extracted by crushing gold-bearing rocks with dolerite ball-hammers and washing the crushed material in troughs lined with the fleeces of sheep, the gold becoming attached to the wool fibres in the process. Gold-bearing river sands and gravels were similarly separated.

Copper was also used extensively in Ancient Egypt, either in the pure form or as one of several alloys. It seems probable that copper was first extracted from an ore during the glazing of pottery using glazes rich in azurite (a form of copper carbonate), this occurring as early as 4000 B.C. It should perhaps be pointed out that the Ancient Egyptians possessed a high level of technology in the field of pottery; they produced, during the time of their civilisation, a wide variety of wares, including a "soft-paste porcelain"* which was fine, white, and translucent, and which was fired in their wood-burning kilns to temperatures in the vicinity of 1200° centigrade.

The extraction of iron from its ores began in this period, and while the discovery of this process is generally attributed to the Hittites, iron smelting was certainly being carried out by the Egyptians and Syrians by 1500 B.C. The best quality iron was used to manufacture swords, knives and armour plate, while inferior grades were used to make axes and ploughshares. The Egyptian blacksmiths discovered that by heating the crude iron in the presence of charcoal it became more malleable and ductile when cold (that is, they perfected steelmaking by a process of carburisation). It remained for the Greeks to develop the process of hardening steel by quenching it in water from red heat and for the Romans to perfect the process of tempering by re-heating to lower temperatures and re-quenching the metal. Modern steel technology is still centred around these three processes.

The Ancient Egyptians perfected the art of stonemasonry using various types of stone for constructional purposes and also for carving statues and vases. Soft rock was often cut with a chisel and mallet or saw, while harder rock was pounded with dolerite hand-hammers or abraded with a hard material like quartz. Trepanning was carried out on blocks of stone in order to hollow them out, the tool used being a hollow copper drill through which a sand and water slurry was passed as the tool rotated. Massive blocks of granite and limestone were quarried by using wedges, levers, rollers and rafts, and were then used in the construction of temples, pyramids and tombs. The surfaces of some pieces of stone were ground to such a degree of accuracy that no more than a 200th of an inch error of flatness or parallelism existed, a remarkable achievement by the craftsmen of the day.

Other materials used extensively during the period of Ancient Egypt

*A "soft-paste" porcelain is a clay-body ceramic which can be vitrified at temperatures lower than those which must be used for ordinary porcelains.

included leather, which was used to make rope, bellows, buckets, chair seats, sheaths, harnesses and the like; bone, horn and ivory for tool handles, pins and rings; glue and hair from animal origins to make brushes; emery, pumice and sand as abrasives; and gypsum plaster, lime mortar, clay and clay bricks for building purposes. Even naturally-occurring bitumen was collected and used for waterproofing and as a matrix into which mosaics could be set. Truly, man was coming to master his environment now that such a wide variety of materials were available to him.

Many of the processes of modern metallurgy originated in the shops of the metalworkers of this period, with such processes as swaging, wire drawing, twisting, stamping and casting being everyday techniques for the goldsmith and the jewellery maker. It is significant to note that the "lost wax" casting technique, a most important industrial technique of today, originates from this period.

The Greek and Roman periods, covering the centuries from 600 B.C. to about A.D. 400 saw many great advances in materials technology, but most of these were lost with the decline of civilisation in the period known as the Dark Ages, which lasted in Western Europe from about A.D. 400 to A.D. 1000.

Wood was a most important material during the period of Greece and Rome, being used as a fuel, for furniture, ship building, container construction, and even for pipes. Seasoning techniques were well advanced and thus greater precision in the woodworking trades became possible. Turnery was introduced by the Greeks, the symmetrical forms produced on the lathe having a particular appeal in this culture. Wood bending, using heat and steam, was widely practised for furniture manufacture. Metal fittings, such as corner pieces, ferrules, iron straps and metal caps on wooden ploughs and spades were in common use, as were iron and bronze nails of various sizes. Forge welding of iron was discovered in Greece in the seventh century B.C., but no cast iron was produced since furnace technology had not advanced sufficiently to allow iron to be fully melted.

The Chinese, however, had perfected the art of making cast iron by the fourth century B.C., this achievement being possible because they had well-developed kilns; they used coal as a fuel, they had better ores and they possessed the necessary refractory clays with which to line the kilns. It is interesting to note that the Chinese were using a process of direct decarburisation of molten cast iron to make steel in the second century B.C. This method known as the "hundred refinings" involved blowing controlled amounts of cold air on to the surface of the molten iron. Clearly this is the predecessor of the method patented by Bessemer in 1856 in England.

Lead found many applications in Greece and Rome. It was produced in great quantities and was used as a roofing material, in the manufacture of pipes, and as a lining for water tanks. Tin was also in common use, and ordinary "soft" solder was used to join lead and lead alloys when watertight seams were required. Clearly, though little was understood about the basic natures of these metals, they were being effectively utilised even at this early stage of man's developing technology.

Concrete, one of the most vital engineering materials of modern times, began its history in the Roman period. A material known as pozzolana (pozzolans), a volcanic earth found in huge quantities near Naples, was mixed with lime, water, and crushed and broken stone and allowed to set. This form of cement is resistant to fire and sets under water and is as strong as its aggregate. It is interesting to note that a patent was not taken out on Portland cement until 1824, and that up to this time cements of pozzolanic origins were the only forms available for all types of building construction.

The Dark Ages, lasting up to about A.D. 1000 in Europe, saw the neglect of all forms of public works, and the concentration of skilled craftsmen either in the castles of kings and other noblemen or in the thousands of monasteries scattered throughout Europe. In fact many of the engineering skills of the Classical World were preserved by the diligence of monks working in the seclusion of their monasteries.

However, by A.D. 1500 the more advanced Western civilisations were far ahead of the old Classical civilisations in the field of engineering technology. The main reasons for this advance were the applications of the scientific method in technology, the development of universities as centres for the dissemination of specialised knowledge, the invention of printing, and the invention of the mechanical clock, the latter being the first automatic machine to be made in quantity. In the field of engineering materials this period saw great developments in casting techniques; for instance, a $4\frac{1}{2}$ ton cannon was cast in Austria in 1404. By the fifteenth century attempts were being made to study the strengths and properties of engineering materials in an objective and scientific manner. Leonardo da Vinci, a pioneer in this and many other fields, developed a machine to test the tensile strength of wire, and also studied the behaviour of simply-supported beams, correctly concluding that the strength of such a beam varies inversely as its length and directly as its width.

The birth of engineering science occurred during the sixteenth and seventeenth centuries, when the art of the craftsman began its long and successful marriage to the experimentation of the scientist. England began to emerge as a great industrial power, a position it held well into the nineteenth century. Robert Hooke expounded his now classical law governing the behaviour of materials within their elastic ranges in 1662 whilst he held the position of curator of the Royal Society in England. A Frenchman, Marriotte, carried out detailed experiments on stress distributions in beams and also in thin-walled cylinders (pipes) under internal pressure. The first full-scale tensile testing machine made its appearance in the University of Leyden, Holland, but such machines did not come into general use for another hundred years.

The microscope was invented, becoming a useful tool of the scientists of the day. Iron and steel technology made fairly rapid advances during this pre-industrial revolution period, and the trend to replace wood by various ferrous alloys for almost all types of construction and machine building began.

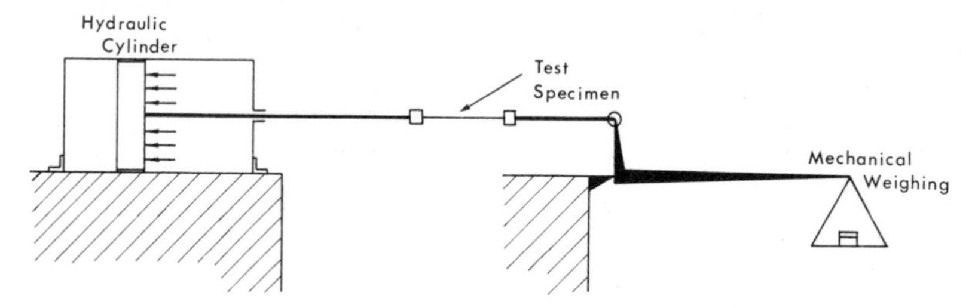

Figure 1.1 *The principle of the tensile testing machine developed by Lamé in 1824.*

The Industrial Revolution with its explosion of industrial production saw many advances in the production of materials, in the development of new forming and working techniques for metals, and in the production of new types of machines, many of which were designed to replace the hand-skills of the craftsman. It is true to say that most if not all of the problems of industrialisation can be traced back to this relatively short time span of about one hundred years.

Perhaps the most spectacular achievement of this period, as far as materials technology was concerned, was the tremendous impetus given to the iron and steel industry, particularly in England. Iron was now being smelted with coke or anthracite instead of charcoal, with the air-blasts necessary for refining operations being provided by steam-driven blowers instead of leather bellows; the "puddling process" came into its own for the manufacture of wrought iron, and pre-heating stoves were developed to warm the air before it entered the furnaces. Cast iron production was high, and special "crucible steel" was made for springs, tools and cutlery where good hardenability was required. During the 1780's Priestly, together with other scientists on the Continent, demonstrated that it is the combination of carbon with iron that makes steel. Bessemer and Siemens both belong to this period; they pioneered the two most important steelmaking processes of their times, the Bessemer Converter and the Open Hearth furnace. Great advances were made in the field of metalworking processes; the steam hammer, usually double-acting, made considerable advances possible in forging; iron tubing was made by rolling strips around a former and forge-welding the seam; while lead was actually extruded into seamless pipe in 1820 by a plumber, Thomas Burr of Shrewsbury.

Thomas Young demonstrated the value of determining the Elastic

Modulus of metals, this modulus now bearing his name; Poisson evaluated the importance of the relationship* between axial and lateral strains in a material under a tensile load; and a great deal of applied research was evaluated mathematically, thus allowing fairly exact design calculations to be made. Research on fatigue in metals began in the early 1840's, with the immediate result that, in 1848, a design safety factor of eight was used in setting the sizes of iron beams used in the construction of railway bridges, and large radii were formed between the shafts and flanges of railway axles to prevent stress concentration causing premature failure.

The foundations of that branch of metallurgy known as metallography were laid by Professor Henry C. Sorby of Sheffield when, in 1861, he demonstrated that the internal structures of metals could be successfully examined by studying prepared sections under a reflecting microscope. Detailed research was carried out on metals and alloys to determine their physical and mechanical properties—strength, density, hardness, melting point and the like. Also the effects of various alloying elements such as nickel, chromium and manganese in steel were evaluated experimentally. Now, at last, man was coming to grips with the real problems of the effective utilisation of metals, and the scene was set for the development of a completely different type of material which was to become known as plastic.

The first commercial plastic was created in 1868 in the United States when John Hyatt, an inventive printer, combined pyroxylin, itself made from cotton linters and nitric acid, with camphor, thus synthesising cellulose nitrate or, as it was to become more commonly called, celluloid. Coloured pink, this material was first used to make billiard balls during the time when ivory was in short supply, but was later employed in the manufacture of denture plates, "wipe-clean" collars, cuffs and shirt fronts, and film. The second vital step in the development of the American plastics industry occurred in 1909 when Dr. Leo H. Baekeland successfully controlled the reaction between phenol and formaldehyde to produce phenol-formaldehyde, or to give it its common trade-name, Bakelite. From 1909 to 1926 only two more plastics were produced, but from 1926 to the present time many different types have been produced by chemically combining organic substances in particular ways. Because of the multitude of uses to which plastics have been put, many people would call the present period the "age of plastics", but in truth these materials are but one small group of the total array available to the engineer and scientist.

Perhaps the most exciting group of modern-day engineering materials are those falling under the generic name "ceramics", for without these materials rocketry and space exploration would still be in the realms of dreams. Try to imagine the intense heat of the exhaust gases of an I.C.B.M. as they pass out of the venturi of the rocket motor; this is an atmosphere in which no

*Poisson's Ratio is discussed in Chapter 5.

known metal can survive, and in which the effective working life of the best ceramic may be only four to five seconds, for, after this short interval of time the ceramic is blasted apart by the heat and abrasion of the escaping gases. Yet ceramics are one of the oldest groups of materials known to man. It has already been said that the Ancient Egyptians were well skilled in the art of making clay-body ceramic articles. The Chinese, in fact, perfected the highly-skilled art of making porcelain and stoneware by A.D. 900, thereby anticipating Western developments in this field by about 700 years. However, many of our modern ceramics are not based upon clays at all, being produced by the carefully controlled reactions between pure compounds of such elements as boron, aluminium, silicon and magnesium.

This scientific period in which we live has much to offer the student of materials science, for with our sophisticated research tools such as the electron microscope and the X-ray machine we are able to look into the very heart of a material and determine its internal structure. Other equally sophisticated tools enable us to assess accurately the mechanical, physical, thermal and electrical properties of the same material, and then scientific reasoning enables conclusions regarding the relationships between structure and properties to be drawn. Since the applications of any material depend upon its properties, and the properties depend upon the structure, then one most important aspect of modern materials science must be that dealing with the internal makeup of materials.

Technologists of today, by virtue of the fact that they can analyse a material from all aspects, are much better equipped to utilise engineering materials effectively than were the craftsmen of earlier times, but complacency cannot be permitted, for there is still a great deal more to be discovered. Thus, the science of engineering materials is a living and vital driving force in our present technology.

GLOSSARY OF TERMS

Metallography: the scientific study of the constitution and structure of metals and alloys as revealed by microscopic examination.

Metallurgy: that branch of applied science concerned with the detailed study of metals—their ores, extraction, working, alloying effects, joining, and physical, chemical and mechanical properties.

Materials science: that branch of applied science concerned with investigating the relationships existing between the structures of materials and their properties.

2

The Classification and Grouping of Engineering Materials

EVERY science requires a comprehensive classificatory scheme so that its fundamental knowledge can be ordered and structured in some logical manner. Very often this classification arises partly from the existing body of knowledge and partly from the various theories and speculations of persons working in the field.

The science of engineering materials, or materials science, is no exception to this rule, and several very important classifications are in use today. The following factors, either taken singly or in combination, form the bases of the various systems of classification of engineering materials:

(1) the chemical composition of the material
(2) its mode of occurrence in nature
(3) the refining and manufacturing processes to which it must be subjected before it gains economic importance
(4) its internal (atomic and crystalline) structure.

ELEMENTS, COMPOUNDS AND MIXTURES

All materials, whether solid, liquid or gaseous, can be classified as elements, compounds or mixtures on the basis of chemical composition and internal structure.

Elements are simple substances which cannot be broken down into chemically simpler substances. Of the known elements only a very small number have importance as engineering materials in their elemental (uncombined) states.

Compounds are formed when two or more elements combine chemically in fixed proportions by weight; all compounds can be broken down, by one means or another, into their component elements. Many important engineering materials exist in the form of chemical compounds.

Mixtures are formed when two or more pure substances (elements or compounds) are mechanically mixed together in any proportion; thus, a mixture differs from a compound in that it is, in theory anyway, mechanically or physically separable. Some metal alloys are of this type; for instance, when cadmium and bismuth are melted together in any proportion, the final solid alloy consists of an intimate mixture of fine particles (grains) of both metals.

METALS AND NON-METALS

The elements may be roughly divided* according to their properties into two groups, the metals and the non-metals. The characteristic properties of these two groups are summarised in Table 2.1:

TABLE 2.1

THE CHARACTERISTIC PROPERTIES OF METALS AND NON-METALS

Metals	*Non-Metals*
Usually solid at ordinary temperatures	May be solids, liquids or gases
Are lustrous on a freshly-cut surface	Usually dull if solid
Usually malleable and ductile to some degree	Usually brittle if solid
Usually fair to good thermal and electrical conductors	Usually non-conductors (insulators) of heat and electricity
Usually form alloys	Do not alloy, but may chemically combine to form compounds

An illustration of these properties may be obtained by comparing copper and sulphur. Copper, a reddish-pink metal, has excellent malleability and ductility, it melts at 1083°C, it forms a wide range of alloys, and both it and its alloys have excellent conductivities. Sulphur, on the other hand, is a yellow vitreous non-metallic solid or powder, it is brittle, it melts at 119°C it is a non-conductor, and, although it combines with a wide range of metals to form sulphides, it does not form any alloys.

Laboratory (1): *Using samples of several pure metals, several common alloys, and several non-metals, verify the properties set out in Table 2.1.*

The characteristics of metals as outlined in Table 2.1 are really those of pure metals and, while many also apply to alloys and other "impure" metals, there are significant differences between pure metals and alloys which must be considered.

The engineer looks at metals from a very different viewpoint from that of the pure scientist, being more concerned with their properties and applications than with their behaviour in chemical reactions. A typical classification of metals as used in engineering would be as follows:

*For a much more exact division see Figure 3.3.

METALS

Pure | Alloyed | Coated | Clad | Ferrous | Non-Ferrous | Sintered

Laboratory (2): *Take small samples of copper, brass, zinc, galvanised iron, terneplate, mild steel, tool steel, cast iron and a "porous" bronze bearing. Classify each as ferrous—non-ferrous; pure-alloyed; clad-coated; sintered. List significant features of each material.*

Pure metals are those obtained when an ore is refined to yield a metallic element. In practice it is very difficult to obtain a perfectly pure metal, and indeed many metals are worthless to the engineer when in this state, but by using specialised and, incidentally, often very expensive techniques some metals can be obtained 99.999 % pure. Copper and aluminium are common examples of this.

An alloy is formed when two or more relatively pure metals are melted together to form a new metal having, in many instances, properties quite different from those of the two metals used in its manufacture. An outstanding example of this is the alloy formed when 18 % chromium and 8 % nickel are added to low-carbon steel, forming a steel which will not rust or corrode, this being quite different from the behaviour of the original low-carbon steel.

Coated metals, including such metals as tinplate, terneplate, galvanised iron, and zincanneal, are manufactured so that the desirable properties of both metals can be "married together". For instance, tinplate (mild steel sheet coated with a very thin layer of tin) enables steel to be used for food containers, while galvanised iron (mild steel coated with a thin layer of zinc) enables strong thin sheets of steel to be used for constructional purposes without fear of rusting.

Clad metals are also made to take advantage of the properties of both materials used, but in this case a "sandwich" is made of the metals concerned. For instance duralumin is clad with thin sheets of pure aluminium for use on exposed surfaces, since the surface layers of aluminium resist corrosion while the centre layer of duralumin provides high strength. Stainless steel is often backed by a relatively thick layer of mild steel by rolling the two metals together while they are red hot. This gives a clad metal which will not rust or corrode on one surface, and which is still relatively inexpensive to manufacture. A typical application of such a clad metal is in the construction of oil storage tanks, where rust and corrosion cannot be tolerated inside the tank, but where the cost of fabrication using thick sheets of stainless steel would be prohibitive.

A ferrous metal is one composed principally of iron, while *non-ferrous metals* are those composed principally of metals other than iron. It should be noted from the outset that a non-ferrous alloy may contain iron in some small proportion (for instance, high-tensile brass often contains about 1% iron by weight), and that ferrous alloys may contain quite significant proportions of non-ferrous metals.

Ferrous alloys are of the utmost importance in engineering, and are classified on the basis of the percentage of carbon and other alloying elements present. Some of the more important groupings are given below:

(1) *Mild Steels*: containing between about 0.15% and 0.25% carbon, are used where low cost, moderate strength and good weldability are required.
(2) *Medium Carbon Steels*: containing between about 0.3% and 0.6% carbon, these steels have higher strengths but reduced weldability.
(3) *High Carbon Steels and Carbon Tool Steels*: contain up to 1.5% carbon; they have poor weldability but can be heat-treated to become hard and tough.
(4) *Cast Irons*: contain between 2% and 4% carbon, and are excellent and inexpensive general purpose ferrous casting alloys.

Of the vast number of non-ferrous metals available to man, only seven are available at a low enough cost and in sufficient quantity to serve as bases for common engineering metals. Three of these (copper, lead and tin) have been used for thousands of years, while the remaining four (aluminium, magnesium, nickel and zinc) are the more recent additions, achieving commercial importance only in the latter part of the nineteenth century. At least fourteen other metals—antimony, beryllium, cadmium, chromium, cobalt, manganese, mercury, tantalum, molybdenum, niobium, titanium, tungsten, vanadium and zirconium—are vital to modern industry although they are only produced in relatively small quantities.

Sintered metals are those produced by the techniques of powder metallurgy, and they possess very different properties and structures from those of metals that have been cast and worked using conventional techniques. Metals to be sintered must first of all be powdered, a process which is sometimes quite complex. The next step is to mix the required powdered metals in correct proportions, then press them into shape in a die, and finally sinter them in a furnace. This produces a metal which is not a true alloy, but which may possess some or many of the properties of typical alloys. Metals formed into shape by powdering and sintering can be made with some degree of porosity, this being an advantage for such things as bronze bearings which can be soaked in oil prior to assembly and become "self-lubricating" during their working life.

ORGANIC, BIOLOGICAL AND INORGANIC MATERIALS

The range of organic compounds available is very extensive owing to the many thousands of hydrocarbon compounds and their derivatives in existence. Furthermore, the importance of the study of organic materials becomes apparent when it is understood that all biological systems are composed of compounds of carbon. However, while all biological systems are organic, there are some materials of biological origins, such as limestone, which are not organic in composition.

Organic materials are those materials derived directly from carbon; they usually consist of carbon chemically combined with hydrogen, oxygen or other non-metallic substances, and their structures are, in many instances, fairly complex. Most plastics are organic, as are petroleum derivatives, many waxes, and wood substance.

Plastics and synthetic rubbers are common organic engineering materials that do not have biolgocal origins. Termed "polymers" because they are formed by polymerization reactions in which relatively simple molecules are chemically combined into massive long-chain molecules or "three-dimensional" structures, these materials have lower specific gravities than even the lightest of metals and yet exhibit good strength-to-weight ratios. Polymers may be said to be more or less solid materials, consisting wholly or in part of chemical combinations of carbon with oxygen, hydrogen, nitrogen and other metallic or non-metallic elements. All polymers are liquid or semi-liquid during some stage of their manufacture from such substances as ethylene, phenol, formaldehyde, butadiene and acetylene, and are generally formed into shape by moulding, rolling or extrusion while in this state.

Two major classes of polymers are known to exist: (1) the thermoplastics, (or thermosoftening plastics) which soften after they have been formed by the application of heat, and reharden upon subsequent cooling; and (2) the thermosetting plastics, which cannot be resoftened once complete polymerization has occurred. Polythene, polystyrene and nylon are examples of thermoplastics, while the phenolics (phenol-formaldehyde, urea formaldehyde and melamine formaldehyde) and the silicones are examples of thermosetting plastics.

Laboratory (3): *Take thin strips of a thermosetting plastic such as 'bakelite' and a thermosoftening plastic such as 'perspex'. Immerse each in boiling water for one minute, and test for plasticity by bending and twisting.*

Natural rubber is an organic material of biological origin being composed in the pure state entirely of a hydrocarbon having the formula $(C_5H_8)n$. It is a thermosoftening material. Plantations of the rubber tree *Hevea Brasiliensis* in Ceylon, Malaysia and Indonesia provide over 90% of the world's supply of latex rubber which, when cured, filled and perhaps reinforced and mixed with certain synthetics, keeps the civilised world in automobile tyres, rubber insulations, rubberised raincoats, mattresses, shoe soles and the many other rubber products so essential in today's technology.

Biological materials may be organic, as in the case of natural rubber, but in every instance they result from the life cycle of some organism. A wide variety of biological materials are of interest to the engineer, some common examples being wood, bone, horn, waxes, leather, diatomite and corralline limestone.

Wood, one of the earliest materials to be used by man, has a complex cellular structure principally composed of fibres of the complex hydrocarbon, cellulose, together with lignin, the latter surrounding the cellulose fibres in a thin layer, and thereby "cementing" them together. Starch, resins, gums, fats, waxes, tannins and traces of organic acids are also present but only in very small amounts since cellulose and lignin account for between 70% to 85% of the wood substance. The sometimes complex and always interesting structure of wood can be seen by using a 10x magnifying lens, the different types of cells making up the structure of the wood appearing quite clearly. Each different type of cell has a specialised function in the living tree, and it is on the basis of the types of cells present that wood is classified as either hardwood or softwood. All hardwoods are porous timbers, possessing very large cells known as vessels while softwoods are non-porous, possessing only small cells. These terms are misleading, since some timbers classified as hardwoods are quite soft, while some softwoods or non-pored

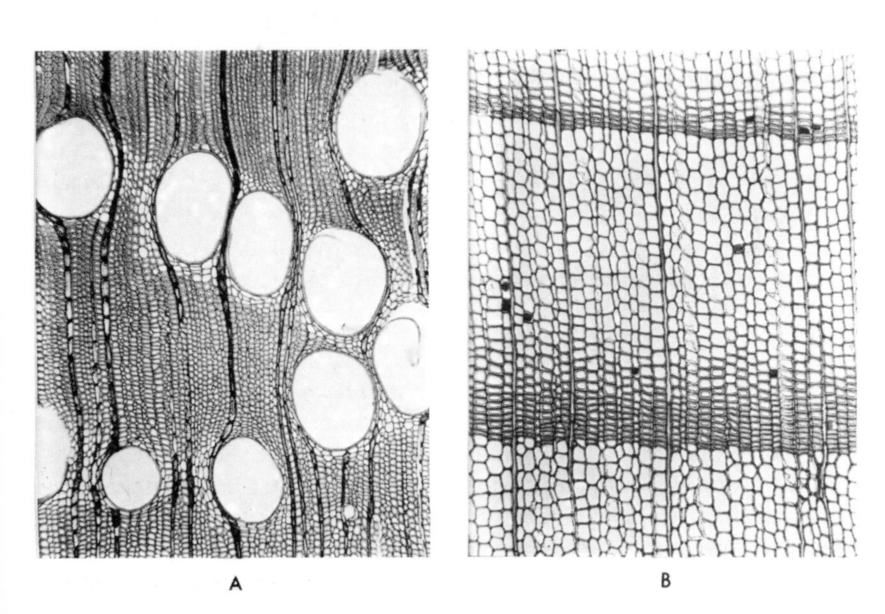

Figure 2.1 *Photomicrographs showing typical hardwood and softwood cellular structures: (A) is the hardwood Flooded Gum showing large vessels; (B) is the softwood King William Pine showing a non-pored structure. Magnification x80. (Photographs courtesy of the New South Wales Division of Wood Technology.)*

woods are quite hard. The classic example of this is balsa, which is porous and thus classified as a hardwood.

Freshly-felled wood always contains a high percentage of water present either in the cell cavities or chemically combined within the structure of the cell walls. Seasoning involves the controlled drying out of the wood, and it is when this process is not properly controlled that sawn timber tends to warp, twist, crack, check and collapse, the latter defect occurring if a too rapid rate of water removal causes the cells to cave in.

Laboratory (4): *Examine the end grain of a piece of softwood (e.g. oregon, pines) and a piece of hardwood (e.g. oak, blackwood) and sketch the cellular structure. Label the important parts of each structure.*

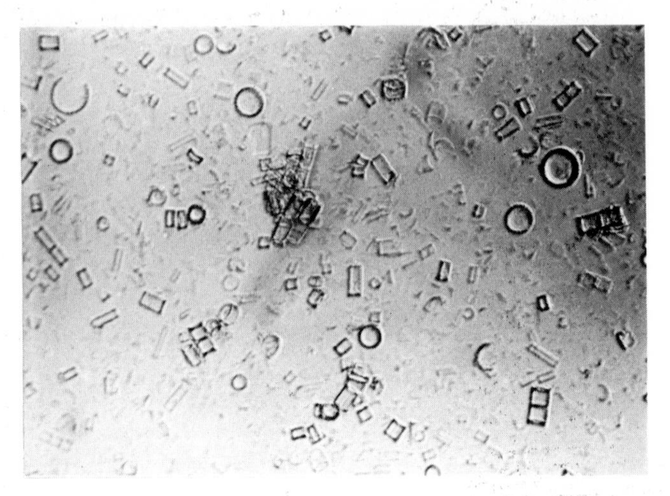

Figure 2.2 *Photomicrograph showing individual skeletal remains of diatoms in powdered diatomite. Magnification x60. (From a slide courtesy of the New South Wales Mines Department.)*

Bone, long ago used by man for tools and weapons, may be considered to be a "laminate" of a complex organic substance called collagen and the phosphate and carbonate of calcium. Living bones possess many living cells; they act as levers, they give form to the animal, they protect vital organs, and they grow strongest where stress concentrations are greatest. Bone is slightly stronger in compression than in tension. Animal bones today find many uses, gelatin and glue manufacture being two fairly important ones.

Leather is the skin of animals after it has been cleaned and tanned, the latter preventing natural decomposition. Early man used the skins and hides of the animals he killed for clothing and weapons and today the skins and hides of many animals (for example, those from cattle, sheep, oxen, goats, kangaroos, horses, pigs and even seals) are economically important. Early tanning techniques probably involved rubbing the skin with fat and drying it in the sun, while modern techniques include "vegetable tanning" and "chrome tanning". The former involves soaking the skin in a tanning liquor such as Wattle bark tannin for several weeks, after which it is partially dried and then rolled to condition the leather. Chrome tanning, used for lighter and more delicate skins, involves pickling the skin in an acid saline solution and then tumbling it in a revolving drum containing a basic solution of chromium salts, after which drying and rolling are carried out. It is interesting to note that perfectly preserved leather articles have been recovered from Egyptian tombs known to be over 3,000 years old.

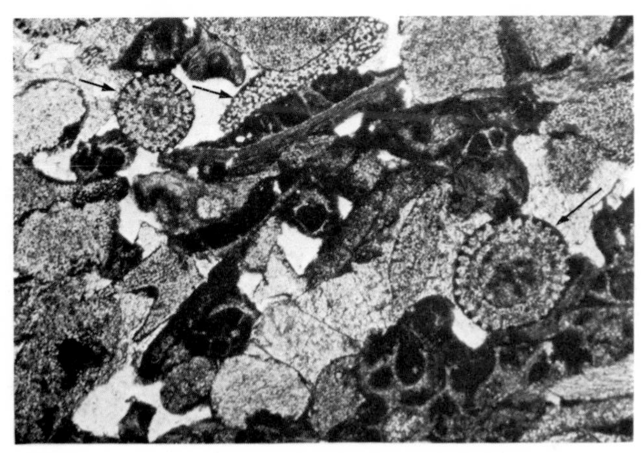

Figure 2.3 *Photomicrograph showing the structure of a corralline limestone; the arrows indicate skeletal remains. Magnification x100.*

Diatomaceous earth, also known as diatomite, kieselguhr, tripolite, and, erroneously as infusorial earth, is a hydrous or opalescent form of silica. It is found in fairly massive deposits as a whitish substance not unlike chalk in appearance, and is composed of the compacted siliceous remains of diatoms, a group of flowerless aquatic plants found in marine or freshwater environments. The relative sizes of diatoms can be appreciated when it is realised that a cubic inch of diatomaceous earth may contain between 40 and 70 million skeletons. Diatomite is light, porous, relatively chemically inert and is a very poor conductor of heat and electricity. It therefore finds applications as an insulating material, as a filtration medium for such things as waste oils, fruit juices, sewage and perfumes, as a neutral filler in paints, varnishes, rubber and plastics, and also as a mild abrasive in metal polishes.

Limestone is also an important material of biological origin which is not organic. Limestone consists mainly of the mineral calcite (calcium carbonate), and limestone laid down as beds in marine environments is made up of the skeletal remains of such animals as corals, worms, crinoids, molluscs and protozoa. The economic importance of limestone is almost beyond estimation; shaly-limestones are a necessary ingredient in the manufacture of cement; crushed limestone is used as a road-base material; and pure limestone is an important flux for many iron and steelmaking processes.

Laboratory (5): *Examine pieces of diatomite and corralline limestone using a 10x lens, and sketch their structures. Can evidence of biological origins be seen using this order of magnification?*

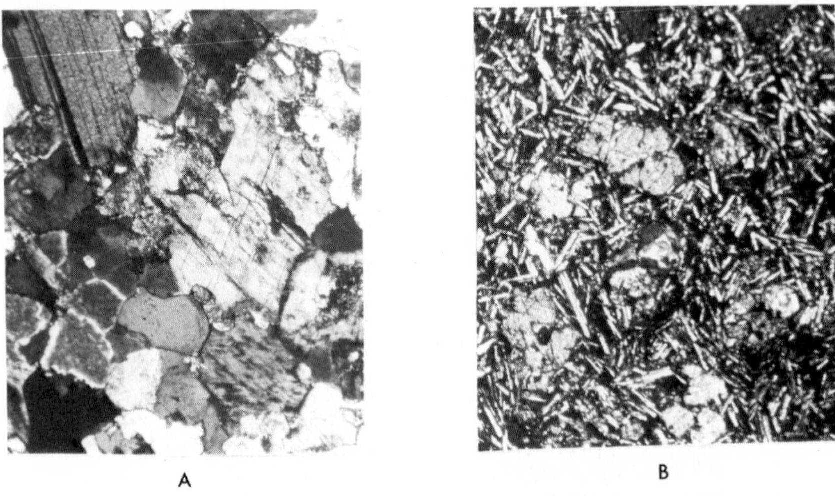

Figure 2.4 *Photomicrographs of (A) granite and (B) basalt. The coarse structure of the granite results from slow cooling; the fine structure of basalt from rapid cooling. Magnifications x 100.*

Inorganic materials are of mineral origin and have not been formed either directly or indirectly as the result of the natural growth and development of living organisms. Thus, it is obvious that no inorganic material can be biological. Common inorganic materials include metals, rocks, minerals, clays, gravels, sands and the multitude of materials grouped under the generic name "ceramic".

Rocks are the more or less definite units forming the crust of the earth and have been used in a variety of ways since man first began his struggle for survival. Petrology, the scientific study of rocks, is a relatively young science and is complicated by the vast array of rocks in existence. Nevertheless, rocks are classified into three major groups based upon modes of origin and formation: igneous, sedimentary and metamorphic. Igneous rocks are those formed by the consolidation of a liquid or semi-liquid material known as magma, the resulting rock material being termed plutonic if this consolidation took place deep in the earth, or volcanic if the magma or lava solidified on the earth's surface. Thus, granite is igneous plutonic while basalt and dolerite are igneous volcanic.

Sedimentary rocks are formed when the broken-down remains of existing rocks are consolidated by pressure; sandstone and shale are good examples, the sediments making up shale being much finer than those making up sandstone. The necessary pressure is supplied by the overlying rocky burden, itself laid down as layers of sediment during some subsequent period of geological history.

Metamorphic rocks are sedimentary rocks altered by heat and great pressure into new rocks. These have structures somewhere in between those of igneous rocks and those of the sedimentary rocks from which they were formed. Slate and marble are typical metamorphic rocks which have had important applications over the centuries.

Figure 2.5 *Photomicrograph of a sandstone. The rounded grains result from abrasion during the formation of the sediments. Magnification x100.*

Building blocks or dimension stone cut from such rocks as granite, marble, limestone and sandstone have been used to construct monuments, arches, bridges, houses, tombs and buildings ever since the period of Ancient Egypt, while shale and more particularly slate have found applications as roofing materials over the same period. Slate, with its superior cleavage and hardness is still used as a roofing material for some classes of domestic dwellings today. Crushed stone, particularly from types like basalt, dolerite, rhyolite and other fine-grained rocks, has many applications including filling, concrete aggregate and as a base material ("blue metal") for road construction.

Laboratory (6): *Examine, using a 10x lens, pieces of the following—granite, basalt, dolerite, marble, sandstone, shale, slate, gypsum. Classify each rock type and describe its appearance.*

Pozzolanic materials are of particular interest since they are naturally-occurring or synthetic siliceous materials which, although not cementitious in themselves, will hydrate in the presence of water and lime to form a cement. The volcanic ash known as tuff, diatomaceous earth, pumice, cherts, some shales, blast furnace slag and fly-ash are all pozzolanic. Pozzolans-Portland cement, containing up to 15% of, say, ground blast furnace slag by weight, sets more slowly than ordinary Portland cement and develops greater resistance to sea water and sulphate solutions.

Ceramics are materials consisting of phases* which are themselves compounds of metallic and non-metallic elements, and thus there is a multitude of materials which fall into this group. For instance, all metallic compounds, many minerals, glass, glass-fibre, many abrasives and all fired clays are ceramics.

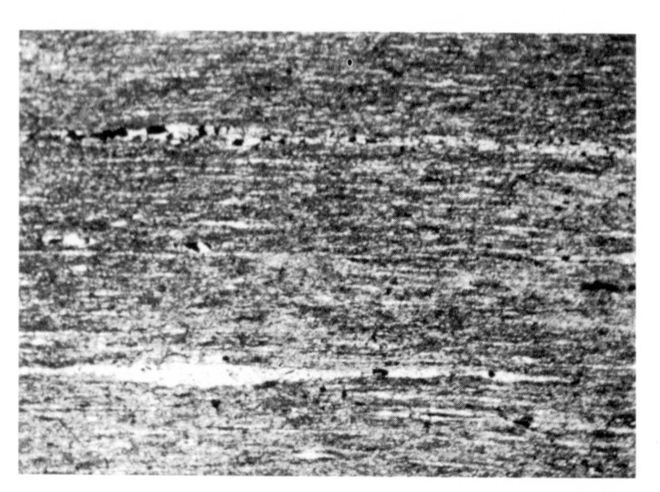

Figure 2.6 *Photomicrographs of slate. Slate is fine grained and has a characteristic cleavage (the horizontal pattern in the photomicrograph). Magnification x100.*

NATURALLY-OCCURRING AND MANUFACTURED MATERIALS

This classification, although cutting across those already discussed, is useful in that it separates engineering materials on the basis of how much processing is required before they gain economic importance.

*A phase is a physically separable and chemically homogeneous constituent of a material.

A naturally-occurring material is one that exists in nature in the form in which it is to be used. Obviously this group of materials is limited in size, but looked at historically includes such things as sticks, wood, rocks, stone and the less active metals; gold, silver and copper. The multitude of materials used today are manufactured; that is, they owe their particular forms and properties largely to the processes by which they have been extracted and treated. Thus an important aspect of the study of engineering materials is concerned with those manufacturing processes by which materials are produced and altered to suit modern needs.

Laboratory (7): *Examine some of the more important manufactured materials derived from wood, such as paper, composition board and plywood. Discuss their structures and economic importance.*

It is significant to note that only those metals that are fairly chemically inert occur in their free or "native" states in nature, while the more reactive metals tend to be found as sulphide, carbonate or sulphate ores. Gold, silver and copper, the three earliest-used metals, are naturally-occurring, possess extremely high corrosion resistance, are extremely malleable and ductile, and alloy readily, forming alloys harder than the parent metals themselves.

TABLE 2.2

SOME IMPORTANT GROUPINGS OF MATERIALS

Class of Material	*Naturally-Occurring Materials*	*Manufactured Materials*
Metals	The less active metals—gold, silver, copper, meteoric iron, mercury*	All metals and alloys including those metals that are naturally-occurring
Polymers	Shellac†, natural rubber, cellulose fibres	All other polymers, e.g. synthetic rubbers, plastics, linen
Ceramics	All minerals that are compounds of metals and non-metals; clays, shales, sands, many rocks, obsidian, emery, corundum	All forms that require purification, mixing, firing or other treatment before gaining economic importance, e.g. clay-body ceramics and glass, carborundum
Biological	Wood, diatomite, bitumen, limestone	Seasoned timber, paper, composition board, lime, diatomite bricks**

*Mercury found as tiny droplets associated with some ores
†The wings of certain beetles
**Pressed diatomite

GLOSSARY OF TERMS

Alloy: a material exhibiting metallic properties made by combining several metals, or metals and non-metals.
Biological material: a material resulting directly from the life cycle of a plant or an animal.
Ceramic: a material containing phases made up of chemical combinations of metals and non-metals; they are typically brittle and high melting-point solids.
Compound: a compound is formed when two or more elements chemically combine in fixed proportions by weight.
Diatomite: (also known as diatomaceous earth or kieselguhr). A material formed by the compaction of the skeletal remains of diatoms.
Dimension stone: rock pieces cut or hewn to regular rectangular shapes suitable for building purposes.
Element: a chemically simple substance made up of atoms all of the same kind.
Ferrous: a metal or alloy having iron as its principal constituent.
Igneous: a primary rock-type formed by the consolidation of mineral solutions known as magma or lava.
Metal: a solid material which is typically ductile, lustrous on a cut surface, and a good conductor.
Metamorphic rock: a secondary rock formed by the alteration of existing rocks by heat and pressure in the earth.
Organic: a material whose structure is based upon carbon atoms held together in molecules.
Polymer: a non-metallic material in which the molecular structure consists of long-chain molecules or macro-molecular structures formed by the combination of many simple molecules.
Pozzolans: naturally-occurring or synthetic silicious materials that will hydrate to form cements.
Sedimentary rock: rock formed by the compaction of the broken-down remains of younger rocks.

REVIEW QUESTIONS

1. Why is the systematic classification of materials of importance to the materials scientist?
2. List and briefly discuss the significance of three factors that are used as bases of systems of materials classifications.
3. Distinguish between elements, compounds, and mixtures.
4. Distinguish between metals and non-metals in terms of their characteristic properties and uses.

5. Briefly explain the meanings of the following terms as they relate to metals: alloy; clad metal; coated metal; non-ferrous; sintered metal. Give examples to illustrate your answer.
6. Distinguish between organic and biological materials in terms of their characteristic structures and properties.
7. Classify the following materials as either naturally-occurring or manufactured: polythene, coal, native copper, mild steel, timber, dimension stone, natural rubber, bone, leather, limestone, diatomaceous earth. Justify your answers.
8. What is an inorganic material? Briefly discuss the important features of any two groups of inorganic materials.
9. Outline the effects that tanning has on animal hide.
10. Explain the important differences between igneous and sedimentary rocks in terms of their modes of formation and characteristic structures. Use examples to illustrate your answer.
11. What are metamorphic rocks and why are they often termed secondary rocks?
12. What are pozzolanic materials?
13. What is the essential feature of a ceramic material?
14. Gold and silver often occur in their free states in nature while iron has to be extracted from its ores. How is this explainable in terms of the fundamental properties of these metals?

3

Structure of the Solid State

MATTER may exist in solid form, or as a liquid, or in the gaseous state, each of these particular forms being characterised by the arrangements and energies of the atomic particles present. When a gas is cooled, the atomic particles present coalesce into the random arrangement characteristic of liquids and upon further cooling the liquid solidifies, the atomic units arranging themselves in the ordered pattern characteristic of that particular solid.

In a gas or vapour the atomic particles are virtually free to move without hindrance since the kinetic energy is sufficiently large to overcome the short-range attractive forces between the atoms. The atomic particles are spaced widely apart and hence gases may be compressed readily since the actual volume of the atomic particles makes up a very small part of the actual volume of the gas. Diffusion rates are very high within gases since the atomic particles of the introduced gas can move freely through the huge spaces between the atomic particles of the gas already present.

In a liquid the atomic particles are closer together than they are in a gas, but this difference is one of degree rather than kind. Diffusion rates between liquids are lower than those between gases since the atomic particles are closer together and collisions become more frequent while the kinetic energies of the particles are themselves lower. However, since both liquids and gases are "mobile" in the sense that their atomic particles possess great freedom of movement they are collectively known as fluids.

The solid state is quite different from either the gaseous or the liquid state, mainly because the atomic particles are packed closely together in some orderly three-dimensional network in space. The kinetic energies are lowered to the extent that movements are restricted to regular vibrations about fixed mean positions and diffusion rates are extremely slow. Nevertheless, diffusion processes in solid metals are of great significance since many metallurgical phenomena depend upon the rates of diffusion of the various atoms within the metal.

Many engineering materials are solid, and since all solids are made up of atoms, molecules, or other atomic particles, the study of the solid state must logically begin at the level of atomic structure.

TABLE 3.1
CHARACTERISTICS OF SOLIDS, LIQUIDS AND GASES

Solids	*Liquids*	*Gases*
Possess definite ordered arrangements	Random atomic arrangements	Random atomic arrangements
Atomic particles at low energies	Energy levels higher	Energy levels very high
Practically incompressible	Can be compressed to some extent	Readily compressible
Diffusion rates very slow	Faster diffusion rates	Diffusion rates extremely rapid
Possess definite shape and volume	Take shape of container but volume remains constant	Take shape and volume of container
All true solids are crystalline	Non-crystalline	Non-crystalline

ATOMIC STRUCTURE

Speculations as to the nature of matter can be traced back to ancient times, but the theory proposed in 1808 by John Dalton, a Scottish chemist, can rightfully be termed the beginnings of the atomic model of matter. Dalton saw the atom as the ultimate particle, as a hard, spherical and indivisible mass exhibiting all of the characteristics and properties of the material which was formed when a large number of similar atoms were bundled together. This theory has been greatly modified and extended by such scientists as Thomson, Rutherford, Bohr and Chadwick so that today the concept of the atom as an indivisible particle is known to be false. Many of the properties of materials are best understood in terms of the structure of the atom and the ways in which atoms join together to form molecules, crystals and macro-units of matter.

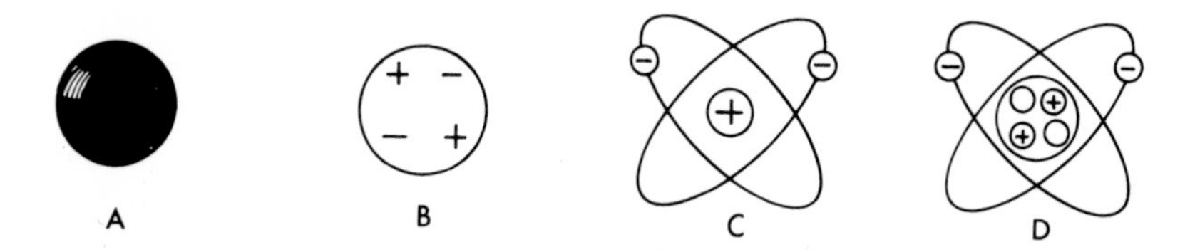

Figure 3.1 *The four figures represent successive stages in the development of the theory of the atom. (A) Dalton's concept of the atom (1808), a hard indivisible ball; (B) Thompson's model (1900), a number of electrons embedded in a sphere of positive charges; (C) Rutherford's model (1912), negative electrons orbiting a positive nucleus; (D) the Bohr-Chadwick model (1932), electrons orbiting a nucleus of protons and neutrons.*

The modern concept of the atom is that of an essentially electrical structure having a diameter of the order of one Angstrom unit ($1\text{Å} = 10^{-8}$ cm) made up of smaller particles, the principal ones being electrons, protons and neutrons. For our purposes the atom can be seen as consisting of two main parts, the heavy nucleus composed of protons and neutrons and the surrounding, highly-structured configuration of electrons. Both are equally important in the explanation of the properties of engineering materials.

Protons possess unit positive electrical charges* and have unit mass†, and exist, usually together with neutrons, in the nucleus of the atom.

Neutrons possess no electrical charge but are 1.008 times as heavy as protons; generally speaking, the larger the size of an atom, the higher the neutron-proton ratio in its nucleus. Both protons and neutrons may be termed nucleons since they comprise the nucleus of an atom.

Electrons exist in shells or orbitals surrounding the nucleus and at relatively great distances from it; an electron possesses only 1/1836th the mass of a proton and has a negative electrical charge equal in magnitude to the positive charge of a proton.

Since the normal atom is electrically neutral, the number of protons in the nucleus must equal the number of electrons in the surrounding orbitals. The nucleus of an atom accounts for almost all of the mass of that atom, while the diameter of the atom tends to be of the order of 10,000 times greater than that of its nucleus. The Atomic Number (Z) of an element is given by the number of protons (or electrons) possessed by a normal atom of that element, while the Mass Number (A) equals the sum of the number of protons plus neutrons. The Mass Number should not be confused with the Atomic Weight of an element, the latter being the weight of an atom of an element as compared to the weight of the oxygen atom which is taken as sixteen. However, there is usually very little difference between the Mass Number and the Atomic Weight of an atom, since the total number of protons plus neutrons accounts for practically all of the mass present.

The electrons, particularly those in the outermost orbitals (the valency electrons) determine many engineering properties of materials, such as chemical reactivity, binding patterns established with other atoms and thus strength characteristics, electrical properties, and optical characteristics.

The Bohr Atom

Bohr pictured the electrons of an atom rotating in circular orbits about the nucleus, each electron neither gaining nor losing energy while it remained in a particular orbit. He further postulated that if an electron jumped from one orbit to another, it would lose or gain a certain amount of energy. This

*1.602×10^{-19} coulombs
†1.673×10^{-27} kilograms

theory, providing as it does a very simplified model of the atom, suffers from many inaccuracies, most of which are brought about by the mechanical nature of the atomic model itself. The modern view is to regard electron orbits as somewhat vague regions in the space surrounding the nucleus in which there is a certain probability of finding a certain electron.

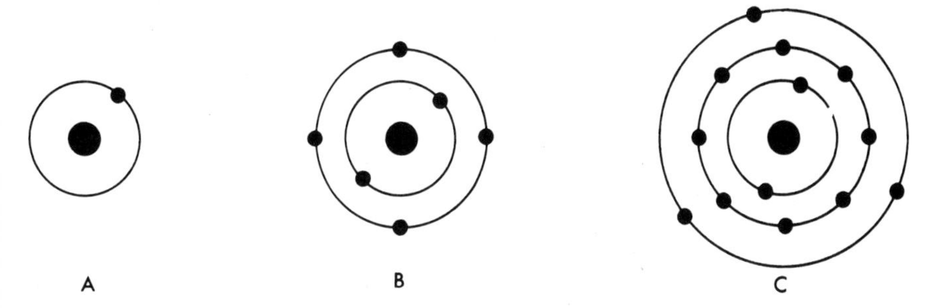

Figure 3.2 *Simplified two-dimensional representations of the atoms of (A) hydrogen, (B) carbon, and (C) sodium, according to the Bohr theory.*

Quantum Numbers

Each electron in a given atom possesses a definite level of energy, this being defined by its four quantum numbers: (1) the principal quantum number; (2) the azimuthal quantum number; (3) the magnetic quantum number; (4) the spin quantum number. Two of these, the principal quantum number and the azimuthal quantum number, serve to locate the position of the electron in the atom, the principal quantum number indicating the electron shell and the azimuthal quantum number the sub-orbital position within the shell. Electron shells are designated by the letters K, L, M, N, O, P and Q, these having principal quantum numbers of 1, 2, 3, 4, 5, 6 and 7 respectively, while the sub-orbital are designated by the letters s, p, d and f which correspond to the azimuthal quantum numbers 0, 1, 2 and 3 respectively. Table 3.2 shows the electronic distributions for periods I to IV of the Periodic Table.

CHEMICAL PERIODICITY

A repetitive pattern or periodicity exists with respect to the chemical and physical properties of all known elements and on this basis the known elements have been arranged in the Periodic Table in order of increasing atomic numbers. (The Periodic Table is printed in the front of the book.) Each vertical column within this table is numbered from one to eight and is called a Group, while each of the seven horizontal rows is known as a Period. Note that sub-Groups exist, for instance Groups Ia and Ib. The

TABLE 3.2

ELECTRONIC CONFIGURATIONS OF ELEMENTS IN PERIODS I TO IV

	n		*K* 1	*L* 2		*M* 3			*N* 4	
	l		*s*	*s*	*p*	*s*	*p*	*d*	*s*	*p*
1st PERIOD	1	H	1							
	2	He	2							
2nd PERIOD	3	Li	2	1						
	4	Be	2	2						
	5	B	2	2	1					
	6	C	2	2	2					
	7	N	2	2	3					
	8	O	2	2	4					
	9	F	2	2	5					
	10	Ne	2	2	6					
3rd PERIOD	11	Na	2	2	6	1				
	12	Mg	2	2	6	2				
	13	Al	2	2	6	2	1			
	14	Si	2	2	6	2	2			
	15	P	2	2	6	2	3			
	16	S	2	2	6	2	4			
	17	Cl	2	2	6	2	5			
	18	A	2	2	6	2	6			
4th PERIOD	19	K	2	2	6	2	6		1	
	20	Ca	2	2	6	2	6		2	
Transition Elements	21	Sc	2	2	6	2	6	1	2	
	22	Ti	2	2	6	2	6	2	2	
	23	V	2	2	6	2	6	3	2	
	24	Cr	2	2	6	2	6	5	1	
	25	Mn	2	2	6	2	6	5	2	
	26	Fe	2	2	6	2	6	6	2	
	27	Co	2	2	6	2	6	7	2	
	28	Ni	2	2	6	2	6	8	2	
	29	Cu	2	2	6	2	6	10	1	
	30	Zn	2	2	6	2	6	10	2	
	31	Ga	2	2	6	2	6	10	2	1
	32	Ge	2	2	6	2	6	10	2	2
	33	As	2	2	6	2	6	10	2	3
	34	Se	2	2	6	2	6	10	2	4
	35	Br	2	2	6	2	6	10	2	5
	36	Kr	2	2	6	2	6	10	2	6

Periodic Table can be conveniently divided into four distinct zones as shown in Figure 3.3.

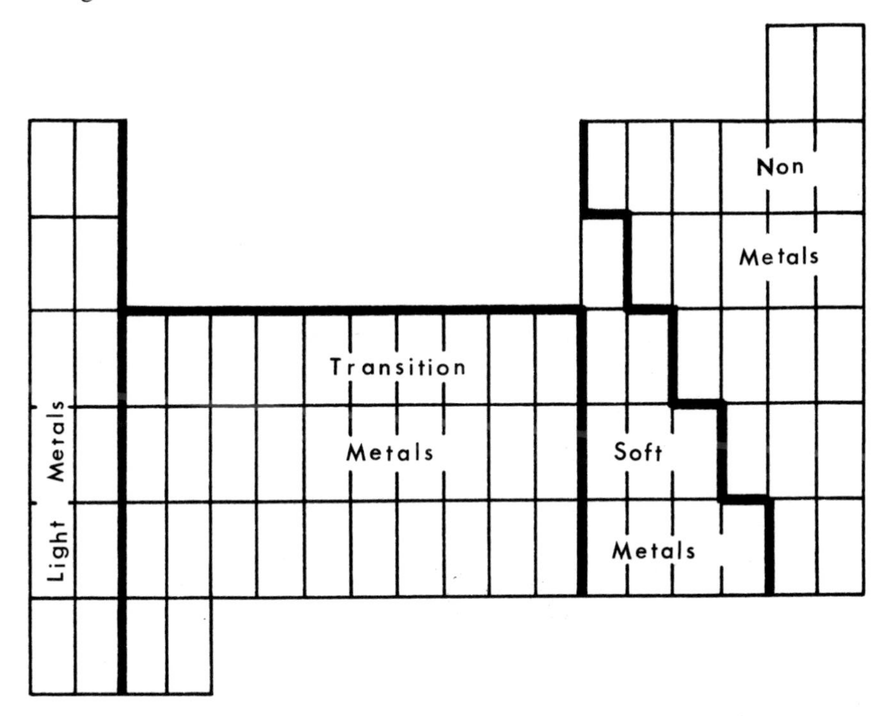

Figure 3.3 *The elements arranged in zones within the Periodic Table.*

(1) *The Non-Metals* are the extreme right-hand group of elements in the Table and include the inert gases. Silicon, arsenic and tellurium belong to that sub-group known as the "half-metals" since they exhibit some of the properties of metals and some of the properties of non-metals.

(2) *The Light Metals* occupy the two groups on the extreme left of the Table and are characterised by low densities, low melting points, high reactivity in chemical reactions and a tendency to form ionic compounds. This zone is made up of the alkali metals (Group I) and the alkaline earths (Group II), these having valencies of +1 and +2 respectively.

(3) *The Transition Metals* occupy the central portion of the Table and possess three general characteristics that distinguish them as a group. They are all high melting-point materials, they usually form coloured compounds and most of them are capable of forming compounds in several different valencies (iron +2 and +3 forming *Fe*II and *Fe*III compounds). Most of the traditional engineering metals come from this zone in the Table, and they all possess relatively high densities.

(4) *The Soft Metals* fall between the transition metals and the non-metals, and the properties of some grade out into those of these two classes of metals. Generally speaking, the metals of this zone are comparatively soft and unreactive, possess low melting points, and some form groups of compounds in valency groups separated by 2 (tin, for instance, forms *Sn*II and *Sn*IV compounds).

CHEMICAL BONDING

The arrangement of atoms in any solid material is determined by the character, strength and directionality of the binding forces, cohesive forces, or chemical bonds present within the material. It should be noted that bonding exists not only between atoms but also between other atomic units such as ions, radicles, and molecules if such are present in the material. Since the particular bonding pattern within a material plays a large part in determining the physical, chemical, and electrical properties of that material, the engineer and metallurgist must possess a sound working knowledge of the types of bonding patterns found in engineering materials. This study can be pursued either qualitatively or quantitatively, both approaches making significant contributions.

Primary Bonds

At this qualitative or descriptive level the nature, strength, and directionality of chemical bonds can be explained by reference to the movements and locations of the electrons involved. Chemical bonds are classified as either primary or secondary depending upon the degree of electron involvement, a primary bond being formed only if there is a pronounced lowering of the energies of the electrons involved during the formation of the bond. Since a primary bond does involve a substantial lowering of the energies of the bond electrons, it follows that such a bond must be stronger and more stable than secondary bonds.

Three limiting types of primary bonds are distinguished on the basis of the positions assumed by the bond electrons during the formation of the bond. These three limiting types are the ionic, covalent and metallic bonds, and are much more clear-cut than the different types of secondary bonds, most of which result from intermolecular (dipole) attractions

An ionic bond is really the attractive force existing between a positive and a negative ion when they are brought into close proximity, these ions being formed when the atoms involved lose or gain electrons in order to stabilise their outer shell electron configurations. Elements are classified as either electropositive or electronegative, depending upon whether they tend to lose or gain electrons in order to achieve this stable outer-shell configuration. Ionisation energies or potentials measure the amounts of energy required to successively remove the outer electrons from an atom; the first ionisation potential being the amount of energy required to remove the outermost electron, the second ionisation potential the second outermost electron, and so on. Reference to the first ionisation potential of various elements shows that in general it requires less energy to strip one electron from a metal atom than from a non-metallic atom, this rule also applying to the removal of the second and subsequent electrons in most cases. Thus, metals are termed electropositive while non-metals are termed electronegative, these terms referring to the abilities of the elements to form positive and negative ions respectively under favourable conditions during chemical reactions.

TABLE 3.3

FIRST IONISATION ENERGIES OF SOME ELEMENTS (kcal g-atom^{-1})

Metals		*Non-Metals*	
Sodium	118	Oxygen	314
Aluminium	138	Nitrogen	335
Nickel	176	Hydrogen	313
Iron	181	Carbon	260
Copper	178		
Zinc	217		

The ionic bond formed in sodium chloride is typical; the bond itself is non-directional since each positive ion attracts all neighbouring negative ions and vice versa. There is no actual molecule $NaCl$; the solid is an aggregate of Na^+ and Cl^- ions and is electrically neutral since the numbers of each ion are equal. The bond formed is strong since the bond electrons have their energies lowered considerably.

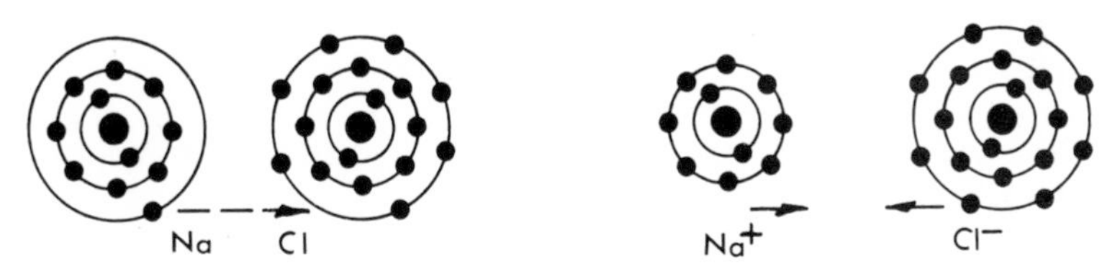

Figure 3.4 *Electron transfer precedes the formation of the ionic bond within sodium chloride.*

A covalent bond is formed when pairs of electrons are shared by several atoms and have their energies lowered as a result of this. Stable covalent bonds are formed between many non-metallic elements since the atoms of these elements usually possess half-filled outer electron shells which resist the direct electron transfer required for the formation of an ionic bond.

In line with the modern ideas involving probability distributions of electrons, the covalent bond may be considered to be a region of high electron density existing between two atoms.

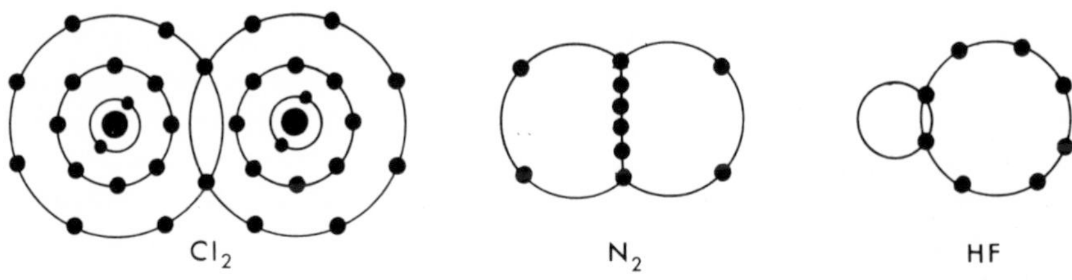

Figure 3.5 *The natures of the covalent bonds in the molecules of chlorine, nitrogen, and hydrogen fluoride. The bonds in the Cl_2 and HF are single bonds; the bond in the N_2 is a triple bond.*

Because pairs of electrons are shared in covalent bonds there is some degree of orbital overlap between the bonded atoms. This overlap is limited by the electrostatic repulsion of the positively charged nuclei, the resultant shapes of the bond orbitals determining whether the bond will be directional or non-directional in space with respect to the nuclei (or ion cores). The greater the degree of orbital overlap the stronger the bond since extensive overlap means greatly lowered energy levels of the bond electrons.

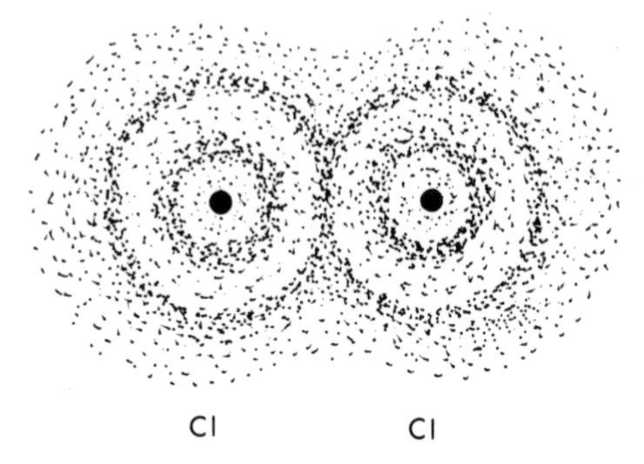

Figure 3.6 *The covalent bond in the Cl_2 molecule seen as a region of high electron density between the two atoms.*

The metallic bond is a far more complex model than either the ionic or covalent bonds, but has great importance in materials science since it predominates in metals.

At the simplest level this bond may be seen as the result of the individual metallic atoms losing their outer valency electrons which then exist in the metal as a cloud or "electron gas". The metallic bond is the electrostatic attraction between the fixed positive ion cores of the atoms and this electron cloud. A metallic bond, thus conceived, can exist only between a large aggregate of metallic atoms and must therefore be non-directional. This type of bond causes a significant lowering of the energy levels of the electrons involved since they are in close proximity to several ion cores at all times. Another view of the metallic bond is to regard it as a time-averaged, fluctuating covalent bond brought about by atoms in close proximity alternately sharing and losing electrons in their outer orbitals.

In general, metals possessing a few loosely held valency electrons exhibit a strong degree of metallic bonding; for example, sodium, potassium, silver and gold. The transition metals, however, such as iron, nickel, and tungsten exhibit a fair degree of covalency due in part to their incomplete 'd' electron orbitals.

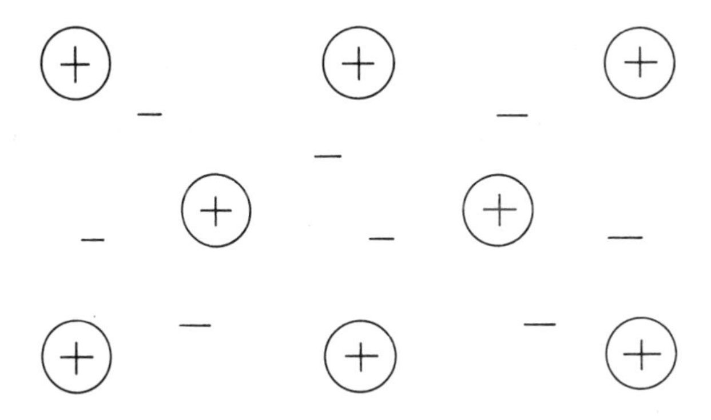

Figure 3.7 *A simple representation of the metallic bond. The positive ion cores of the metal atoms are fixed within a "sea" or "cloud" of electrons.*

Secondary Bonds

While primary bonds result from either a transfer of electrons (metallic, ionic) or from a sharing of electrons (covalent), secondary bonds do not generally involve the valency electrons at all. Secondary bonds are those comparatively weak intermolecular bonds formed as the result of dipole attractions, the dipoles forming as a result of the unbalanced distributions of electrons in asymetrical molecules. Dipoles may be permanent as in the hydrogen fluoride and water molecules, or they may be temporary, in which case the bonds are known as *van der Waals forces*. Although weak in comparison to primary bonds, secondary bonds often play a significant part in determining the structures and properties of many non-metallic materials.

The quantitative approach to chemical bonding involves the direct measurement of the strengths of bonds. This is usually done by breaking the bond, the amount of energy required for this being termed the bond energy. This procedure is not always feasible since bond energy determinations are done more readily when the material is in the gaseous state, and thus this approach to the study of chemical bonds has its limitations.

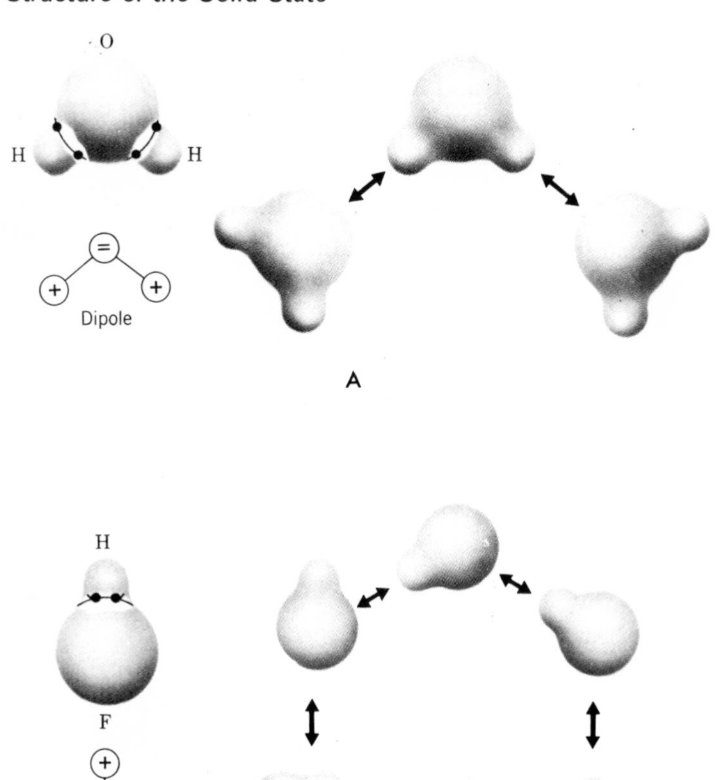

Figure 3.8 *Dipole forces originating in the asymmetry of molecules. (A) the water molecule, showing the dipole and the direction of attraction between the molecules; (B) the hydrogen fluoride molecule, showing the dipole and the direction of attraction between molecules. (Reproduced with permission from Volume I of "The Structure and Properties of Materials" by Moffatt, Pearsall, and Wulff. John Wiley and Sons Inc., N.Y. 1964.)*

THE RELATIONSHIP BETWEEN CHEMICAL BONDING AND THE PROPERTIES OF SOLID MATERIALS

It is useful to classify solid materials according to the types of basic units from which they are built and the nature of the chemical bonds which bind these units together. This classification reveals that all solid materials can be termed ionic, molecular, covalent (giant molecular) or metallic, the characteristics and properties of these four groups being set out in Table 3.4.

TABLE 3.4

Type of Solid	*Crystal* Units*	*Binding Force*	*Optical*	*Electrical*	*Thermal*	*Mechanical*	*Examples*
Ionic	Simple and complex ions	Electrostatic attraction of oppositely charged ions (the ionic bond)	Transparent or coloured by characteristic absorption by ions	Insulators, forming conducting solutions in ionising solvents	Fairly high melting to form ions	Hardness increases with ionic charge; break by cleavage	sodium chloride, calcite, magnesia
Molecular	Rare gas atoms; molecules	Dispersion and multi-pole forces (secondary bonds)	Transparent and like its molten form	Insulators; dissolve in non-ionising solvents	Fairly low melting points	Soft and plastically deformable	argon, paraffins, calomel ice solid CO_2
Covalent (giant molecular)	Group IV Elements; III-V and II-VI compounds	Covalent, sometimes with some ionic character	Transparent, high refractive index; or opaque	Semiconductors except diamond; insoluble	Very high melting	Very hard; break by cleavage	diamond carborundum rutile
Metallic	Positive ions and "free" electrons	Attraction between ions and electron gas (the metallic bond)	Opaque and reflecting	Electronic conductors; soluble in acids to form salts	Moderately high melting; good heat conductors	Tough and ductile except tungsten	copper iron sodium

*These "crystal units" are the "basic units" referred to above, and the particular arrangements of such units within crystalline solids is discussed later in this chapter.

Materials bonded ionically or covalently are non-conductors since their electrons are bound up tightly in the bonds themselves. Metals, however, are excellent conductors since the "electron cloud" within the metal can be readily directed by an applied potential difference. A material bonded entirely covalently will have high melting and boiling points and will be hard and strong. The outstanding example of this is diamond which is the hardest of all materials and which must be heated to above 6,000°C before vaporisation will occur. Materials bonded predominantly in a metallic manner are always opaque since impinging radiant energy is readily absorbed by the outer-shell electrons which move to higher energy levels as a result of this. As these electrons fall back to lower energy levels the excess energy is emitted as radiant energy, and thus metallic materials are lustrous unless obscured by a surface layer of oxide, carbonate, or other shielding material.

It is important to recall that many metals exhibit a significant degree of covalent bonding, thus reducing their conductivities; however, these are still classified as "metallic crystals". Crystals with molecules as their basic units are weak, soft and have low melting points because of the weak secondary bonds holding the molecules together, while both ionically and covalently bonded crystalline materials are non-conductors, and are hard, strong, high melting-point solids. Metallic crystals possess ductility because of the ability of the bond electrons to make and break bonds readily, while properties such as conductivity, lustre, strength and hardness depend upon the configurations of the atoms involved and particularly upon the numbers of electrons involved in bonding.

Laboratory (8): *Samples of solid sodium chloride, sodium chloride solution, carbon tetrachloride, copper, aluminium, lead, nylon, phenol-formaldehyde, alumina rod and a fired-clay ceramic are to be tested for—*

(a) electrical conductivity

(b) hardness, brittleness, lustre

(c) tendency of the solids to soften in boiling water or in a bunsen flame.

Results are to be tabulated and correlated with Table 3.4

THE CRYSTALLINE STATE

The vast majority of engineering materials, for instance all metals, many ceramics, some plastics, and over 99% of all minerals are crystalline, which means that they are characterised by particular three-dimensional atomic arrangements or repeating patterns known as lattices. Crystallography, the science of crystals and crystalline materials, offers two different approaches which are termed the "external" and the "internal" viewpoints. While the external viewpoint studies individual crystals with regard to their external shape and symmetry, the internal viewpoint is more fundamental, since it examines the arrangements of atomic units within crystals and crystalline materials, these largely determining external shape and symmetry.

A crystal may be thought of as an homogeneous solid of definite chemical composition which is bounded by plane faces, the arrangement of which is an expression of the internal order of the atomic units within the crystal. However, other factors, such as the physical and chemical environments in which the crystal was formed, play an important part in determining external shape and symmetry, so that the one material may exist in several different crystal forms or modifications.

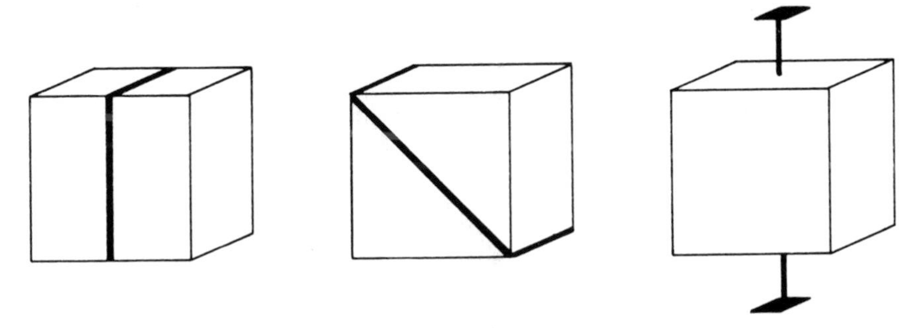

Figure 3.9 *Two planes of symmetry and a tetrad axis of symmetry in a simple cubic crystal.*

Crystal Symmetry

The detailed study of individual crystals relies very heavily on symmetry, which may be defined as the regular arrangement or repetition in distinct numbers of equivalent constituent parts of a whole. In describing the symmetry of crystals three separate and distinct symmetry operations need to be examined, these being:

(a) the symmetry plane, which divides a crystal into two similar and similarly situated halves or sections, so that one becomes the "mirror image" of the other. It is, however, only in an "ideal" crystal that these faces are of exactly the same size.

(b) the symmetry axis, which causes the crystal to occupy more than one congruent position during rotation about that axis during a 360° rotation. Crystals may possess diad, triad, tetrad or hexad axes of symmetry, these giving two, three, four and six congruent positions respectively.

(c) the centre of symmetry, which is some point within the crystal about which crystallographically similar faces are arranged in parallel and corresponding positions.

While the centre of a cube is a centre of symmetry, a tetrahedron has no such centre. The vertical or c-axis of a hexagonal bipyramid is a hexad (six-fold) axis of symmetry; the principal axes of a cube are tetrad (four-fold) axes, but the body diagonal axes are triad (three-fold). In fact, the cube has very high symmetry, possessing a centre of symmetry, nine planes of symmetry, three tetrad axes, four triad axes, and six diad axes; twenty-three symmetry elements in all.

Crystal Systems and Classes

Long before X-ray analysis made possible the detailed study of internal structure or symmetry, crystallographers had classified all known crystalline substances into systems and classes according to the external symmetry of individual well-formed crystals. There are thirty-two individual crystal classes into which the vast array of all known crystalline materials are classified, one class possessing no external symmetry at all. These thirty-two crystal classes are grouped into the seven crystal systems, with external symmetry again being the basis of classification.

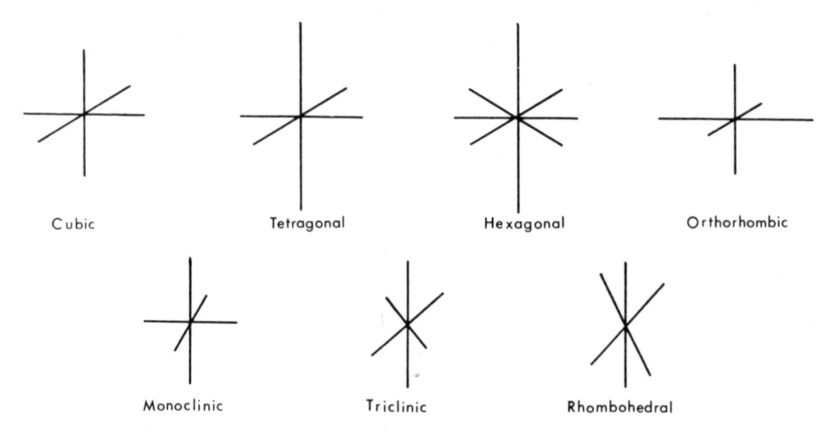

Figure 3.10 *The axial crosses of the seven crystal systems.*

The seven crystal systems are best defined in terms of their patterns of crystallographic axes, the two parameters involved being axial lengths and interaxial angles. The crystallographic axes are chosen so as to be closely related to symmetry planes or axes, being usually parallel with or normal to symmetry planes or axes, or so constructed that they bisect the angles between several symmetry planes or axes.

It should be noted that several possible variations can occur within this classification. Firstly, it is possible to designate the Rhombohedral system as the Trigonal system and to regard it as having a set of four axes similar to those of the Hexagonal system. However, this offers little if any advantage over the system as proposed in Table 3.5, while the second variation, which is to regard this system as a special class within the Hexagonal system, is also unworthy of serious consideration because of differences in symmetry patterns. The third point to be noted is that the name Orthorhombic is often shortened to Rhombic, but this offers no real advantage and is undesirable.

TABLE 3.5

GEOMETRY OF THE CRYSTAL SYSTEMS

	Axial Parameters	
Systems	*Lengths*	*Angles*
Cubic	Three equal axes $a_1 = a_2 = a_3$	All angles $= 90°$
Tetragonal	Three axes, two of equal length $a = b \neq c$	All angles $= 90°$
Orthorhombic	Three axes, all unequal $a \neq b \neq c$	All angles $= 90°$
Monoclinic	Three axes of any lengths $a \neq b \neq c$	Two angles $= 90°$ One angle $\neq 90°$
Triclinic	Three axes of any lengths $a \neq b \neq c$	All angles different and none equal 90°
Hexagonal	Three coplanar equal axes, and a fourth axis (at 90° to these three) of different length $a_1 = a_2 = a_3 \neq c$	Coplanar axes spaced 120° apart, c-axis perpendicular to them
Rhombohedral	Three equal axes $a = b = c$	All angles equal but not 90°

Laboratory (9): *Determine the symmetry elements present in simple crystal models such as the cube, hexagonal prism, tetrahedron, and square prism. Tabulate your result and sketch the crystal models so that the position of the symmetry elements are shown.*

Miller Indices

Crystallographers and others interested in crystal structure need some method whereby the angular position of any given crystal face or plane can be described with respect to the particular crystallographic axes of the crystal. Miller indices provide one means of achieving this end, but the introduction of a reference or parametral plane for each crystal is at first glance a complicating factor. A parametral face or plane must be selected so that it is not parallel to any crystallographic axis; in actual fact the parametral plane must intersect all crystallographic axes.

Suppose that the selected parametral plane makes intercepts of x, y, and z units on the three crystallographic axes a, b, and c (the hexagonal system will not be treated at this stage). Miller indices may now be determined by carrying out the following operations:

(1) determine the intercepts of the face or plane in question in terms of whole numbers which are multiples or submultiples of the parametral intercepts x, y and z;

(2) take reciprocals of these numbers and reduce them to their smallest possible values by dividing throughout by the largest possible common factor;

(3) Enclose these numbers in brackets without commas in between e.g. (hkl)

Thus for the parametral plane the procedure becomes:

	a-axis	b-axis	c-axis
Expression in terms of parametral intercepts	$\frac{x}{x}$	$\frac{y}{y}$	$\frac{z}{z}$
Simplifying fractions	1	1	1
Reciprocals	$\frac{1}{1}$	$\frac{1}{1}$	$\frac{1}{1}$
Miller indices	1	1	1

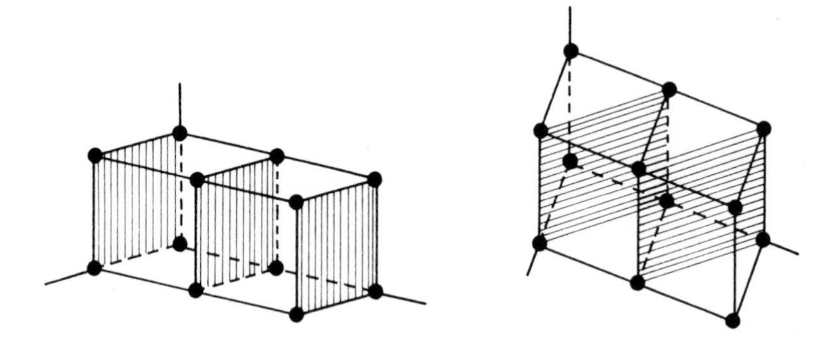

Figure 3.11 *The (010) and (110) planes in the primitive cubic lattice.*

Thus, it is quite obvious that the parametral plane for a crystal belonging to any system having three crystallographic axes will be (111). This procedure will also assign the same indices to any number of parallel planes within a crystal, and negative indices, written with the negative sign above the indices (hkl) are possible. The use of reciprocals means that a plane parallel to a crystallographic axis, and thus making an indefinite intercept on it, has an index of zero with respect to that particular axis.

A modified form of Miller indices, known as Miller-Bravais indices, are used to designate orientations of crystal planes in the hexagonal system; four indices are given, with the parametral plane being (1111). The basal plane of a normal hexagonal prism is (0001), while the six sides are the open form* {1010} .

Laboratory (10): *Solve the following problem: A crystal fragment of a cubic mineral was studied and the intercepts of four well-defined crystal faces were found to be as follows—*

Axial Intercepts

a_1	a_2	a_3
0.257	1.0	0.211
–0.257	1.0	∞
∞	3.0	0.105
0.257	∞	∞

Determine Miller indices for these planes, given that the first plane is to be considered as the parametral plane.

SPACE LATTICES—THE INTERNAL APPROACH

A space lattice depicts the internal symmetry existing within a crystal and and may be said to be a network of straight lines drawn in space such that space is divided into equal volumes. The lattice points formed by the intersections of these lines all possess identical surroundings, and are more significant than the lattice itself. The lattice points may be occupied by single atoms as in the case of metals, by ions as in the case of ionic crystals, or by molecules as in molecular crystals such as solid carbon dioxide.

At first glance it may appear that many space lattices could be constructed but in fact, as Bravais demonstrated in 1848, only fourteen are possible, these being now known as the fourteen Bravais Space Lattices. All lattices possess unit cells, a unit cell being defined as the smallest unit of a lattice which, when translated along the axes of the lattice, forms the lattice. A given lattice could have a number of different unit cells, but generally the one having the simplest structure and the least number of lattice points is selected.

Lattice parameters distinguish different crystalline solids having the same Bravais space lattice from one another since the lattice parameters, the unit lengths on the unit cells and the angles between the lattice lines, are different for each crystalline substance. In other words, each and every crystalline substance will have its own characteristic set of lattice parameters.

*A crystallographic form is a group of faces having like relationships to the crystallographic axes; for example the cube is the form {100}, the like faces being (100), (010), (001), ($\bar{1}$00), (0$\bar{1}$0), and (00$\bar{1}$).

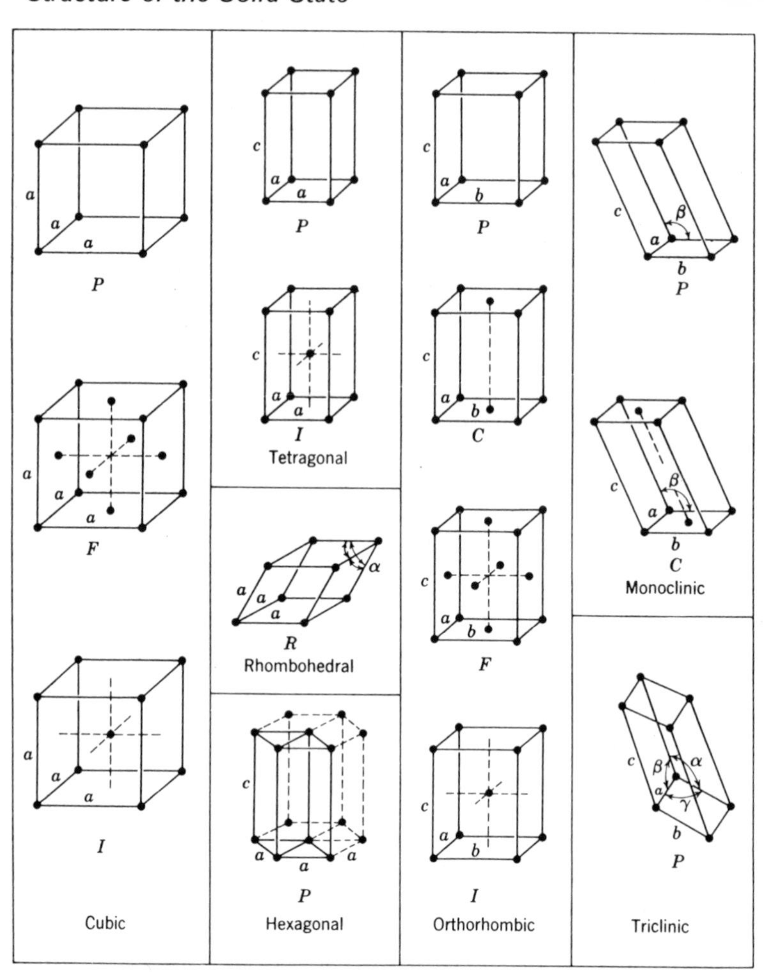

Figure 3.12 *Conventional unit cells of the fourteen Bravais space lattices. The capital letters refer to the type of cell: P, primitive cell; C, cell with a lattice point in the centre of two parallel faces; F, cell with a lattice point in the centre of each face; I, cell with a lattice point in the centre of the interior; R, rhombohedral primitive cell. All points indicated are lattice points. There is no general agreement on the unit cell to use for the hexagonal lattice; some prefer the P cell shown with solid lines, and others prefer the C cell shown with dashed lines. (Reproduced with permission from Volume I of "The Structure and Properties of Materials" by Moffatt, Pearsall, and Wulff. John Wiley and Sons Inc., N.Y. 1964.)*

Lattice co-ordinates describe the positions of points on or within a unit cell, the origin being taken as one corner of the cell. For instance, the corner lattice points of a body-centred cubic cell are designated (000) while the central lattice point becomes ($\frac{1}{2}\,\frac{1}{2}\,\frac{1}{2}$) if the unit lengths along the edges of the cell are taken as 1. The face-centred cubic lattice has the following points within its unit cell — (000), ($\frac{1}{2}$ $\frac{1}{2}$ 0), ($\frac{1}{2}$ 0 $\frac{1}{2}$) and (0 $\frac{1}{2}$ $\frac{1}{2}$), and contains four atoms per cell as compared to two for the body-centred cubic cell.

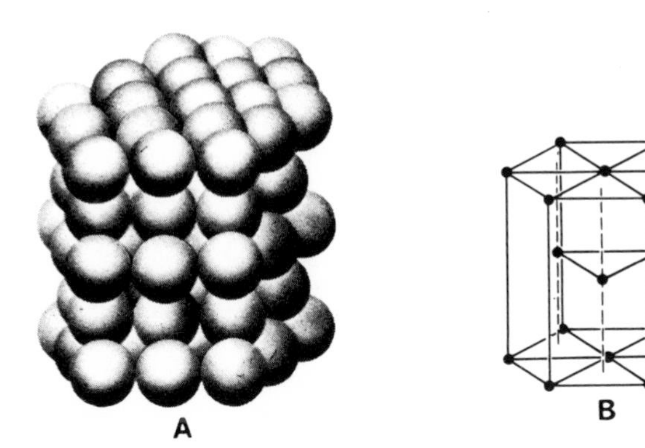

Figure 3.13 *The HCP structure. (A) shows the structure as an arrangement of the atoms in close-packed atomic planes; (B) shows the conventional unit cell.*

Co-ordination Numbers and Close-Packed Structures

Many metals have crystalline structures that are best termed close-packed or closest-packed, since the atoms are packed together in the lattice so that a minimum of "free space" occurs in the lattice. The co-ordination number is an indication of the closeness of atomic packing in crystals since it is found by counting the number of equivalent equidistant near-neighbours possessed by any atomic particle (atom, ion, molecule) occupying a lattice point. If atoms are represented by hard solid spheres then "layers" of spheres can be built in which each sphere touches six others. If several of these layers are then fitted together in various ways, it is found that there are two possible arrangements that allow any one atom to touch twelve others. These two close-packed structures, known as the face-centred cubic (FCC) and hexagonal close-packed (HCP) lattices, have co-ordination numbers of twelve. It should be noted that the internal atom sites in the hexagonal close-packed unit cell are not Bravais lattice points; thus this structure is not included with the Bravais space lattices.

The body-centred cubic structure is not closest-packed since it has a co-ordination number of eight, and consequently this crystal structure contains more "free space" than either of the previously discussed closest-packed structures. Taken collectively these three structures account for about three quarters of all metals. Some common metals crystallising in these systems at room temperature are:

BCC: chromium, iron, molybdenum, tantalum, tungsten, vanadium, niobium.

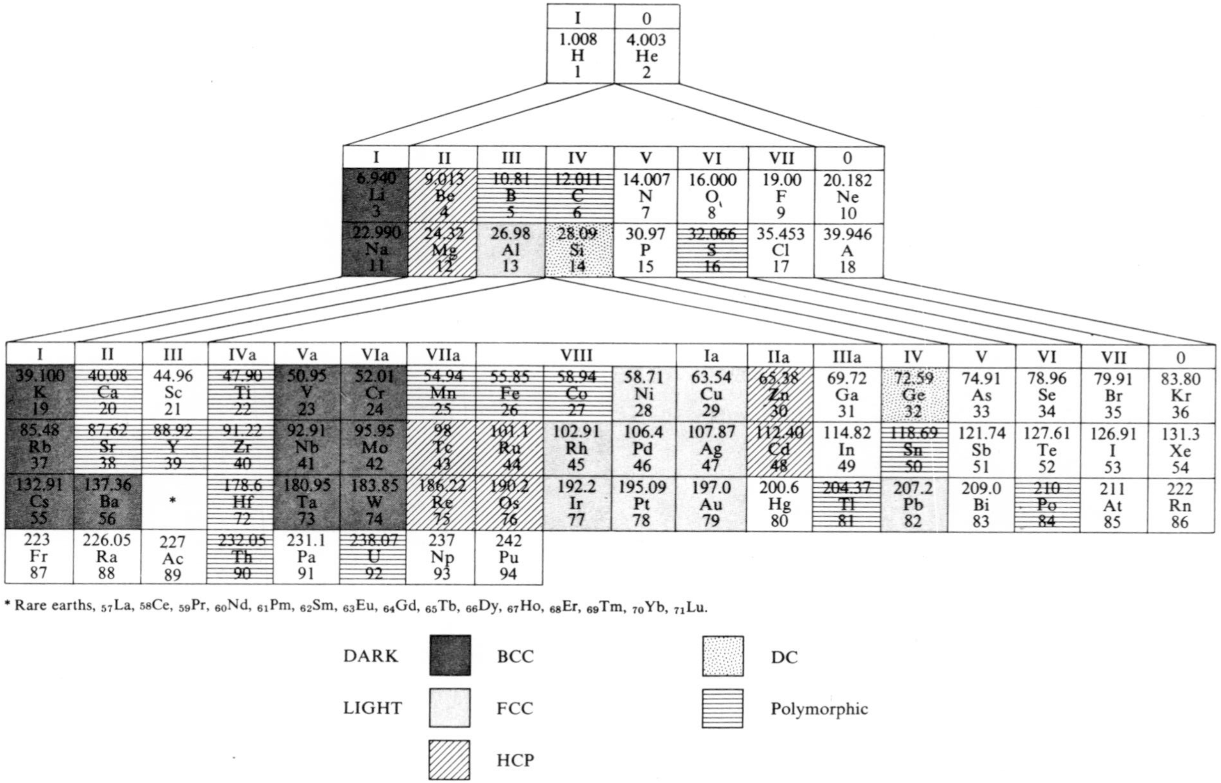

I	0
1.008 H 1	4.003 He 2

I	II	III	IV	V	VI	VII	0
6.940 Li 3	9.013 Be 4	10.81 B 5	12.011 C 6	14.007 N 7	16.000 O 8	19.00 F 9	20.182 Ne 10
22.990 Na 11	24.32 Mg 12	26.98 Al 13	28.09 Si 14	30.97 P 15	32.066 S 16	35.453 Cl 17	39.946 A 18

I	II	III	IVa	Va	VIa	VIIa	VIII			Ia	IIa	IIIa	IV	V	VI	VII	0
39.100 K 19	40.08 Ca 20	44.96 Sc 21	47.90 Ti 22	50.95 V 23	52.01 Cr 24	54.94 Mn 25	55.85 Fe 26	58.94 Co 27	58.71 Ni 28	63.54 Cu 29	65.38 Zn 30	69.72 Ga 31	72.59 Ge 32	74.91 As 33	78.96 Se 34	79.91 Br 35	83.80 Kr 36
85.48 Rb 37	87.62 Sr 38	88.92 Y 39	91.22 Zr 40	92.91 Nb 41	95.95 Mo 42	98 Tc 43	101.1 Ru 44	102.91 Rh 45	106.4 Pd 46	107.87 Ag 47	112.40 Cd 48	114.82 In 49	118.69 Sn 50	121.74 Sb 51	127.61 Te 52	126.91 I 53	131.3 Xe 54
132.91 Cs 55	137.36 Ba 56	*	178.6 Hf 72	180.95 Ta 73	183.85 W 74	186.22 Re 75	190.2 Os 76	192.2 Ir 77	195.09 Pt 78	197.0 Au 79	200.6 Hg 80	204.37 Tl 81	207.2 Pb 82	209.0 Bi 83	210 Po 84	211 At 85	222 Rn 86
223 Fr 87	226.05 Ra 88	227 Ac 89	232.05 Th 90	231.1 Pa 91	238.07 U 92	237 Np 93	242 Pu 94										

* Rare earths, $_{57}$La, $_{58}$Ce, $_{59}$Pr, $_{60}$Nd, $_{61}$Pm, $_{62}$Sm, $_{63}$Eu, $_{64}$Gd, $_{65}$Tb, $_{66}$Dy, $_{67}$Ho, $_{68}$Er, $_{69}$Tm, $_{70}$Yb, $_{71}$Lu.

DARK BCC
LIGHT FCC
HCP
DC
Polymorphic

Figure 3.14 *Crystal structures of the elements. (Reproduced with permission from Volume I of "The Structure and Properties of Materials" by Moffatt, Pearsall, and Wulff. John Wiley and Sons Inc., N.Y. 1964.)*

FCC: aluminium, copper, gold, lead, nickel, palladium, platinum, rhodium, silver.
HCP: beryllium, cadmium, cobalt, magnesium, titanium, zinc, zirconium and (solid) mercury.

NON-CRYSTALLINE MATERIALS

Many apparently solid substances are non-crystalline by virtue of the fact that detailed X-ray analysis reveals that their internal structures are not based upon some regular lattice pattern. The crystalline state is the lowest energy-level state attainable by a material and hence non-crystalline materials are often unstable and will revert to crystalline forms under favourable conditions. An example of this is the reversion of rapidly-cooled "glassy" fused felspar to a crystalline form over a fairly long period at room temperature, the felspar becoming cloudy and finally opaque as this change goes to completion.

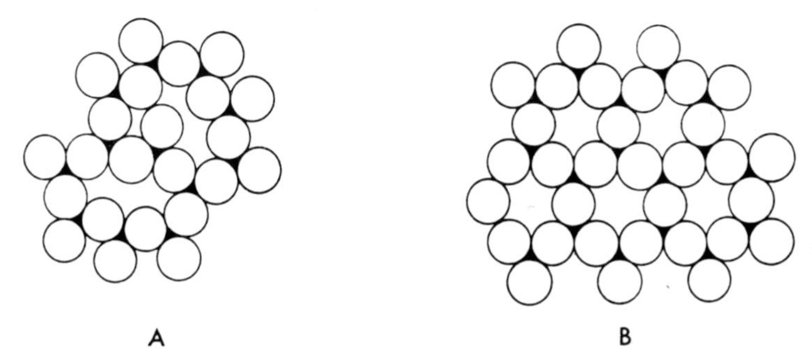

Figure 3.15 *Crystalline and non-crystalline forms of B_2O_3. (A) the glassy form shows only short range order within its sub-units; (B) the crystalline form reveals long range order within its crystalline structure. (Reproduced with permission from "Elements of Materials Science" 2nd Edition, by L. H. Van Vlack. Addison-Wesley Publishing Co., Reading, Mass. 1964.)*

Non-crystalline materials include all liquids and gases, glass and related materials, and most plastics. Some materials contain both crystalline and non-crystalline phases, clay-body ceramics being an example of considerable interest. All non-crystalline materials may be termed amorphous, this meaning literally "without form", and must be regarded as existing in disordered states. A significant feature of non-crystalline materials is that they do not reveal definite melting or freezing points or ranges; instead, they progressively soften and harden over an extremely wide temperature range. Such materials may be naturally-occurring or manufactured; obsidian (volcanic glass) being an instance of the former, and soda-lime ("window") glass an instance of the latter.

There has been a tendency to call non-crystalline, apparently solid materials (such as glass) supercooled liquids, but this is a misleading term since many supercooled liquids do not form glassy phases at all. Glassy materials are also termed vitreous. Non-crystalline plastics are formed because the long-chain molecules in their structure exist in tangled and disordered patterns. Those few plastics which can crystallise are usually prevented from doing so by the addition of fillers and plasticisers (see Chapter 15) since plasticity is lost to some extent when a polymer crystallises.

Many non-crystalline materials possess local order within their structural sub-units. Silica glass, for instance, contains a network of silicon-oxygen tetrahedra (SiO_4^{4-}), each tetrahedron being completely ordered in its structure. Similarly, plastics contain local order within their macro-molecules, but the molecular arrangements themselves are disordered and tangled. Graphite, the softer form of carbon, will be either crystalline or non-crystalline depending upon the arrangement and alignment existing between the atom layers.

Laboratory (11): *Examine the fracture of non-crystalline materials such as obsidian and ordinary soda-lime glass. Take a small sample of glass and attempt to melt it in a small crucible (you will need to get about 1250 °C at least). Note how the glass softens gradually and does not melt at a definite temperature.*

TABLE 3.6

STRUCTURE-PROPERTY RELATIONSHIPS IN METALS, CERAMICS, AND PLASTICS

Material	*Crystallinity*	*Bonding*	*Important Properties*
Metals	Crystalline	Predominantly metallic	Excellent conductors, high to moderate strengths, usually malleable and ductile to some degree, opaque, lustrous, moderate to high melting points.
Ceramics	(a) Crystalline	Ionic and covalent	Non-conductors, high melting points, brittle, dull to lustrous, hard.
	(b) Non-crystalline (vitreous)	Ionic and covalent	Non-conductors, brittle, no definite melting points, transparent to translucent, hard.
	(c) Crystalline and non-crystalline phases	Ionic and covalent	Very poor conductors or insulators, brittle, hard, low strengths, opaque, high melting ranges.
Polymers (plastics and rubbers)	Predominantly non-crystalline	Covalent plus secondary	Very poor conductors or insulators, dull and usually translucent to opaque, low strengths, deform plastically with little applied load, low softening temperatures.

STRUCTURE-PROPERTY RELATIONSHIPS

Definite relationships exist between the structure of a material and its properties, and some of the more important relationships are listed in Table 3.6.

It must be realised that there are many exceptions to the above table because of the extremely wide range of the properties encountered within each group of materials. There are, for instance, some extremely soft, low melting-point metals, some transparent plastics, some shock-resistant glasses, and some ceramics which are not particularly hard and which do not possess very high melting points.

Studying Structure at Various Levels

When considering structures at various levels it is useful to have some concept of relative sizes, and the so-called "logarithmic scale of the universe" provides one method of comparison. The dimensions given are approximations made by taking into account all of the principal dimensions of the entities in question; man, for instance, is certainly usually higher than one metre, but averaging his height, "thickness" and "width" give the approximation of one metre.

TABLE 3.7

RELATIVE SIZES WITHIN THE UNIVERSE—LOGARITHMIC SCALE

Object	*Average Size*	*Logarithmic Scale*
A galaxy	1 million light years	10^{21}
Distance to nearest star	1 light year	10^{15}
The sun	1 million kilometres	10^{9}
The earth	1000 kilometres	10^{6}
Man	1 metre	10^{1}
Sand grain	1 millimetre	10^{-3}
Dust particle	1 micron	10^{-6}
Molecule	1 Angstrom unit	10^{-9}
Atom	1000 Fermis	10^{-12}

Studies of the structures of materials are usually made at the following levels:

Level	*Unit Equivalent from Table* 3.7
(1) Macrostructure	millimetre (10^{-3})
(2) Microstructure	micron (10^{-6})
(3) Crystalline (Internal Studies)	Angstrom (10^{-9})
(4) Atomic	Fermi (10^{-15})

Macroscopic Investigation: Macroscopic investigation is concerned with the overall constitution of the material as seen with the naked eye or a hand lens, or by using a low-power microscope. Only gross features can be examined in this manner but such investigations are often very rewarding and may obviate the need, in particular instances, to use any of the more sophisticated techniques which require expensive and time-consuming material preparation. The fibrous structure of wood can readily be examined using a 10x lens, as can the porous structure of firebrick. With metals, macro-

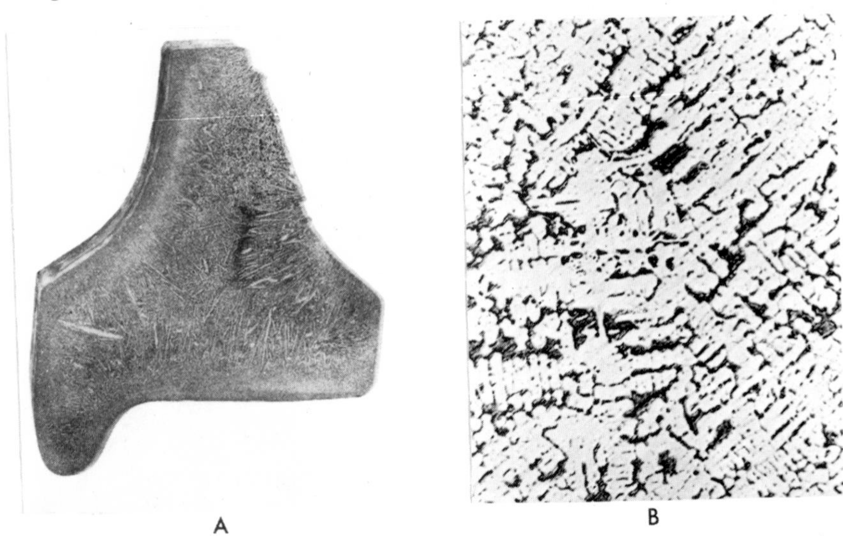

Figure 3.16 *Macro and micro examination. (A) shows a section of a porous casting that has been polished and heavily etched; dendrites are clearly visible. Magnification* $x\frac{1}{2}$. *(B) is a photomicrograph of an alloy of 70% copper and 30% lead. Since copper and lead are mutually insoluble in the solid state, they both appear in the micrograph, the copper being light and lead dark. ((A) Courtesy of the New South Wales Department of Railways. (B) Courtesy of The Copper and Brass Information Centre.)*

scopic studies can be made of such features as structural changes brought about by methods of forming and working, fractures, slag inclusions, porosity, non-metallic inclusions such as sulphide globules in cast steel, oversize or unusually large grains, and segregations. Rocks and minerals lend themselves to macro investigation with regard to their crystalline forms and constitution. Special preparation may be necessary in some instances but a lot of materials can be examined without this at the macro-level.

Laboratory (12): *Carry out macroscopic studies on the zinc coating on galvanised iron, various rocks, and some fracture surfaces in metals. Record your observations, and outline the limitations of this method of structural investigation.*

Microscopic Investigation: Microscopic examination reveals greater detail of the structure of a material but in almost every instance some sort of special specimen preparation and mounting is necessary. The range of magnification generally used is from fifty to one thousand diameters, with two thousand diameters being the effective upper limit since the wavelength of white light is too long to allow higher magnifications. Microscopes themselves are either of the "transmitted light" or "reflected light" types and may be monocular or stereoscopic. Thin sections of rocks, minerals and wood are studied under a transmitted light microscope after being mounted on glass slides in Canada balsam, while metals, because of their opacity, must be examined under reflected light.

Laboratory (13): *Etch the tin coating on tinplate and examine the surface macroscopically and microscopically. What features are visible at these levels of investigation?*

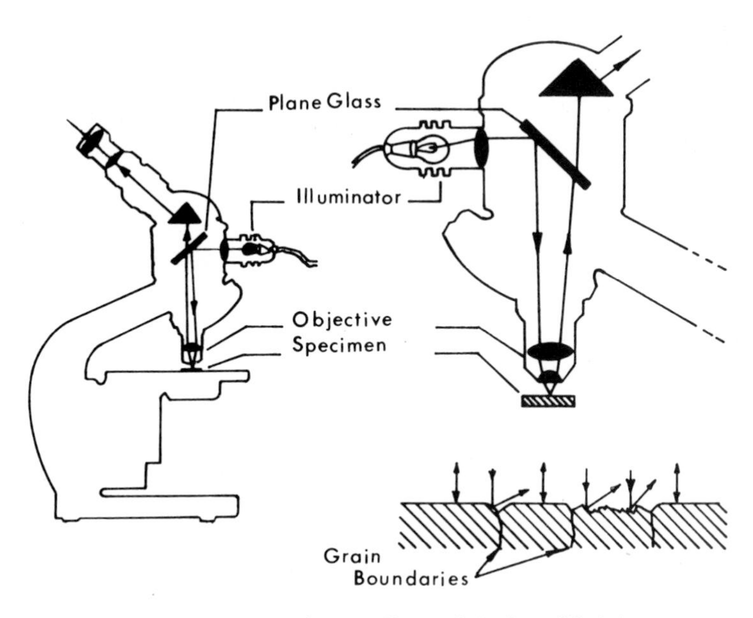

Figure 3.17 *The principle of the metallurgical (reflected light) microscope. Grain boundaries appear dark because of internal reflection of light rays.*

The Electron Microscope

It has already been stated that the optical microscope is limited to magnifications of some 2,000 diameters since detail size becomes comparable with the wavelength of light at just above this upper limit. The development of the electron microscope in the 1930's extended this range to about an upper limit of 75,000 diameters but presented some new problems as well. The principal of the electron microscope is that a "beam" of electrons of wavelength about 40,000 times less than that of white light is directed on to the surface of the specimen by an electro-magnetic focusing mechanism and then either passes through the specimen or is reflected if the incident beam is not normal to the surface. If an oblique beam is used, a reflection photograph is obtained when the reflected electrons impinge on a photographic plate, but such a photograph suffers from severe distortions except along its central area.

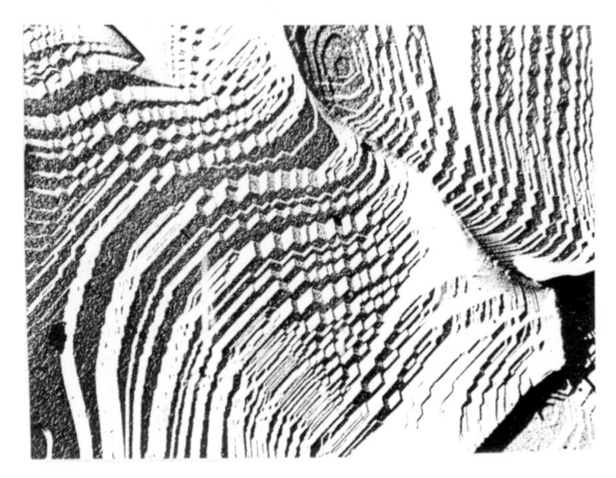

Figure 3.18 *An electron photomicrograph of a deeply etched grain of iron. The cubic crystal structure is clearly revealed. Magnification x10,000.*

Early researchers used the "replica technique" which overcame the problem of obtaining the actual specimens in thicknesses not greater than about 10^{-5} centimetres, the maximum thickness of metal that the electron beams could pass through. Replicas, faithful reproductions of the surfaces to be studied, are made of a thin layer of a suitable plastic, and transmission photographs can be reaily obtained of surface detail using this process. However, this was not altogether satisfactory from a number of viewpoints, and non-metallic inclusions were usually "lost" in the process of making the replica. Direct electron beam studies of extremely thin metal sections made by machining and chemical attack were begun in the 1950's, and using this technique internal structure could be studied directly.

X-Ray Analysis

Crystalline solids, because of their regularity of internal structures, are susceptible to structural analysis by means of various types of radiation. Practically our entire knowledge of crystal structure has evolved from such studies, which were begun in a very crude manner at the beginning of this century very shortly after Rontgen discovered X-rays. Electron beams are also used for structural analysis but electrons, being charged particles, can easily be deflected by other charged particles within the material and hence are commonly restricted to surface studies.

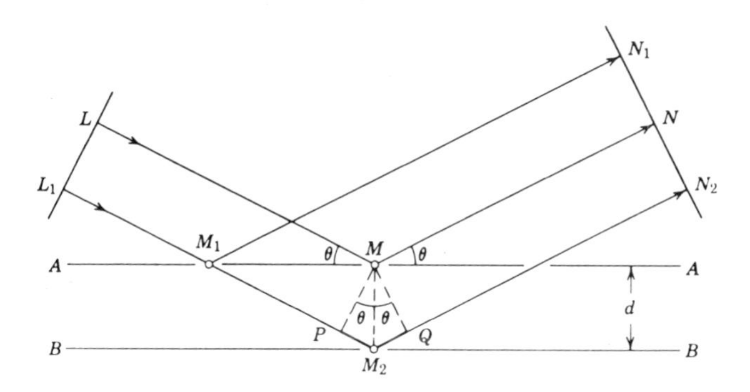

Figure 3.19 *X-ray diffraction—deriving the Bragg equation. (Reproduced with permission from "The Physics of Engineering Solids" 2nd Edition, by Hutchison and Baird, John Wiley and Sons Inc., N.Y. 1968.)*

X-rays are a form of electro-magnetic radiation, being produced when the rays emanating from the cathode in a cathode-ray tube are allowed to impinge on to a metal plate known as the anti-cathode. While the character of the X-rays so produced depends to some extent upon the composition of this metal plate, all X-rays behave somewhat similarly to light rays in that they can be reflected, refracted, diffracted and transmitted under suitable conditions. X-ray beams possess extremely short wavelengths, the range extending from about 0.1 to 1,000 Angstrom units. They are classified as "hard" or "soft" depending upon wavelength, "hard" X-rays having wavelengths between about 0.7 to 3.0 Angstrom units. "Hard" X-rays are most useful when studying crystal structure.

Diffraction studies are used to analyse crystalline structures. Since atoms are essentially electrical structures they will oscillate when irradiated with electro-magnetic waves, the atoms in effect absorbing some of this radiation and re-emitting it at the same frequency. This is the process of diffraction, and analysis of the patterns of diffracted radiation is usually based upon the Bragg equation which is derived from the assumption that particular atomic planes within a crystalline material will act as "reflecting" planes under favourable circumstances.

The parallel lines AA and BB in Figure 3.19 represent planes of atoms that "reflect" incident X-rays. The path difference for rays LMN and $L_1M_2N_2$, reflected from adjacent planes, is the length PM_2Q which is equal to $2d\sin\theta$, where d is the atomic spacing and θ the angle of incidence of the radiation. If this path difference is equal to an integral number of wavelengths (λ) the "reflected" rays from adjacent planes will be in phase and will reinforce one another. This relationship is expressed in the Bragg equation:

$$n\lambda = 2d \sin\theta$$

when: n = the order of the diffracted image (1, 2, 3, etc.)
λ = wavelength of incident radiation
d = interplanar spacing
θ = glancing angle or incident angle of X-radiation.

Problem: *A parallel beam of X-radiation of wavelength* 0.669 *Angstrom units is diffracted by a crystal, giving the first-order maximum when the incident angle is* 6°50′. *Calculate the interplanar spacings in the crystal.*

Solution: $n\lambda = 2d\sin\theta$

$$d = \frac{n\lambda}{2\sin\theta}$$

$$= \frac{1 \times 0.669}{2 \times 0.119}$$

$$= 2.81\ \text{Å}$$

X-Ray Diffraction Techniques

Many variations of basic techniques are possible, but generally speaking they can all be classified as either belonging to the Laue method, the rotating crystal method, or the powder technique. Of these, the latter is preferred in many instances since it reveals the lattice parameters accurately without requiring a single crystal of the material.

In the powder method, a sample of the material is ground into a fine powder which is then pressed and cemented into a thin spindle. While a narrow beam of monochromatic X-rays impinges upon the spindle, it is slowly revolved inside a specially constructed circular "camera". The strip of film used in this camera shows a series of curved lines, each of which represents "reflecting" plane within the crystalline material.

The distance "S" on the strip of film as shown in Figure 3.20 is related to the angle θ required for the Bragg equation by the formula

$$S = 4\theta R$$

where R is the known radius of the camera. Thus values for θ can be worked out, and then by substituting all known data into the Bragg equation the atomic spacings can be calculated quite readily.

X-ray diffraction techniques enable—

(a) The identification of elements and compounds, since the "powder patterns" are uniquely characteristic for every substance.

(b) The estimation of grainsize in polycrystalline aggregates, particularly when the average grainsize is less than 10^{-2} cm and consequently beyond the working range of the optical microscope.

(c) The determination of the orientation of a single crystal, the Laue technique being preferable for this. This can be important since many mechanical, optical and electrical properties are different in different crystallographic directions.

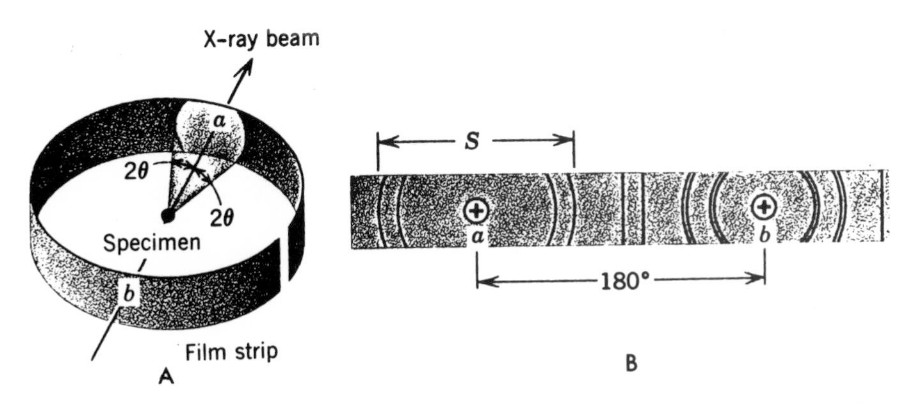

Figure 3.20 *Schematic diagram showing the method of obtaining a powder photograph; (A) shows the specimen and film set-up; (B) shows the resulting diffraction pattern recorded on the strip of film. (Reproduced with permission from "The Physics of Engineering Solids" 2nd Edition, by Hutchison and Baird. John Wiley and Sons Inc., N.Y. 1968.)*

Calculating Interplanar Spacings Using Lattice Parameters and Miller Indices

If the lattice parameters are known, it is possible to calculate the spacings of any given set of parallel planes within a crystal lattice given their Miller indices. For instance, if the length along the edges of the cubic unit cell is taken as "a" the spacing of planes of indices (hkl) would be

$$d = \frac{a}{\sqrt{h^2 + k^2 + l^2}}$$

Similar relationships exist for all other crystal systems.

Appendix to Chapter 3

PREPARATION OF METAL SPECIMENS FOR MACRO AND MICRO EXAMINATION

Mounting and Cutting of Specimens

Since very large or very small specimens are difficult to manipulate, metal specimens for micro examination should be kept within the limits of from $\frac{1}{4}''$ to $1''$ square or diameter with the thickness always less than the linear dimensions of the face to be polished. Small or very thin specimens must be mounted prior to preparation in some sort of thermosetting plastic which is usually made $\frac{3}{4}''$ to $1''$ in diameter and $\frac{1}{2}''$ thick. This may be done using a hot press, or a cold-setting resin may be employed. Specimens should be so selected that they are representative of the material; for example, both longitudinal and transverse sections will be required from rolled, extruded or drawn stock, and samples must be taken from many different parts of a casting.

Where possible, specimens should be cut out using a hacksaw which is lubricated with soap rather than oil, and minimum pressure should be exerted. Some specimens may require cutting or trimming by the lathe, shaping machine, or band saw, but care must be exercised to avoid extreme mechanical action which may distort or alter the structure within the metal. Cutting by a flame technique must never be used, and grinding, which may be necessary on hardened steels, must be carried out at as low a speed as possible using a copious supply of coolant. It must be remembered that even slight rises in temperature may cause alterations in structure, particularly in hardened steels.

Grinding and Polishing Specimens for Micro Examination

After the specimen has been cut or ground to size and mounted if necessary, a level surface must be prepared by careful filing. A fine file should be used and pressure kept to a minimum to prevent the formation of a deep "flow layer" which will resist the actions of both fine grinding and etching. A linisher may be employed for hard metals in which case a fine belt must be used together with copious amounts of water. The sharp corners should then be rounded off to prevent them digging into the abrasive papers used in the next stages of polishing.

Fine grinding is now begun using emery papers of various grades laid flat on $\frac{1}{4}''$ plate glass or on a surface plate. The specimen is held face downwards on the emery paper and rubbed backwards and forwards with as little pressure as possible until the surface is covered with an even pattern of fine scratches. The process is repeated on finer grades of paper, the specimen being turned through 90° at each change of paper and the polishing plate, paper, and specimen dusted free of grit. Wet grinding is now carried out using a similar technique, successively finer grades of silicon carbide papers of the "wet-and-dry" type being employed, with water or kerosene as the wetting agent. During each change to a finer grade the specimen, glass plate, and surrounding bench should be washed to remove grit from the previous grinding operation. A successful fine grinding sequence would be to use grades 1 and 0 emery paper, followed by 240, 320, 400 and 600 silicon carbide papers. For very soft metals, such as lead, glycerine should be used as the wetting agent on all grinding papers to prevent scratching.

The specimen is now ready for polishing which may be done by holding it against a slowly-rotating cloth pad impregnated with a suitable polishing powder in the form of a fine suspension in water. Hand polishing may be employed when a machine is not available, strips of leather (using the reverse side), serge, or other cloths stretched over wooden bases being suitable when impregnated with a polishing powder. Alumina, 600 grit or finer, is the most successful polishing powder for most metals; tin oxide is also good for most of the harder non-ferrous metals; while the grade of magnesium oxide known as "heavy calcined magnesia" is suitable for irons and steels. Diamond powder is most useful for the harder metals, particularly when machine polishing is carried out. A fine metal polish such as "Brasso" is satisfactory for copper alloys, while soap solution is used to polish lead and aluminium. Buffing must never be used since it develops a "flow layer" of metal that distorts the surface structure. After polishing has been completed the polishing cloth should be washed clean, and the specimen should be washed in water until all traces of the polishing agent are removed and then dried carefully by dabbing (*not rubbing*) with a very soft cloth. If commercial metal polish has been used, the specimen should be washed in acetone or benzene prior to washing in water.

Etching and its Effects

Etching involves the controlled attack by a mild reagent such as an acid solution which, by preferentially attacking different phases and their grain boundaries, reveals many details of the crystalline structure of the metal. If a normal acid etchant is to be used the most satisfactory procedure is to immerse the specimen, polished face downwards, in the etchant and gently agitate for the required time, after which the specimen should be quickly removed, washed thoroughly first under running water and then in a beaker of alcohol (methylated spirits is satisfactory), and finally dabbed dry on a soft cloth. The main difficulties associated with etching are those of (1) uneven attack, often caused by a greasy surface, or (2) staining, which is usually caused by a delay in the final washing and drying procedure, or by etchant oozing from pores and fissures after washing has been carried out.

Etching may also be done electrolytically if the specimen is made the anode in a suitable bath of electrolyte, the cathode being either a platinum or carbon electrode. Low voltages are used, the common range being from two to six volts. This gives a more even etch which tends to reveal more detail than a chemical etch.

It must be stressed that all specimens, whether being prepared for macro or micro examination, should be examined after the final grinding or polishing stage and prior to etching since many features, notably non-metallic inclusions such as graphite flakes in grey cast iron, are better studied while the metal is in its unetched condition.

Preparation of Specimens for Macro Examination

Specimens for macro examination are frequently polished and etched in order to bring out gross detail, the fine grinding, polishing and etching procedures being similar to those employed for micro examination. The degree of surface finish required is often less, however, and in many instances the final polishing processes can be omitted. Sectioned forgings and extrusions are commonly etched immediately after fine grinding on a diamond lap, the gross detail of grain structure being quite clearly revealed by this process.

Etchants for Macro Examination

(1) To reveal dendritic structures in steels

Copper ammonium chloride	= 9 g
Water	= 91 ml

Agitation is necessary, and the etching time may be up to several hours. A small amount of hydrochloric acid may be added after etching has begun, this increasing the contrast obtained. Deposited copper may be removed by light rubbing with cotton wool.

(2) Deep etch for steel

Hydrochloric acid (conc.)	= 140 ml
Sulphuric acid (conc.)	= 3 ml
Water	= 50 ml

Effective etching is obtained by from $\frac{1}{4}$ to $\frac{1}{2}$ hour etch at a temperature of 90°C. Polishing is not important, defects being revealed on fractured and sheared surfaces.

(3) Aluminium and alloys

Hydrofluoric acid	=	10 ml
Nitric acid	=	1 ml
Water	=	200 ml

Etching time is short, and the black film is easily removed by dipping into nitric acid.

(4) Copper and alloys

Nitric acid	=	45 ml
Potassium bichromate	=	50 ml
Water	=	0.2 g

Etching Reagents for Micro Examination

(1) Ferrous metals

(a) Nital: Nitric acid = 2 ml
Methyl Alcohol = 98 ml

(b) Picral: Saturated solution of picric acid in absolute alcohol

Etching time varies from 10 to 30 seconds, the nital etching grain boundaries clearly. A general etchant for hardened steels is a 10% solution of hydrochloric acid in absolute alcohol.

(2) Copper and alloys

(a) 10% solution of ammonium persulphate in water
(b) Saturated solution of chromic acid in water
(c) 25% solution of ammonium hydroxide in water with a few drops of hydrogen peroxide

(3) Aluminium and alloys

(a) Nitric acid = 5 ml
Hydrofluoric acid = 2 ml
Water = 100 ml

Etching time 15–60 seconds

(b) Sodium hydroxide (solid) = 1 g
Water = 99 ml

Etchant swabbed on for 10 seconds

(4) Lead, solders

Immerse in a 5% ammonium molybdate to remove scratches, then etch in—

Glycerol	=	40 ml
Acetic acid	=	10 ml
Nitric acid	=	10 ml

(5) Tin coatings on steel

Ammonium persulphate	=	5 g
Water	=	95 ml

Unless specified otherwise, all acids are used in their concentrated forms, and distilled water should be used. Extreme caution should be observed when using concentrated hydrofluoric acid and medical attention must be obtained if contact is made with skin. All etchants should be used fresh.

GLOSSARY OF TERMS

Amorphous: a material lacking orderliness in its atomic or molecular arrangement; there is some evidence that certain so-called amorphous materials do possess very short-range order; that is, order within their molecular groups.

Bond energy: the energy associated with a chemical bond holding two or more atomic particles together.

Closest-packed structure: crystal structures having the highest possible co-ordination (12); these are the FCC and HCP crystal lattices.

Co-ordination number: the number of equivalent, equidistant near-neighbours that an atom, ion, or molecule has in a particular crystal lattice; the higher the co-ordination number, the closer the atomic packing within the structure.

Covalent bond: a primary bond arising from the reduction in energy associated with the sharing of pairs of electrons by several atoms.

Crystal: a homogeneous solid possessing a regular three-dimensional lattice arrangement of atomic particles.

Crystalline state: a low-energy state characterised by a high degree of internal order which is expressed in the lattice structure; all true solids are crystalline.

Electron: a small almost weightless particle of unit negative charge that orbits the nucleus of an atom.

Ion core: that part of a (metallic) atom left after the valency electrons are removed.

Ionic bond: a primary bond arising from the electrostatic attraction between two oppositely charged ions.

Macrostructure: the gross features of a material or component as revealed by examination using the naked eye or a low-power lens or microscope.

Microstructure: fine details of structure as revealed by microscopic examination of fractured or polished and etched specimens.

Metallic bond: a primary bond arising from the attraction between the positive ion cores and the negatively charged electrons of the "electron cloud" within an aggregate of metallic atoms.

Miller indices: indices used to specify the (angular) position of any crystallographic face or plane within a given crystal structure.

Neutron: an atomic particle having no electrical charge but 1840 × the mass of an electron.
Non-crystalline solid: an apparently solid material possessing no internal order.
Proton: an atomic particle of unit positive charge having 1836 × the mass of an electron.
Space lattice: a network of straight lines drawn in space such that space is divided into equal volumes; the lattice points, the intersections of these lines, have identical surroundings.
van der Waals forces: a secondary bond which is in effect the result of fluctuating dipole attractions.
X-ray diffraction: a technique used to study crystal structures in which parallel beams of monochromatic X-rays are passed through the crystalline solid to give a diffraction pattern.

REVIEW QUESTIONS

1. List the important characteristics of a solid material and contrast a solid with a liquid.
2. Why are diffusion rates in gases and liquids significantly greater than those in solid materials?
3. Explain the following terms: electron, proton, neutron, nucleus, electron shell, electron orbital.
4. Discuss the important features of the Bohr theory of the atom, and give one reason why this theory is inadequate.
5. Briefly discuss the characteristics of the elements placed in the four zones of the Periodic Table.
6. Distinguish between primary and secondary chemical bonds in terms of bond energies and electron involvement.
7. Distinguish between electropositive and electronegative elements and give examples to illustrate your answer.
8. Compare the bonds existing in sodium chloride with those in hydrogen fluoride, and explain why a molecule of sodium chloride cannot exist.
9. Describe the essential features of the metallic bond, referring particularly to bond directionality and electron involvement.
10. Why do the transition metals possess significant degrees of covalent bonding? What effect does this have on their properties?
11. What are van der Waals forces and why are they important?
12. What are the important features of a "giant molecular" material? Use an example to illustrate your answer.
13. How does the metallic bond explain the opacity, lustre, ductility, and high conductivity of metals?

14. Define the terms "crystal" and "crystalline".
15. What is meant by crystal symmetry? Sketch the symmetry elements of a primitive cubic crystal.
16. What is the significance of Miller indices to the materials scientist?
17. Distinguish between the terms "crystal lattice" and "crystal structure".
18. Sketch the unit cells of the two closest-packed crystal structures and state their co-ordination numbers. Give examples of substances crystallising in these structures.
19. What are the essential features of non-crystalline materials?
20. Distinguish between the terms "amorphous" and "vitreous", using examples to illustrate your answer.
21. Compare and contrast the macroscopic and microscopic levels of investigation.
22. Why is X-ray diffraction an important technique of the crystallographer?
23. Briefly discuss the important steps in the preparation of specimens for micro examination.
24. What is the effect of chemical etching on metallographic specimens?

4

Properties of Engineering Materials

THE INTELLIGENT use of all types of engineering materials depends upon a thorough knowledge of their particular properties. This knowledge is obtained in a cumulative manner from scientifically-designed tests and from the previous experiences of engineers, designers and architects with materials. It is possible to approach the study of the properties of engineering materials from both a qualitative or descriptive viewpoint and from a quantitative viewpoint, the latter belonging to the field of scientific testing. The range of properties found in different classes of materials is very large and Table 4.1 provides a partial classification.

TABLE 4.1

A CLASSIFICATION OF THE PROPERTIES OF ENGINEERING MATERIALS

Class	*Property*	*Class*	*Property*
Physical	Dimensions, shape Density or specific gravity Porosity Moisture content Macrostructure Microstructure	Mechanical	Strength: Tension, compression shear, and flexure (under static, impact, or fatigue conditions) Stiffness Elasticity, Plasticity Ductility, Brittleness Hardness, Wear resistance
Chemical	Oxide or compound composition Acidity or alkalinity Resistance to corrosion or weathering, etc.	Thermal	Specific heat Expansion Conductivity
Physico-chemical	Water-absorptive or water-repellent action Shrinkage and swell due to moisture changes	Electrical and magnetic	Conductivity Magnetic permeability Galvanic action
		Acoustical	Sound transmission Sound reflection
		Optical	Colour Light transmission Light reflection

SOME IMPORTANT PHYSICAL, THERMAL AND ELECTRICAL PROPERTIES

The density of a material is its mass per unit volume, and is directly related to (1) the atomic weights of the atoms or other atomic particles present in the material, and (2) the number of atoms of each type per unit volume which is in turn determined by the method of packing and the interatomic distances. Heavy atoms packed closely together in a structure having a high co-ordination number give a material of high density, while a large amount of "free space" in the structure makes for low density. The density of a solid is usually found by weighing it in air and then determining its volume by a displacement technique. The density of an alloy or composite material may be assessed with some accuracy if the densities and proportions of the individual metals or constituents are known, but it must be clearly understood that this is an approximation to be used when it would be impractical or impossible to measure density directly. *Specific gravity* is sometimes used instead of density, the specific gravity of a solid material being defined as the ratio of the density of that substance to the density of water. This means that the specific gravity of water is 1.

The density and the specific gravity of a material are convenient methods of expressing the weight of that material. They are useful to the design engineer because they allow the weights of structures and machines to be readily calculated when appropriate volumes are known.

The porosity of a substance reveals the volume of voids or "pores" present in the material, and is usually expressed as a percentage of unit volume. Porosity can be measured as *apparent porosity* or as *true porosity*. Apparent porosity is measured by the amount of water that can be absorbed by the material and therefore does not include those voids sealed off from the outside surfaces. Porosity may arise out of processes of natural growth, as in wood; it may be deliberately introduced into a material by processing, as in a firebrick; or it may occur accidentally, as in some metals because of imperfect casting techniques, or in some ceramics because of a lack of control in the firing cycle. It has considerable influence, in particular instances, upon such properties as heat conductivity, refractoriness, strength, brittleness and absorptive capacity. The true porosity of a piece of material can be readily assessed if its true density is known.

Laboratory (14): *Take a perfectly dry rectangular kitchen sponge and calculate its volume geometrically. Immerse the sponge in water, remove, and squeeze out the absorbed water into a measuring cylinder. The ratio of volume of absorbed water to volume of sponge gives the apparent porosity of the sponge.*

Moisture content can have considerable influence upon the strength properties of many porous materials. Wood is the most outstanding example in this field. Freshly-felled "green" wood often contains water in amounts two to three times the actual amount of wood substance present. Seasoning reduces this down to 7 to 16% moisture, this being accompanied by a tremendous increase in strength properties. Furthermore, the loss of moisture is accompanied by a corresponding decrease in thermal and electrical conductivities. Many plastics can absorb quite significant quantities of moisture, this bringing about considerable modifications to such properties as strength and electrical insulation.

Very often alterations in moisture content cause change in the overall volume or "bulk" of the material, wood being the outstanding example. Changes in atmospheric conditions cause increases or decreases in the moisture content of unprotected seasoned timber resulting in swelling or shrinkage. This is a problem since unprotected wooden structures alter their shapes with atmospheric and seasonal changes.

Laboratory (15): *Cut and plane a rectangular block of wood to the sizes 6″ × 3″ × 1″ approx., measure its external dimensions accurately, and weigh it. Dry the block in an oven at 105°C for several hours, remeasure and reweigh it. Now immerse the block in water for several days, remove, blot dry with absorbent paper, and again remeasure and reweigh. Compare the three sets of results.*

The *specific heat* of a substance is the amount of heat required to cause unit mass of that substance to rise through one degree of temperature. The specific heat of a substance determines the rate at which it may be heated and cooled since it is a measure of the amount of heat required for this type of change to take place. Water, because of its extremely high specific heat, finds widespread application as a coolant since it is cheap, readily available and can absorb relatively large amounts of heat without undergoing very great changes in temperature.

The ability of a material to transmit heat energy is expressed in terms of its *thermal conductivity*, metals being by far the best conductors of heat. One common method used in engineering is to "rate" different materials in terms of their relative thermal conductivities.

TABLE 4.2

THERMAL CONDUCTIVITIES COMPARED, WITH SILVER TAKEN AS THE STANDARD OF 100

Metals		*Non-Metals*	
Silver	= 100	Glass	= 0.15
Copper	= 90	Asbestos	= 0.06
Aluminium	= 48	Wood (seasoned)	= 0.05
Brass	= 27	Cork	= 0.01
Iron	= 12	Felt	= 0.01
Mercury	= 0.15		

Thermal conductivity is high in metals because of the large number of electrons available to carry heat energy throughout the metal. Ionic and covalent compounds are poor conductors, but some conductivity is present in materials with a significant degree of secondary bonding.

Laboratory (16): *Cut six-inch lengths of 1/8" diameter steel, copper and glass rod. Hold one end of each rod in your hand so that each is just separated from the other, and hold the other ends together in a bunsen flame. Which rod is the best conductor? Why?*

Substances commonly expand and contract with temperature changes. This is seen in all materials, whether solids, liquids or gases, and is greatest in magnitude in gases. As the temperature of a substance is increased the energies of the atomic particles present increase. The particles move further apart in order to retain stability, causing an overall volume expansion of the material. Within crystalline materials, increasing temperature causes an increase in the amplitude of vibration of the atomic particles present at the lattice points, this causing the average distance apart to be effectively increased. The coefficient of thermal expansion is related both to specific heat and to melting point, the most important relationship being that lower melting-point solids tend to have greater coefficients of thermal expansion. Plastics, for example, expand much more than most metals given similar temperature increases, but this can be reduced by the addition of filler materials of lower expansivity. Table 4.3 shows the coefficients of thermal expansions of some metals and their melting points.

TABLE 4.3

Metal	*Melting Point °C*	*Coefficient of Linear Expansion ($\times 10^{-6}$ cm/cm)*
Mercury	−39°	40
Lead	327°	29
Aluminium	660°	25
Copper	1083°	17
Iron	1539°	12
Tungsten	3410°	4.2

Heat Resistance is largely determined by the melting point of a material together with its chemical stability, oxidation resistance, and ability to retain strength at high temperatures. Refractories are those materials that give satisfactory service at elevated temperatures, but they often possess low resistance to thermal and mechanical shock. The refractory metals, such as molybdenum, titanium and tungsten, retain good strength properties at elevated temperatures. However, most refractories are ceramics such as alumina, silica, zirconia, magnesia, tungsten carbide, and zircon. Alloys such as Nimonic and Incoloy used in jet turbines are very refractory. Organic materials such as wood, plastics and rubbers have low heat resistance; however, thermosetting plastics such as the silicones and the phenolics have greater heat resistance.

TABLE 4.4

MELTING POINTS OF SOME COMMON REFRACTORY MATERIALS

Silica Brick	1700°C
Magnesia Brick	1900°C (softens at about 1700°C)
Graphite Brick	3500°C
Titanium	1812°C
Tungsten	3380°C
Tungsten Carbide	2870°C
Zircon	1852°C

Electrical conductivity, the ability of a material to conduct an electric current, depends upon internal structure. Conduction in solid materials is the result of the movement of electrons throughout the atomic lattice, while liquids conduct if they possess ions which can move in accordance with an applied potential difference. Electrical conductivity of solid materials is largely dependent upon the type of chemical bonding present, covalent and ionic solids being non-conductors, while metals, because of the large number of valency electrons involved in the metallic bond, are extremely good conductors. Conductors have their resistivities lying between 1.6×10^{-6} to 1.4×10^{-4} ohm-cm, while the values for insulators lie between 10^9 to 10^{20} ohm-cm.

TABLE 4.5

GOOD CONDUCTORS, POOR CONDUCTORS AND INSULATORS

Good Conductors	*Poor Conductors*	*Insulators*
The Metals	Water (pure)	Air Glass
Carbon	Stone	Silk Rubber (pure)
	Wood	Oils Paper
	Cotton	Many plastics Many ceramics

Electrical insulating materials can be solids, liquids, or gases and include some polymers and many ceramics. Electrical conductivity is probably the most widely variant property of all, some materials differing by as much as twenty orders of magnitude from one another.

MAGNETIC BEHAVIOUR

Magnetic properties arise out of the behaviour of spinning and orbiting electrons and are thus closely related to electronic configuration. If two electrons within the one energy level (sub-orbital) are "paired" (rotating in opposite directions), their magnetic effects cancel out. However, when there are unpaired electrons magnetic effects become possible. Solid materials are classified as diamagnetic, paramagnetic, or ferromagnetic. Ionic and molecular solids belong to the first group; metals which have one electron in their outer orbital (such as sodium) belong to the second; only iron, nickel, cobalt, and some of their compounds, belong to the third. These ferromagnetic materials, under suitable conditions, become permanent magnets, large numbers of electrons being "unpaired" in the outer orbitals. The Curie point is the temperature to which a ferromagnetic material must be heated in order for its magnetism to disappear; in the case of pure iron the Curie point is 768°C.

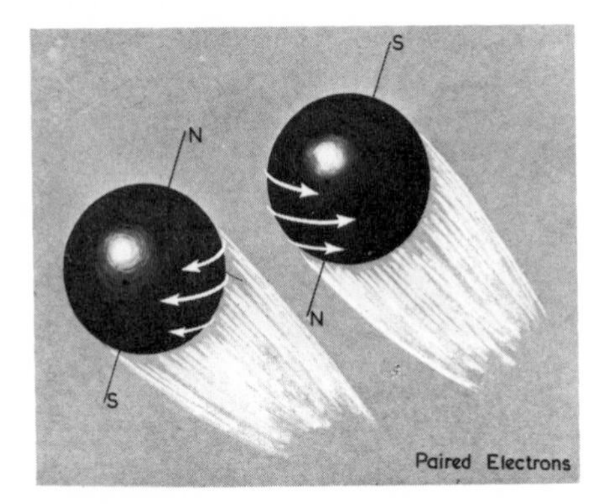

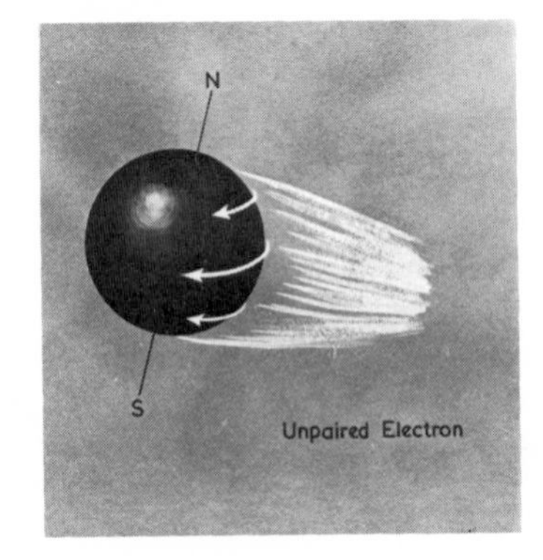

Figure 4.1 *Electrons as determinants of magnetism. Unpaired electrons give rise to residual magnetic effects in metals.*

OPTICAL PROPERTIES

The optical properties of an engineering material may or may not be of importance although such properties as colour and lustre often assist in the identification of a material. When light rays strike a non-luminous body they will be transmitted, reflected, or absorbed according to the nature of the body, the colour of the body depending upon the degree of transmission, reflection or absorption. A transparent material transmits almost all light incident upon it; a translucent material also transmits light but in a distorted manner; an opaque body does not transmit light incident upon it. While all gases and most liquids are transparent, most solids are opaque, with the exception of some vitreous materials and some polymers. Some materials are termed luminous because they emit light; this is the case when metals are heated, the colour of the emissions being determined by the temperature of the body. A distinguishing feature of almost all metals is their lustre which may be said to be the selective reflection of certain wavelengths. Lustre is only seen on a clean surface, as oxides and other surface layers exercise a "blanketing" effect upon incident light.

CORROSION

Perhaps the most important chemical property of engineering materials is the degree of their resistance to corrosion and weathering when left unprotected in the atmosphere. Metals found in nature in the form of oxides, carbonates, and sulphides, must be termed metastable since they tend to alter to these stable chemical states upon continued exposure to the atmosphere. Thus, corrosion may be thought of as nature's answer to the processes of extractive metallurgy. Corrosion is a natural process requiring no external energy, and costs highly industrialised nations thousands of millions of dollars each year.

Chemical corrosion or attack accounts for a small amount of the total damage inflicted each year, and is due to exposure to corrosive liquids such as acids and alkalis. The weathering away of limestone deposits by solutions of carbon dioxide in water (carbonic acid) is a related instance of chemical attack upon a non-metallic material. Many beautiful limestone caves have been formed as a direct result of this, but man-made limestone structures are similarly affected. However, most corrosion occurs as an electro-chemical process resulting from electrical or galvanic cells set up between dissimilar metals, or between dissimilar parts of the same metal.

Stray-Current Corrosion

An example of electrolytic corrosion due to an external current source is stray-current corrosion, which occurs because of leakage from an existing electrical system such as that employed to run tramways and railways. Consider the situation when a tramway is being operated by D.C. current supplied to the tram by means of an overhead conductor, the return flow being through the rails. Suppose that a buried water pipe or gas main runs parallel to, or even across the path of, the tramway. Some of the current will follow the buried pipeline, and electrolytic cells will be set up wherever the current leaves the rails, enters the pipe, leaves the pipe, and re-enters the rails, the moist soil providing the electrolyte for these cells. Severe corrosion will result where the current leaves the rails to enter the pipe and leaves the pipe to re-enter the rails (anodic areas); however, the areas where the current enters the pipe and re-enters the rails are completely protected from rust (cathodic protection), and existing rust may even be reduced back to the metallic form.

Two Metals in Contact

If two dissimilar metals are placed in a solution of an electrolyte and connected externally by a wire, an electric current will flow in the wire and a Voltaic cell will be formed. Conduction in the electrolyte is accomplished by the movement of the ions formed when the more electropositive metal becomes the anode and goes into solution. These positive ions are deposited on the cathode (the more electronegative metal) which may become electroplated. Thus, the anodic metal undergoes corrosion. Metals that corrode readily are known as base metals, and those that do not as noble metals. The galvanic series grades metals in terms of their electropositivity or electronegativity.

TABLE 4.6

ABBREVIATED GALVANIC SERIES OF METALS

NOBLE END (Electronegative)	Lead
	Tin
Platinum	*Pb-Sn* Solders
Gold	Stainless Steel (active)
Silver	Iron
Stainless Steel (passive)	Steel
Copper	Cadmium
Ni-Cu Alloys	Aluminium
Cu-Sn Alloys	Zinc
Cu-Zn Alloys	Magnesium
Nickel	
	BASE END (Electropositive)

Thus, if two dissimilar metals are in contact in a wet situation, the less noble metal will corrode, the rate of corrosion depending upon the separation of the two metals in the galvanic series and the conductivity of the electrolyte present. Consider the case of a bronze union in an iron pipeline; since the iron is less noble than the bronze, it becomes the anodic area, and severe corrosion will take place in the region near the bronze union.

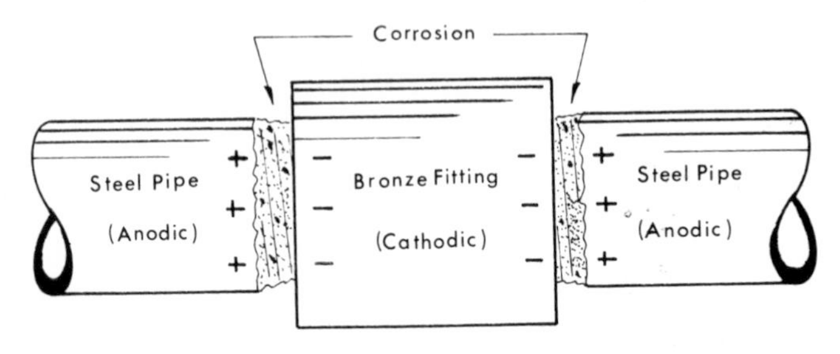

Figure 4.2 *Corrosion around a bronze union in a steel pipe. A bronze union would not in practice be used in this situation since the pipe would corrode while the union itself remained intact.*

Corrosion Within One Metal

If neighbouring areas within a piece of seemingly homogeneous metal develop a potential difference, then corrosion will occur at the anodic area. One of the main reasons why this situation occurs is the inherent inhomogeneity of the metal brought about by:

(1) the presence of impurities
(2) different orientations of neighbouring grains
(3) the presence of several different phases within the metal, a common situation in commercial alloys
(4) the presence of cored grains*, commonly found in castings.

If differential aeration of the metal surface occurs, as in the crevices in metal immersed in water, the metal in contact with the more highly aerated water beomes the cathode while nearby areas become anodic and so corrode quite rapidly. This often results in "pitting", a highly localised and dangerous form of corrosion that ultimately leads to the perforation of the metal.

When a metal is cold worked the grains absorb slightly different amounts of strain energy due to their different orientations within the metal. Because of this some grains become anodic while others become cathodic, and corrosion will proceed if the other conditions, such as the presence of an electrolyte, are fulfilled.

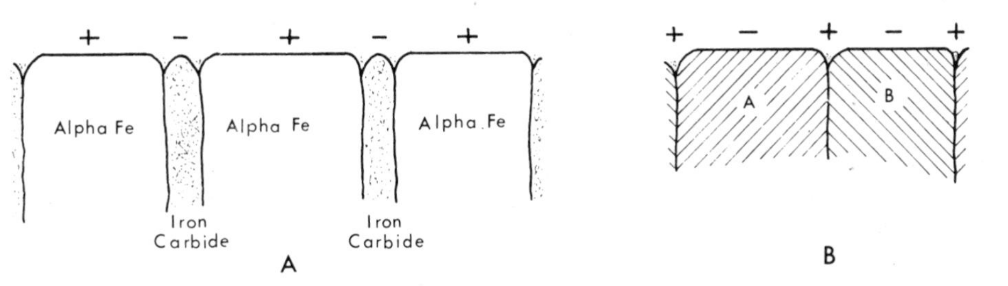

Figure 4.3 *(A) In a two-phase material like pearlite (an important microconstituent of steels—see Chapter 11), one phase becomes cathodic while the other becomes anodic. (B) However, in a pure metal, the grain boundaries themselves become anodic.*

The Rusting of Iron and Steel

Most metals dissolve to a very slight extent in pure water, but as a corrosion mechanism this is of little importance. However, if the water becomes slightly acidic due to the absorption of either CO_2 or SO_3 from the atmosphere a continuing corrosion reaction will begin if the metal is above hydrogen in the activity series. In the case of iron the most usual reaction is

*See Chapter 6 for a description of cored grains.

$$Fe + H_2SO_4 \rightleftharpoons FeSO_4 + H_2\uparrow$$

If oxygen is present, rusting will occur, the reactions being

$$Fe + 2H_2O \longrightarrow Fe(OH)_2 + H_2\uparrow$$

$$4Fe(OH)_2 + 2H_2O + O_2 \longrightarrow 4Fe(OH)_3 \text{ (ferric oxide or rust)}$$

These reactions will occur most vigorously in the zones where oxygen concentrations are high and water or water vapour is present.

Laboratory (17): *Take six identical mild steel nails. Place two in an uncorked dry test tube, a second two in another uncorked test tube which contains a little water, and the last two into an uncorked test tube also containing some water. Boil the water in this third test tube for several minutes until the steam has displaced the air in the test tube, and then cork the test tube securely. Leave all test tubes to stand for a week. Comment on the different degrees of corrosion in the three pairs of nails.*

MECHANICAL PROPERTIES

The mechanical properties of materials are usually of primary interest to the engineer since they determine the ability of a machine or structure to withstand loads without failure. Very often if a material exhibits good strength properties but is deficient in some other direction, then steps will be taken to overcome this deficiency rather than to find a substitute material. Thus, steel, which possesses good strength properties but corrodes badly by rusting, is protected by galvanising or by painting. Many mechanical properties are the subject of standardised tests and are best defined in terms of these tests, but some properties are regarded as fundamental and can be discussed in a qualitative way.

Strength is one of the most important mechanical properties since it determines the ability of a material to withstand stress without failure. Strength varies according to the type of applied loading, and thus it is possible to assess tensile, compressive, shearing, and torsional strengths.

Hardness is a very fundamental property which is closely related to strength. Hardness is usually defined in terms of the ability of a material to resist scratching, abrasion, indentation, or penetration. A very early method of assessing the relative hardnesses of materials was developed for field use in geology, and has become known as Moh's scale of (scratch) hardness. The scale consists of ten minerals arranged in order of ascending hardnesses so that a mineral high on the scale will scratch one lower than itself. Other materials can be rated on this scale; for instance soda-lime glass is about 5.5 while hardened steels are about 6.5.

This scale is not satisfactory for engineering materials, and specialised tests have been developed for metals, wood, and other materials.

TABLE 4.7

MOH'S SCALE OF HARDNESS

1	Talc	6	Orthoclase feldspar
2	Gypsum	7	Quartz
3	Calcite	8	Topaz
4	Fluorite	9	Corundum
5	Apatite	10	Diamond

Laboratory (18): *Take small pieces of copper, mild steel, cast iron, glass, hardened steel, porcelain and rubber and, by using a Moh's hardness determination set of minerals, rank these materials in order of hardness, and give each an approximate hardness number.*

The elasticity of a material is its ability to return to its original shape and dimensions after being subjected to a load that caused, or tended to cause, deformation. The elastic properties of a material are of extreme importance to the structural engineer since materials loaded beyond their elastic range develop a permanent set and may cause a weakening of the whole structure by altering stresses imposed on other members within the structure.

The stiffness of a material or structure is a measure of its ability to resist deformation or deflection under load, and is a property closely associated with elasticity. It is usually defined by the Young's Modulus, also variously termed the Modulus of Stiffness or Modulus of Elasticity, of the material.

Plasticity refers to the ability of a solid material to undergo some degree of permanent deformation without rupture. Plastic deformation will only occur after the elastic range has been exceeded. Many hot and cold working processes such as rolling, extruding, pressing, forging, and spinning depend upon plastic deformation and the ability of a material to readily alter its shape under relatively severe loading. Plasticity usually increases with increasing temperature, many materials being "worked" at elevated temperatures for this reason. Some materials such as glass (at room temperature) and cast iron exhibit no plasticity and are termed *brittle*.

Workability (or formability) denotes the ability of a solid material to undergo all types of deformation processes without failure. It is clearly related to the concept of plastic deformation, and is generally considered to embody the properties of *malleability* and *ductility*. In the strict sense of the word, malleability is that property of a material that allows it to be hammered and rolled out into thin sheets, while ductility refers to the ability to be drawn out into thin wire. Current usage of the word ductility makes it almost synonymous with workability, the older term malleability falling into partial disuse. However, it is important to realise that some materials may be malleable and not ductile; lead, for instance, can be readily rolled and hammered out but cannot be drawn into wire. Also, the workability of a material may vary from one particular situation to the next, a fact exemplified by the excellent rolling properties of low carbon steels as contrasted to their behaviour in the spinning lathe. Malleability is related closely to the compressive properties of a material while ductility relates more to its tensile properties.

Laboratory (19): *Take 4″ discs of annealed aluminium, brass, copper, and lead and attempt to spin them, one after the other, over a simple cup-shaped spinning chuck. What information concerning the formability of these materials do you gain from this exercise? (If two sets of perpendicular lines (as for a noughts and crosses game) are scribed deeply into one side of each disc prior to spinning, then the actual degree of plastic deformation can be assessed visually after the spinning operation has been completed.)*

TABLE 4.8

Metals in Order of Decreasing Malleability	*Metals in Order of Decreasing Ductility*
Gold	Gold
Silver	Silver
Copper	Platinum
Aluminium	Iron
Tin	Nickel
Platinum	Copper
Zinc	Aluminium
Iron	Zinc
Nickel	Tin
	Lead
	Antimony

The fatigue properties of a material determine its behaviour when subjected to many thousands or even millions of cyclic load applications in which the maximum stress developed in each cycle is well within the elastic range of the material. Under these conditions failure may occur after a certain number of load applications, or the material may continue to give service indefinitely. In many instances a component is designed to give a certain length of service under a specified loading cycle; many components of high-speed aero and turbine engines are of this type.

Laboratory (20): *Take $^1/_2$″ wide strips of brass, copper, and lead and bend each one backwards and forwards until it breaks. Count the number of reversals necessary to cause failure in each case. Is this a valid test of the fatigue strength of these metals? Why?*

The strength properties of a material depend upon:

(1) the type of chemical bonding present and the absolute strengths of these bonds
(2) the crystalline lattice present
(3) the number of imperfections or dislocations* present within the crystal structure
(4) the microstructure and macrostructure.

Thus, metals are generally malleable and ductile due to the ability of the bonding electrons to make and break bonds readily, while strength and hardness are closely related to bond strengths. However, the influence of crystal structure upon strength properties is very complex and will be treated more fully in Chapter 7. As far as macrostructure is concerned, flaws, cracks, and other imperfections will act as "stress raisers" by becoming points of stress concentration, this leading to premature failure, particularly when cyclic loading is present.

GLOSSARY OF TERMS

Apparent porosity: porosity taking into account only those pores open to the surface; usually measured by the relative amount of water absorption that can occur.
Corrosion: the destruction of a material by chemical or electrochemical attack; in the case of metals, there is usually a gradual reversion to the more stable oxide, sulphide, or carbonate.
Density: the mass per unit volume of a material.
Ductility: the ability of a material to be plastically deformed by predominantly tensile stresses; for example, as in wire-drawing.
Elasticity: the ability of a material to return to its original dimensions after being subjected to stresses that caused or tended to cause deformation.
Fatigue: the tendency of a metal to break when subjected to conditions of repeated cyclic stressing well below the elastic limit.
Formability: the ability of a metal to undergo all kinds of deformation processes without failure.
Hardness: the ability of a material to resist scratching, abrasion, indentation or penetration.
Lustre: the characteristic surface colour of a metal as seen on a clean surface caused by the selective reflection of certain wavelengths of light.
Malleability: the ability of a metal to be deformed by predominantly compressive stresses; for example, as in rolling and forging.

*A detailed explanation of the dislocation theory is given in Chapter 7.

Plasticity: the ability of a material to undergo permanent deformation without rupture occurring.

Refractoriness: that property of a material which allows it to withstand high temperatures during its service life without spalling, pitting, or deforming.

Specific gravity: the specific gravity of a material is the ratio of its density to the density of water.

Specific heat: the amount of heat needed to raise the temperature of unit mass of a substance through one degree of temperature.

Stiffness: the ability of a material to resist deformation under load. (Compare "Young's Modulus"; see Glossary to Chapter 5.)

Strength: the ability of a material or structure to withstand stress without failure. Alternatively, the stress at which some previously specified end condition occurs.

REVIEW QUESTIONS

1. Explain the meanings of the terms "density" and "specific gravity".

2. Distinguish between true and apparent porosity. Which would be more important in (i) an earthenware agricultural drainpipe and (ii) a firebrick.

3. How can the moisture content of wood be accurately determined?

4. What is the significance of the Curie point of a ferromagnetic metal?

5. What is a luminous material?

6. Define the term "corrosion".

7. A piece of pure aluminium and a piece of a two-phase aluminium-silicon alloy are left unprotected in the atmosphere for one month. What would you expect to find with regard to the corrosion of each piece?

8. What is meant by the term "sacrificial anode"? Give an example of where one could be used.

9. What happens when iron rusts?

10. Why are the mechanical properties of engineering materials of paramount importance to the designer?

11. Outline Moh's scale of hardness and discuss its importance to the materials scientist.

12. What relationship, if any, exists between the elasticity of a material and its stiffness?

13. Define the term "plasticity" in so far as it relates to metals.

14. Distinguish between malleability and ductility, and give an example of an extremely malleable material that lacks ductility.

15. What is meant by the strength properties of a material?

5

The Mechanical Testing and Inspection of Materials

THE MECHANICAL testing laboratory is primarily concerned with the evaluation of all classes of engineering materials in terms of their mechanical properties. It must be established at the outset, however, that it is not the function of the personnel of the testing laboratory to specify the suitability of a material for any particular application; this is the province of the design engineer, who must assess material suitability on the basis of other factors as well as those revealed by mechanical tests. All tests carried out in registered testing laboratories are conducted according to the recognised standards established by the registering bodies or other recognised institutions; for example, standards are laid down in the U.S.A. by the American Society For Testing And Materials (ASTM), in Germany by the Deutsche Industrie Norm (DIN), and in Australia by the Australian Standards Association (ASA). There are national and regional differences between testing procedures and this makes it difficult to compare the results of tests conducted in areas or countries operating under different standards. Fortunately, however, the trend is towards a more or less uniform set of testing standards, so this difficulty should eventually cease to exist.

TYPES OF TESTS

All testing procedures, whether of a mechanical nature or otherwise, are classified into one or more of the following seven groups—

(1) routine tests
(2) exploratory tests
(3) destructive tests
(4) non-destructive tests, including proving tests
(5) tests on specially prepared samples or scaled models
(6) full-scale tests, or tests on the completed article or structure
(7) inspection techniques.

The principal function of any industrial testing laboratory is the *routine testing* of commercial materials in order either to verify manufacturer's specifications or to examine the results of various production or forming techniques upon the material. The continual checking of the purchase specifications of materials ensures that the manufacturer is supplying his material as guaranteed in the agreement made with the purchaser, this in turn protecting the interests of the purchaser since it is less likely that he will, however unwittingly, produce inferior components.

Exploratory tests are the very backbone of materials research. The purposes of such testing may be to extend the amount of knowledge already available on some commercial material or to investigate the properties of a newly-discovered material. Both of these purposes require a higher level of skill and training than routine testing, so that a great deal of this work is carried out in specially-equipped testing laboratories associated with the research section of a large industrial concern. This type of fundamental research leads to continued industrial progress. Exploratory testing may also be conducted in order to evaluate new testing procedures so that their reliability and suitability as routine tests may be accurately assessed. This often results in the modification of long-established testing procedures or their complete replacement by a new type of test.

Destructive tests entail the complete failure of a specimen and thus have their limitations. For instance, such tests obviously cannot be used to check the quality of a completed article, nor, usually, can they be used to test anything other than a fairly small component or structure. Destructive tests are used principally in routine testing where the aim is to check acceptance specifications or alterations in properties resulting from manufacturing techniques. They are usually carried out using small, specially prepared samples, the testing being done strictly according to the appropriate standards.

When it is desired to check empirically the design calculations which have been used to plan a large and expensive structure, such as a bridge or multi-storey building, the *scale-model technique* may be used. This involves the manufacture of a detailed and very accurate scale model of the structure which can then be subjected to scaled design loadings in order to check deflections or distortions occurring when the maximum permissible loadings are applied.

Non-destructive tests are many and varied and are usually designed to test the properties of the finished article or component, or to evaluate its properties at various stages of manufacture. Such tests are therefore the very essence of the manufacturer's quality control. The most common type of non-destructive test is the hardness test, because of the close relationship between hardness and strength properties.

Proving tests belong to the non-destructive group, and are the means whereby the behaviour of an article or component is tested at its maximum permissible design load before it goes into service. Large crane hooks and safety chains are commonly subjected to proving tests before being used because of the hazards that could result from their failure while in service.

Figure 5.1 *A proving tensile test on a crane hook. (Photograph courtesy of the New South Wales Department of Railways.)*

Figure 5.2 *A tensile test on steel strip being conducted on a hydraulic loading universal testing machine. (Photograph courtesy of the New South Wales Department of Railways.)*

Another instance of non-destructive testing is the hardness test carried out on each auto axle before it leaves the manufacturer. This provides a simple, inexpensive, and thoroughly reliable check upon the effectiveness of the heat treatment process which in turn determines the torsional strength of the axle.

Inspection may be considered as a type of non-destructive testing procedure, but some distinction is necessary. Whereas testing is concerned with an exact evaluation of properties, inspection is only qualitative and may involve no more than a visual examination to ensure accuracy of dimensions or correctness of surface finish. However, many inspection procedures are quite complex and involve a detailed and exact evaluation of the quality of the finished article; X-ray examinations, penetrant dye tests, ultrasonic tests, and the like are commonly used for this latter purpose.

It should be clearly understood at the outset that mechanical tests, irrespective of their particular features, do not generally provide absolute answers to the problems of the design engineer. Many tests are designed merely to evaluate behaviour under highly artificial test situations bearing little or no relationship to actual service conditions, and the results of such tests can be applied to design problems only because of the previous experience of the design engineer.

STRESS AND STRAIN

The actual measurements made during mechanical tests which involve the deformation of the specimen are the applied loads and the resultant deformations. Loads, which should be measured to at least an accuracy of one percent, are measured in force units, such as pounds-weight (*lbf*), kips (1,000*lbf*), tons force, or kilograms. However, for the torsion test a measure of the applied moment (in *in-lbf*) replaces the actual load. Deformation is measured in various units, depending upon the type of test. For example, the tensile test involves the measurement of elongation in fractions of a suitable length unit; the compression test the measurement of compression, also in fractions of a suitable length unit; while deformation in the torsion test involves measurement of the angle of twist over a specified gauge length.

The results of mechanical tests are usually expressed in terms of stress and strain, these units being mathematically derived from load and deformation. *Stress* may be defined as the distribution of internal force within the specimen and is calculated from the following relationship:

$$\sigma = \frac{P}{a} \tag{1}$$

where: σ = stress in specimen
P = applied load
a = cross-sectional area of specimen

In routine testing, stress is always calculated using the original cross-sectional area of the specimen. The value thus obtained is termed *engineering stress* to distinguish it from the true stress obtained by taking into account the instantaneous changes of area. Strain may be defined as the change per unit length in a linear dimension; it is really a dimensionless quantity, but is commonly given the rather meaningless units of "inches per inch" when using F.P.S. units. Strain is calculated by taking the original gauge length of the specimen into account:

$$\varepsilon = \frac{e}{L} \tag{2}$$

where: ε = strain
e = deformation (elongation, compression, etc.)
L = gauge length

Measurements of deformation are made using a strainometer, a term which covers an extensometer (tensile test), a compressometer (compression test) and a deflectometer (torsion test). While strain measurements of 0.0001 inch per inch are accurate enough for routine tests, at least 0.00001 inch per inch is required for experimental testing.

Poisson's Ratio

When a specimen is stressed by a uniaxial force it commonly deforms in the direction of that force (axial strain), and there is a corresponding adjustment in dimensions at 90° to the direction of the force (lateral strain). Poisson's ratio expresses the relationship existing between the lateral strain and the axial strain:

$$\nu = \frac{x}{y} \qquad (3)$$

where: ν = Poisson's ratio
x = lateral strain
y = axial strain

For most structural materials the values of Poisson's ratio lie between 0.3 and 0.6.

Stress-Strain Diagrams

The results of many mechanical tests are exemplified in stress-strain diagrams which show graphically the relationships existing between stress and strain throughout the duration of the test. Some machines are capable of plotting stress-strain diagrams directly since they possess the necessary autographic attachment. However, many machines are not so equipped and frequently it is more convenient to plot applied load against deformation (elongation, compression, etc). Load-deformation curves have the same general shapes as stress-strain curves. The values of load or stress are the ordinate scale, and deformation or strain are the abscissa scale.

UNIVERSAL TESTING MACHINES

One of the most important items in any mechanical testing laboratory is the universal testing machine, so named because it is possible to conduct tension, compression, bending, transverse, and shearing tests on this one machine by using different types of attachments. These machines are hydraulic or mechanical in operation, hydraulic systems being preferred due to their simplicity and lower cost. A universal testing machine must be capable of applying a maximum load of at least 60,000 *lbf* with some machines having capacities as high as 180,000 *lbf*. These machines always consist of four main parts.

(1) *The Straining Side*: This part of the machine supplies the load or force by which the specimen is deformed and may consist of an hydraulic system or a mechanical worm drive. Either the load is applied at a constant rate or a constant rate of straining is maintained in the specimen. Constant straining rates for engineering materials vary between 0.01 and 3.0 inches per inch per minute, the relevant standards specifying straining rates for particular materials. Stress-strain curves produced using a constant rate of straining differ slightly from those produced using a constant rate of loading, the latter method being more applicable to non-metallic materials such as plastics or textiles.

(2) *The Weighing Side*: It is equally important that the applied load be measured accurately, at least to an accuracy of $\pm 1\%$. The weighing side performs this function either mechanically, by means of a lever or spring system, or hydraulically by means of a balancing cylinder.

(3) *The Gripping Devices*: These secure the specimen in the machine, their design depending upon the type of specimen to be handled. Grips are usually self-aligning or self-centering as eccentric loading will introduce inaccuracies into the test. For instance, tensile test grips may be of the wedge-grip, split collar or screwed types, depending upon the type of tensile specimen to be held.

(4) *The Recording Unit*: The older type of machine required one or even two operators whose functions were to record corresponding readings of load and deformation. However, most modern machines record both load and deformation graphically, producing the load-deformation curve during the test. Most recording units involve a rotating drum and a moving pen or stylus, the speed of drum rotation being controlled by the strainometer while the pen movements are controlled by the weighing system.

The Hounsfield Tensometer

This is a laboratory-sized universal testing machine which can be operated by hand and is therefore very suitable for student use. It is a horizontal machine capable of applying loads up to two tons using beams of different capacities. The machine gives a basic tensile load which, by the use of suitable attachments, can be used for tensile, compression, flexural, shear, hardness, cupping, and notched-bar tests. The relationships between load and deformation are readily recorded on the autographic attachment which is also partially hand-operated. If an extensometer is fitted, Young's Modulus and

Proof Stress determinations are readily made. Specimens are necessarily very small, but test results are directly comparable to those obtained in full-scale tests as normally conducted in testing laboratories.

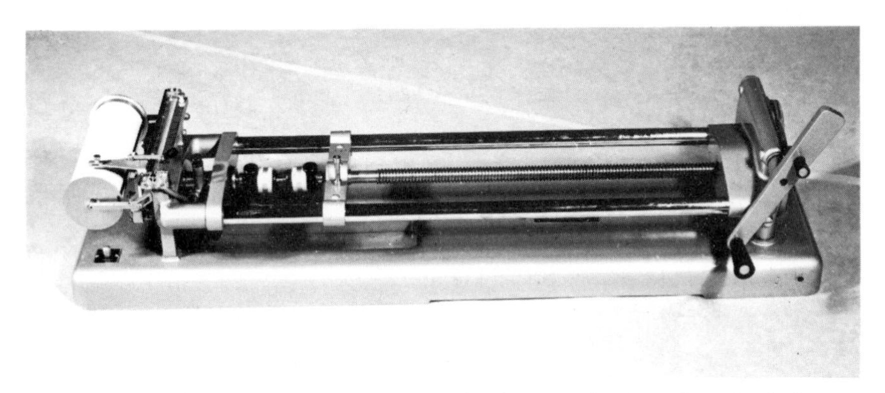

Figure 5.3 *The Hounsfield Tensometer, a hand operated universal mechanical testing machine.*

THE TENSION TEST

The test consists of the gradual application of an increasing uniaxial tensile load to the specimen until rupture occurs. (See Figure 5.2.) Various specimens may be used, either full-sized or in the form of standardised test pieces, the latter being usually of circular cross-section and having shouldered, screwed, or plain round ends depending upon the type of gripping device. The ends

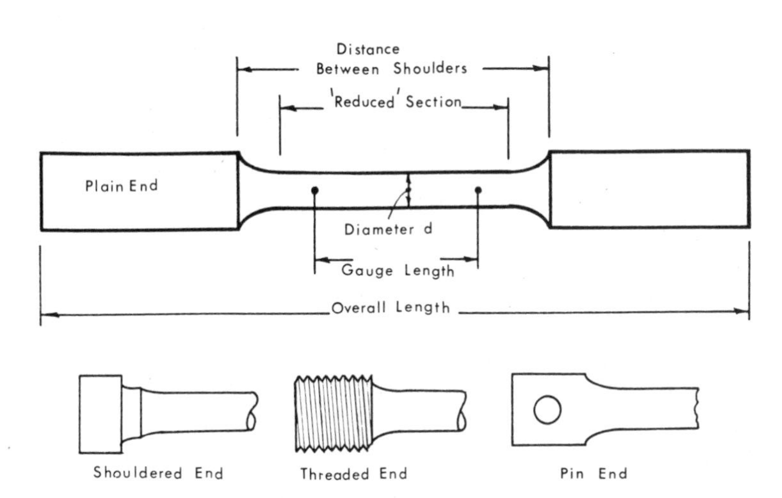

Figure 5.4 *Details of typical tensile test specimens.*

are usually of a larger cross-section than the centre portion so that failure will not occur in those areas affected by gripping stresses. However, this does not apply to tests on wire, reinforcing rod, tube, and the like where the test sample is cut off the end of the material to be tested. Sheet and plate are usually made into flat specimens which also have reduced cross-sectional areas in their centre portions. Tubular specimens must have their ends plugged in some suitable manner to prevent crushing in the wedge grips commonly employed for this type of test.

Round specimens usually have a parallel "reduced section" to give a uniform stress distribution along the gauge length. The following size relationships apply to standard round tensile specimens.

Cross-sectional area	A	$= 0.7854\ d^2$	
Gauge length	L	$= 4\sqrt{A}$	$= 3.54\ d$
Parallel length	P	$= \frac{9}{8}L$	$= 3.98\ d$ (minimum)
Shoulder radius	R	$= \frac{5}{4}L$	for all cast metals,

Where d equals the reduced diameter.

The common tensile specimen has a gauge length of 2.00 inches and a reduced diameter of 0.564 inches in order to have a cross-sectional area of 0.25 square inches.

Before the test is conducted the gauge length must be marked upon the specimen, this being done commonly by using two centre punch marks placed the correct distance apart. The diameter must then be measured to an accuracy of at least 0.001″. The specimen is assembled into the machine and the extensometer fitted. Loading is then begun at a rate suitable for the particular material under test, a loading rate of 100 kips per square inch per minute being the maximum permissible for metallic materials. If the extensometer is attached directly to the specimen it is usually removed after the yield point has been passed in order to prevent damage to its delicate mechanism when the specimen ruptures; however, some extensometers can be left on to failure.

The following data are usually calculated from tensile tests.

(a) For ductile materials showing a definite yield point: proportional limit stress, yield stress, ultimate stress, breaking stress, percentage elongation, and percentage reduction in area.

(b) For ductile materials showing progressive yield: as in (*a*), but a proof stress determination is made since no definite yield point exists.

(c) For brittle materials: only the ultimate stress, percentage elongation, and percentage reduction in area are assessed.

The type of failure is recorded, a sketch often being included.

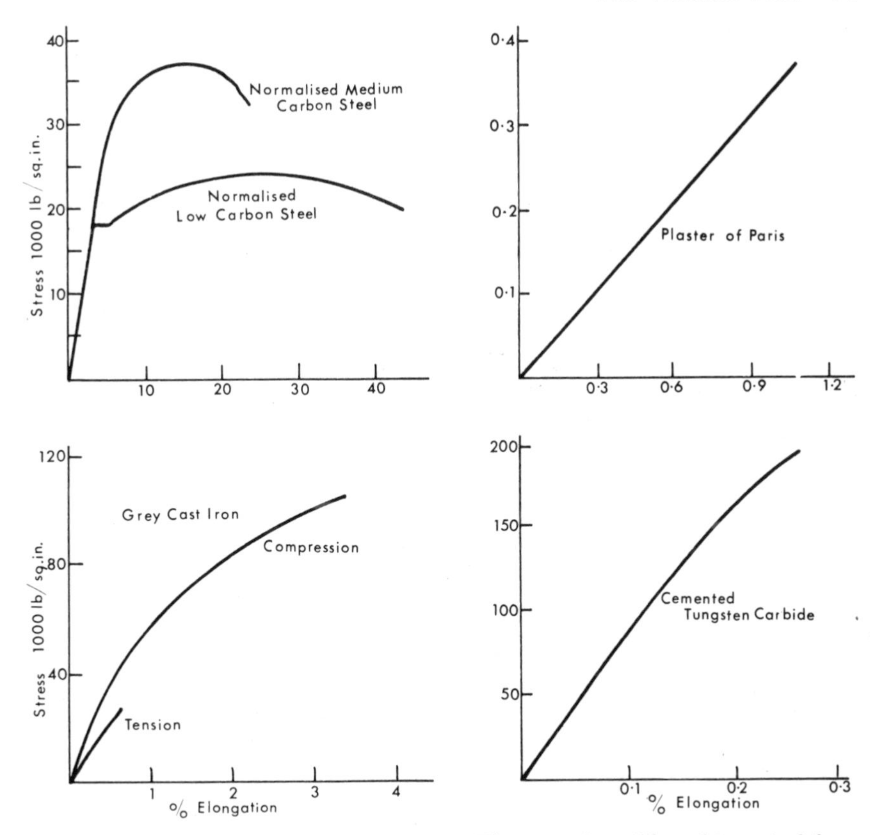

Figure 5.5 *Typical tensile stress-strain curves. The curve for mild steel is typical for a ductile metal showing a definite yield point; Plaster of Paris, cemented tungsten carbide, and cast iron are brittle. The stress-strain curve for cast iron tested in compression is given for comparison.*

Measures of Elastic Strength

Measures of elastic strength are important to the design engineer. Common measures include the elastic limit, the proportional limit, the yield stress (or yield strength), and proof stress, the latter only applying to those materials not showing a definite yield point.

The *elastic limit* is defined as the greatest stress that a material can endure without taking up some permanent set. The only way in which this stress can be determined is to successively load and unload the specimen with progressively greater loads until the extensometer readings reveal a permanent set upon complete removal of the applied load. This procedure is very time consuming, and results depend in large measure upon the accuracy of the extensometer.

The *proportional limit* is defined as the greatest stress that a material can endure without losing straight-line proportionality between stress and strain. In contrast to the elastic limit, the proportional limit is readily determined from the stress-strain curve. Since most materials exhibit straight-line proportionality between stress and strain within the elastic limit, for all practical purposes the proportional limit can be regarded as identical with the elastic limit, and thus is often known as the proportional elastic limit. It is interesting to note that this concept of proportionality between stress and strain within the elastic limit was first expounded by Robert Hooke in 1678, and is known as Hooke's Law.

The *yield stress* is defined as that stress at which some marked increase in strain occurs without a corresponding increase in stress. Some materials exhibit a definite yield point, in which case the yield stress is simply the stress at this point. Mild steel is an instance of this. However, most ductile materials exhibit progressive yield and so another measure of yield stress, usually known as proof stress, must be employed.

Proof stress is defined as that amount of stress required to produce some previously specified amount of permanent set within the specimen, common measures being 0.1% or 0.2% of the original gauge length. This is usually determined upon completion of the test by the "offset method" shown in Figure 5.6.

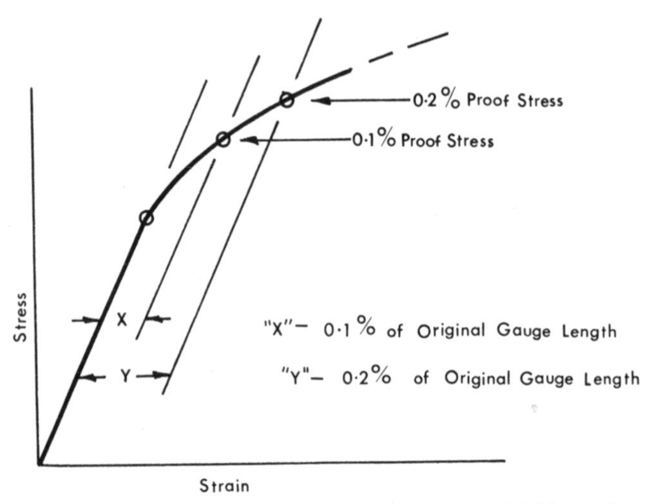

Figure 5.6 *The offset method for determining 0.1% and 0.2% proof stress.*

Ultimate Strength

The ultimate strength of a material is the maximum stress that the material is capable of developing. In practice, the ultimate strength is computed using the original cross-sectional area of the test specimen and must be regarded as the nominal strength. Once the ultimate tensile strength (UTS) is exceeded the load appears to drop away to failure, and the greatest amount of "necking down" occurs in ductile test specimens during this part of the test.

Stiffness

The stiffness of a material is defined as the relationship between the amount of deformation and the applied load and is commonly expressed in terms of Young's Modulus or the Modulus of Elasticity, the latter term being somewhat confusing. Young's Modulus (E) is defined as the relationship existing between stress and strain within the proportional limit, this being determined from the following equation.

$$E = \frac{PL}{ea} \tag{4}$$

where: P = load at proportional limit
L = gauge length
e = extension at proportional limit
a = original cross-sectional area of specimen.

The higher the value of Young's Modulus, the stiffer the material. It is interesting to note that most steels have a value of E equal to about 30×10^6 psi, so that a steel structure cannot be stiffened by changing to a higher grade of steel unless the overall sizes of the members are also increased.

Values of Young's Modulus vary with varying crystal orientations. Tests conducted on single crystals of the same material often reveal wide variations in Young's Modulus. However, tests conducted on polycrystalline materials yield virtually constant values because of the random crystal arrangements present.

TABLE 5.1

VARIATIONS OF YOUNG'S MODULUS FOR METALS (10^6 psi)

Metal	*Single Crystals*		*Polycrystalline Specimens*
	Maximum	*Minimum*	
Aluminium	11.0	9.1	10.0
Copper	27.9	9.7	16.1
Lead	5.6	1.6	2.3
Iron (pure)	41.2	19.2	30.0
Magnesium	7.4	6.3	6.3
Zinc	18.0	5.0	14.5
Tin	12.4	3.8	6.6

The Energy Capacity of a Material

The ability of a material to absorb and store energy is related directly to its shock-resistance and toughness. Two measures of energy capacity are in common use, the Modulus of Resilience and the Modulus of Toughness.

The Modulus of Resilience (R) is defined as the amount of energy required to stress unit volume of a material to its proportional limit, and is represented graphically by the area under the stress-strain curve to the proportional limit. It is calculated from the equation

$$R = \frac{\sigma\varepsilon}{2} = \frac{\sigma^2}{2E} \qquad (5)$$

where: σ = proportional limit stress
ε = strain at proportional limit
E = Young's Modulus of Material.

The Modulus of Toughness (T) is defined as the amount of energy required to cause failure in unit volume of a material, and thus is represented by the total area under the stress-strain curve. This is usually assessed in a qualitative way by visually examining the overall appearance of the stress-strain curve. Figure 5.8 illustrates this principle in relation to various ferrous alloys.

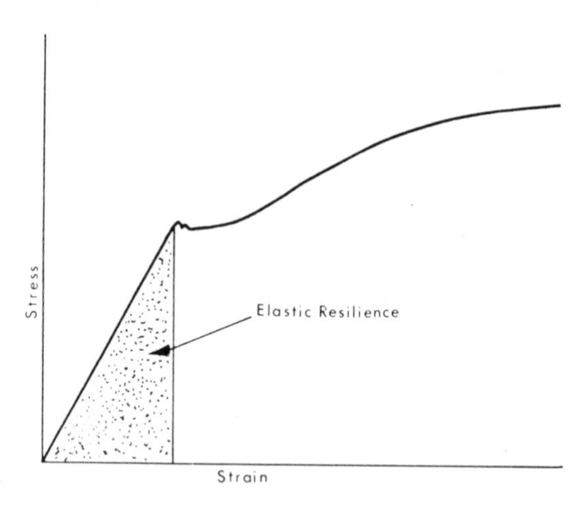

Figure 5.7 *Elastic resilience of a material as the area under that part of the tensile curve up to the proportional limit.*

The Assessment of Ductility

The ductility of a material cannot be assessed exactly from a tensile test. However, excellent indications are given by the overall form of the test curve, and by the percentage elongation and the percentage reduction in area of the test specimen. If the test curve shows that considerable deformation has taken place after the UTS has been exceeded, then the material is clearly ductile, since a great deal of "necking down" occurred before failure.

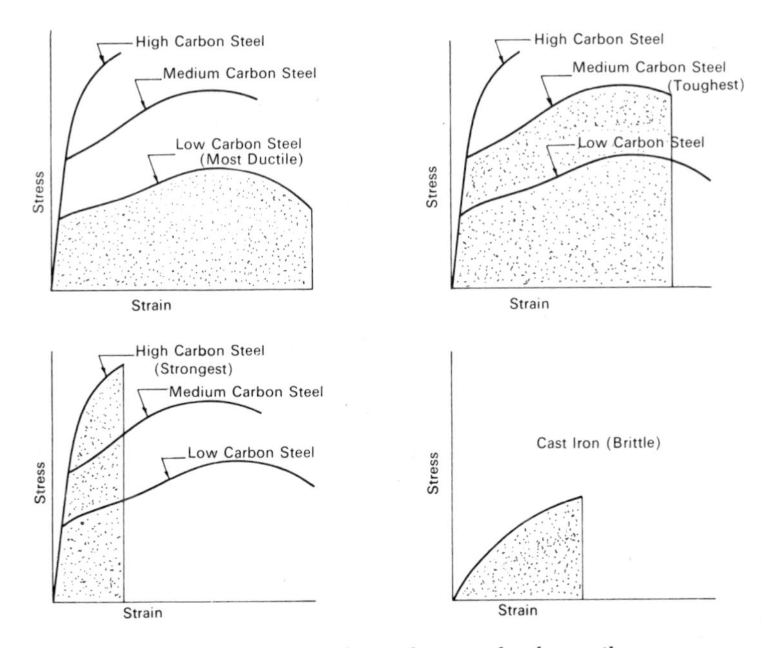

Figure 5.8 *Toughness seen as the total area under the tensile curve.*

The percentage elongation of the test specimen is calculated from the equation

$$\%\ \text{elongation} = \frac{L - L_0}{L_0} \times 100 \qquad (6)$$

where: L_0 = original gauge length
L = length between gauge marks after completion of the test (this is determined when the broken pieces are fitted together by hand after removal from the machine).

The percentage reduction in area is derived from the relationship:

$$\%\ \text{R.A.} = \frac{\quad}{\quad} \times 100 \qquad (7)$$

where: d_0 = original diameter
d = reduced diameter.

It must be noted that the values of percentage elongation depend to some extent on the gauge length used, and accordingly when test figures are stated the gauge length must always be given.

Rupture

Rupture or failure will occur when the material can no longer withstand the applied stress. It is important that the mode of failure be fully described and that any peculiar test conditions that could conceivably relate to premature failure be also noted. The breaking strength of a material is commonly assessed in tests to failure:

$$\text{Breaking Strength} = \frac{P_f}{a} \qquad (8)$$

where: P_f = load at failure
a = original cross sectional area of specimen.

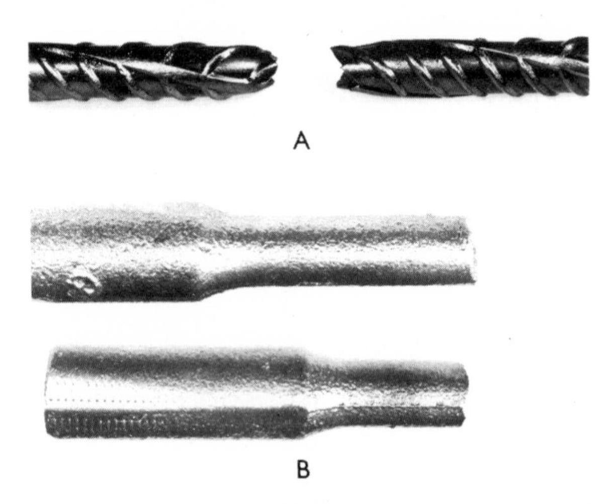

Figure 5.9 *(A) A ductile failure in steel reinforcing rod. (B) A brittle tensile failure in cast iron.*

Laboratory (21): *Using the tensometer, conduct tensile tests on wire samples and standard machined test pieces of mild steel, brass, copper, and aluminium. An extensometer is to be used when the standard specimens are tested. From the load-elongation and stress-strain curves so obtained, calculate*

(1) proportional limit stress
(2) yield stress (either at the yield point or as 0.1% proof stress if no definite yield point occurs).
(3) UTS and breaking strength
(4) percentage elongation
(5) percentage reduction in area.
Tabulate results, sketch modes of failure, and suggest why standard specimens are preferable to plain wire.

THE COMPRESSION TEST

The compression test may be regarded as opposite to the tensile test in so far as the uniaxial load applied is compressive rather than tensile. Although rarely used as an acceptance test for structural metals, it is particularly useful for the evaluation of the strength properties of brittle materials. Such materials are usually used in compression rather than in tension, so that this form of testing is often directly applicable to design problems. Materials commonly tested in compression include cast iron, concrete, mortar, brick, ceramics, and wood, the latter being almost impossible to test in direct tension due to the difficulty of designing suitable gripping devices.

Four common difficulties are encountered when materials are tested in uniaxial compression.

(1) It is extremely difficult to apply a truly axial load because of the difficulties encountered in centering the specimen in the machine.
(2) There is a tendency for bending stresses to be set up in the specimen during the test, but this can be largely prevented by suitable specimen proportions.
(3) The deformation of the specimen is restricted to some extent by the frictional forces existing between the crossheads or plattens of the testing machine and the ends of the specimen and barrelling occurs.
(4) The larger diameters of the test pieces require loads greater than those required by tensile specimens. Thus greater-capacity machines are often necessary.

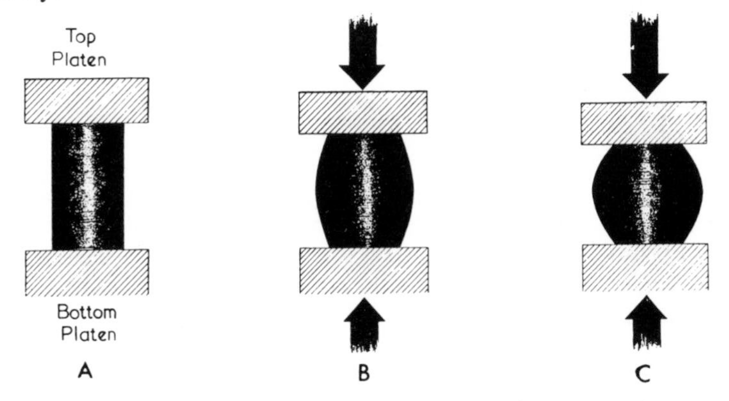

Figure 5.10 *Barrelling in ductile materials tested in compression.*

Compressometers are not usually fitted unless Young's Modulus determinations are to be made, the usual test measures being the proportional limit stress (ductile materials) and the ultimate compressive strength (brittle materials).

Test specimens for metals are usually cylindrical, with the length-to-diameter ratio depending upon the particular metal being tested. Compression tests on wood are conducted using 2″ cubes, or 2″ × 2″ × 8″ specimens with 6″ gauge lengths if elastic modulus determinations are to be carried out. Concrete specimens are usually 6″ in diameter and 12″ in height, although plain cement and mortar tests are conducted on 3″ cubes. Again, the mode of failure has great significance and should be described and sketched. Brittle materials commonly fail in one of three ways, each of which is caused by a concentration of shear stress along some diagonal plane within the material. (See Figure 5.11).

Laboratory (22): *Test samples of brass, copper, cast iron, cement, ebonite, and wood in compression. Obtain load-deformation curves, calculate stress at proportional limit, ultimate stress (if possible), and sketch modes of failure.*

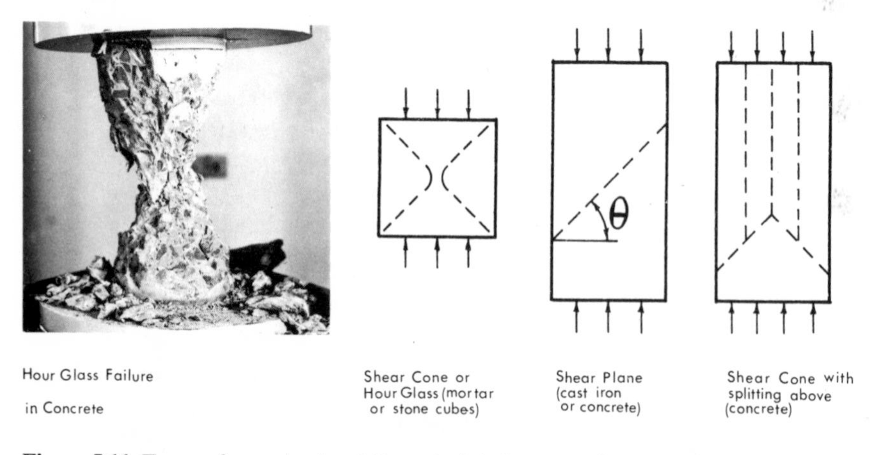

Figure 5.11 *Types of compression failures in brittle materials. θ equals 45° for metals. (Photograph courtesy of New South Wales Department of Public Works.)*

THE TRANSVERSE TEST

Many structural members are used in situations where the applied loads tend to set up tensile stresses in one side of the member and compressive stresses in the other. Such members are said to be "in bending". Transverse, or as they are also called, flexural tests, assess the strength of a material in this type of situation. This test is also useful to assess the tensile strengths of those materials which are difficult to test in direct tension. Wood comes into this category.

The actual test involves placing an appropriate specimen on two fulcrum-type supports set a known distance apart, applying a gradually-increasing load, and measuring deflections using a suitable deflectometer. The test curve resulting from this is known as a load-deflection curve. Various different systems of loading are employed, the two most common systems being central-point loading and loading at the center-thirds. Different methods of load application induce different stress distributions within the test beam, Figure 5.12 showing the bending moment distributions for the two previously-mentioned load systems.

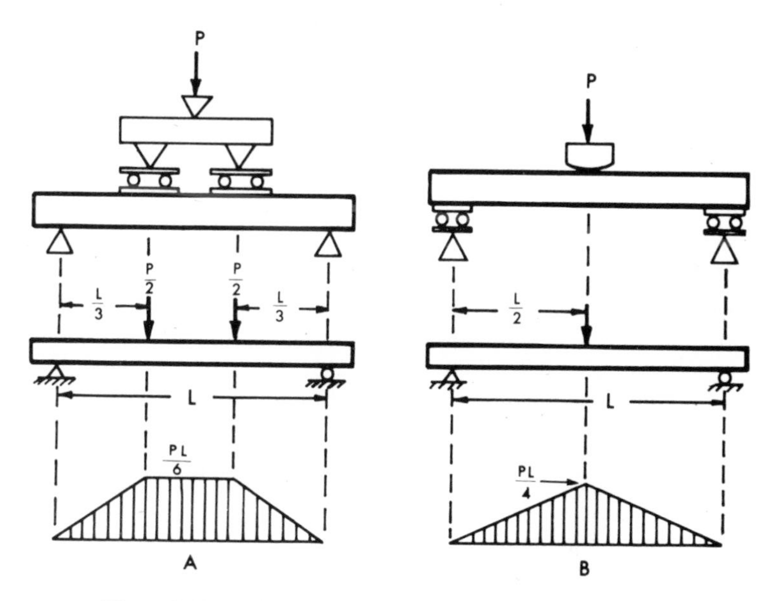

Figure 5.12 *Bending moment distributions in beams subjected to (A) centre-thirds loading and (B) centre-point loading.*

A *neutral plane* exists in a beam subjected to pure bending, the bending stresses along this plane being zero. The neutral plane is usually located along the centroid of the beam. Compressive stresses exist in those parts of the beam above the neutral plane, tensile stresses in those parts below it. Most materials fail in tension rather than in compression under bending loads. Deflection measurements indicate the overall stiffness of the material and are thus most significant in their own right.

The results of flexure tests are usually expressed in terms of the transverse strength at the proportional limit, when

$$T.S. = \frac{\textit{Bending Moment at Proportional Limit}}{\textit{Section Modulus}}$$

Since the section modulus $= \frac{bd^2}{6}$ for rectangular beams, where b = breadth and d = depth of the beam, this becomes

$$(a)\ \text{T.S.} = \frac{\frac{PL}{4}}{\frac{bd}{6}} = \frac{3PL}{2bd^2} \qquad (9)$$

if central-point loading is used, or,

$$(b)\ \text{T.S.} = \frac{\frac{PL}{6}}{\frac{bd^2}{6}} = \frac{PL}{bd^2} \qquad (10)$$

if centre-thirds loading is applied, where P = applied load and L = gauge length.

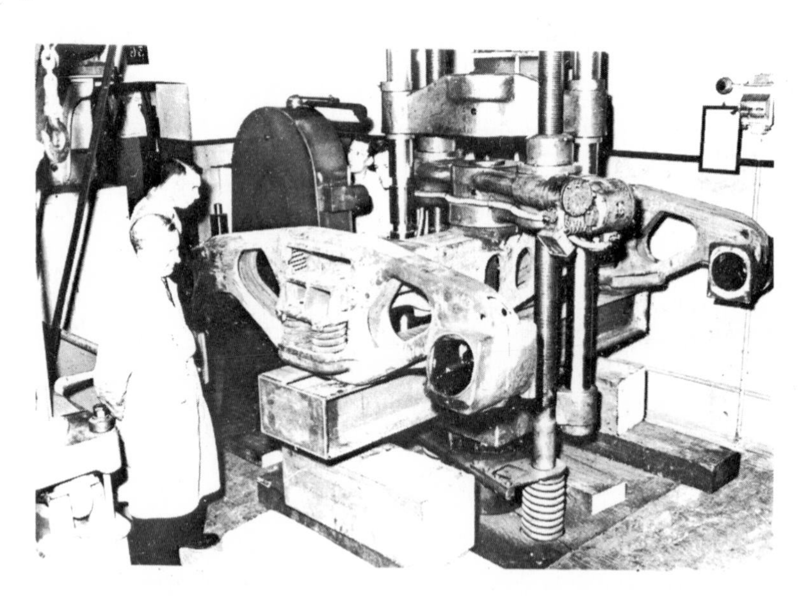

Figure 5.13 *Illustrates a remarkable transverse proving test carried out by the New South Wales Department of Railways on a welded locomotive bogie which had severely fractured in service. The welding proved successful.*

The *Modulus of Rupture* is calculated using similar formulae; however, the maximum bending moment must be substituted for that at the proportional limit.

Transverse tests are commonly conducted on wood using centre-point loading, the specimen, 2″ × 2″ × 30″ and free from defects, being tested over a 28″ span. Concrete specimens are usually 6″ × 6″ × 24″ and are loaded using the centre-thirds method, while round specimens are commonly used for metals such as steel and cast iron.

Laboratory (23): *Perform transverse tests on square specimens of wood (centre-point loading), aluminium, and cement (a suitable test piece for the tensometer is as-cast to ½″ square and 5″ long). In each case calculate the transverse strength.*

SHEAR TESTS

Shearing stresses are those that act parallel to a given plane, and must be distinguished from tensile and compressive stresses which act normal to a given plane. Most materials fail in shear whether they are under tensile, compressive, or shearing loads; however, shear tests measure shear resistance directly.

Three different types of shear tests are in common use, the direct shear test, the punching test, and the torsion test.

Direct Shear Tests

When metals are tested in direct shear a suitable test piece is clamped in some device and sheared across by a special die to which known loads are applied. A special shearing box is used to reduce bending stresses in the specimen, which may be tested in single or double shear. Shear tests are commonly carried out on rivets and bolts using a modified form of double-shear box. Wood is tested in single shear.

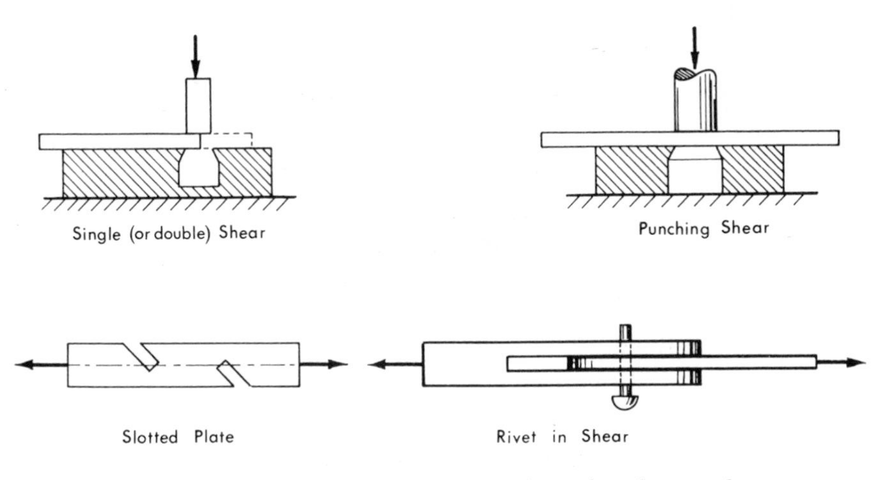

Figure 5.14 *Types of shear tests commonly conducted on metals.*

Punching Shear Tests

Punching shear tests are a special form of direct shearing applicable mainly to sheet metals in which a circular blank is removed by the shearing action of a hardened pin moving into a steel die. Again, bending can occur, this time between the pin and the die, and errors may be introduced into the test results.

The only test measurement made is that of maximum load (P), the ultimate shear strength being calculated according to the relationship

$$\text{Ultimate shear strength} = \frac{\text{maximum load}}{\text{sheared area}}$$

$$= \frac{P}{a} \qquad (11)$$

Laboratory (24): *Conduct punching shear tests on 22 gauge steel, brass, copper, and lead. Calculate ultimate shearing stress in each.*

Torsion Tests

Torsion tests are conducted in specially-designed machines which possess one fixed head and one movable head. The prepared specimen is securely gripped in both heads by means of wedge-grips or serrated cams and the movable head is rotated until failure occurs. Readings of angle of twist (deformation) and applied torque ("load") are recorded during the test, and then a *torque — angle of twist* curve is drawn. If tubes are tested their ends must be securely plugged to prevent crushing. It should be noted that shearing strength is more accurately obtained from tubular specimens. The usual results determined from torsion tests include shearing stresses in extreme "fibres", the modulus of rigidity, and the ultimate shearing stress.

For round solid specimens the following relationships apply.

(a) Shearing Stress at Proportional Limit

$$\tau = \frac{2T}{\pi r^3} \qquad (12)$$

where: τ = shearing stress in outer fibres
T = applied torque at proportional limit
r = radius of test piece.

(b) Modulus of Rigidity

$$E_S = \frac{2TL\theta}{\pi r^4} \qquad (13)$$

where: L = gauge length of specimen
θ = angle of twist measured over L.

Modes of failure in torsion are quite distinct from tensile failures since no "necking down" of the specimen occurs. Ductile materials tend to fail on a plane at right angles to the plane of applied torque; brittle materials develop helicoidal failures; and tubular specimens tend to buckle.

Figure 5.15 *The helicoidal torsional failure typical of solid brittle test specimens. (Photograph courtesy of the New South Wales Department of Railways.)*

DUCTILITY TESTS

Direct tests of ductility are sometimes made in order to check on the ductility of a certain material, to investigate the changes in ductility brought about by various forming processes or heat treatment, or to qualitatively compare several different materials with respect to ductility. Two types of tests are commonly employed, the "bend test" for rod or bar and the "cupping test" for sheet metal.

The Bend Test

The material to be tested is bent around a pin or former of known diameter, and the load required to complete a 180° bend is recorded if the specimen does not break. If fracture does occur, the angle at which it took place is recorded, together with the load at fracture. It is usual to test metals cold but "hot bend" tests are employed when changes in plasticity with temperature are to be investigated.

The Erichsen Cupping Test

The specimen, usually a 3″ square cut from the sheet metal to be tested, is placed between two dies and firmly clamped around its outer surface. A

Figure 5.16 *Three successive stages of a bend test on a piece of $\frac{3}{4}''$ square mild steel stock.*

hemispherical plunger is forced against the centre of the specimen, and loading continues until failure occurs. The index of ductility is the depth of the "cup" so formed, measured to an accuracy of 0.01″. In general, circumferential fractures indicate sound stock while transverse splitting indicates rolling defects. The index obtained from this test gives a good indication of the deep drawing qualities of the sheet material.

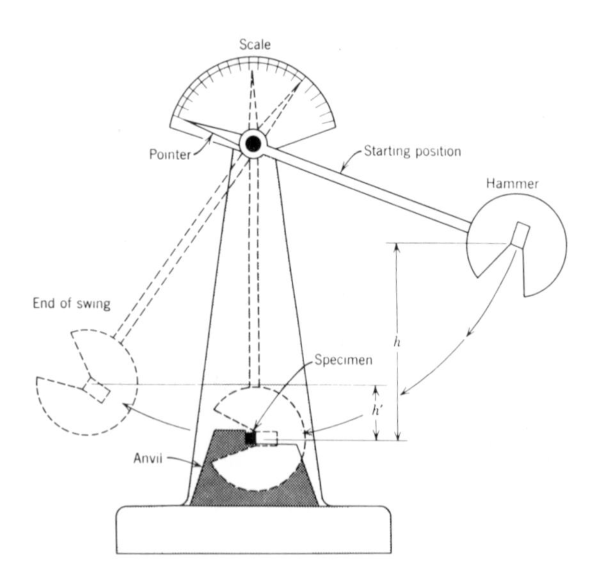

Figure 5.17 *Schematic drawing of an impact testing machine. (Reproduced with permission from Volume 3 of "The Structure and Properties of Engineering Materials" by Hayden, Moffatt, and Wulff. John Wiley and Sons Inc., N.Y. 1965.)*

NOTCHED-BAR IMPACT TESTS

Many materials have to resist suddenly-applied or "impact" loads while in service, and it has been shown that impact strengths are much lower than strengths achieved under slowly-applied loads. The notched-bar impact tests were designed to measure the impact strength of a material, but in actual fact they are unable to do this. Values obtained from notched-bar impact tests are highly arbitrary and are of no direct consequence in design problems. However, they are often used in materials specifications since they do measure "notch toughness", a property quite difficult to define in absolute terms. In practice, the standard specimen with its standard notch is broken or severely deformed by an impact blow, the energy required to propagate the crack from the bottom of the notch across the specimen being measured in *ft lbf*. The size and shape of the notch is of the utmost importance, and very minor variations in notch size or shape will cause large variations in the test figure, which is commonly termed the "energy to fracture" of the material.

Three common types of tests are used.

(1) *The tensile impact test*: uses a notched or un-notched specimen which is broken by a dropping weight. The test is not common because of the large capacity and extreme rigidity required in the testing machine.

(2) *The Izod test*: employs a cantilevered test specimen of 10mm × 10mm section which is broken by means of a swinging pendulum. This is also done using a round specimen of 0.450″ diameter, this type of specimen being more readily manufactured.

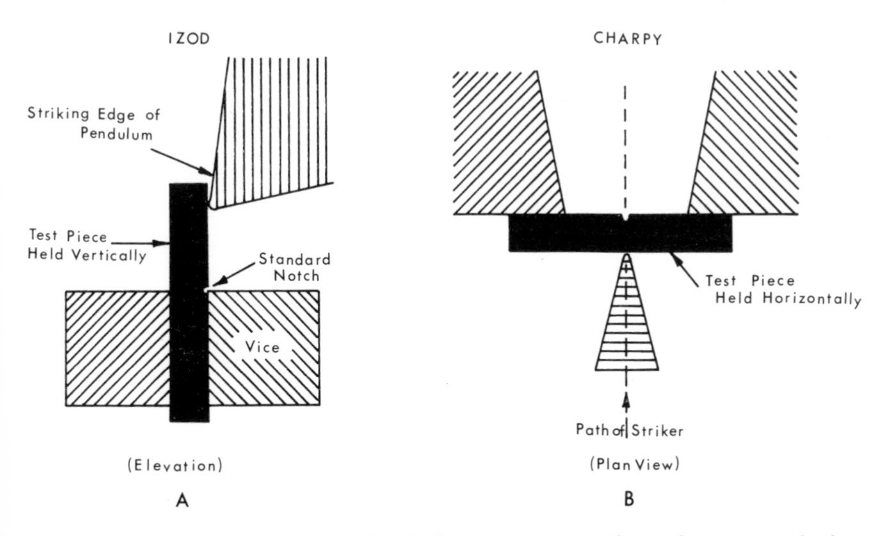

Figure 5.18 *The principles of the Izod and Charpy tests: (A) The Izod specimen is broken as a cantilever, (B) the Charpy specimen as a beam.*

(3) *The Charpy test*: tests the standard specimen as a beam having a span of 40mm, the blow being delivered at a point directly behind the notch.

Ignoring frictional losses, the "energy to fracture" of both Izod and Charpy tests can be calculated as follows. See also Figure 5.19.

$$\begin{aligned}\text{Energy to Rupture} &= \text{Initial Energy of Pendulum} - \text{Residual Energy of Pendulum} \\ &= WH - WH_1 \\ &= WR\,(\text{Cos}B - \text{Cos}A)\ \textit{ft lbf} \qquad (14)\end{aligned}$$

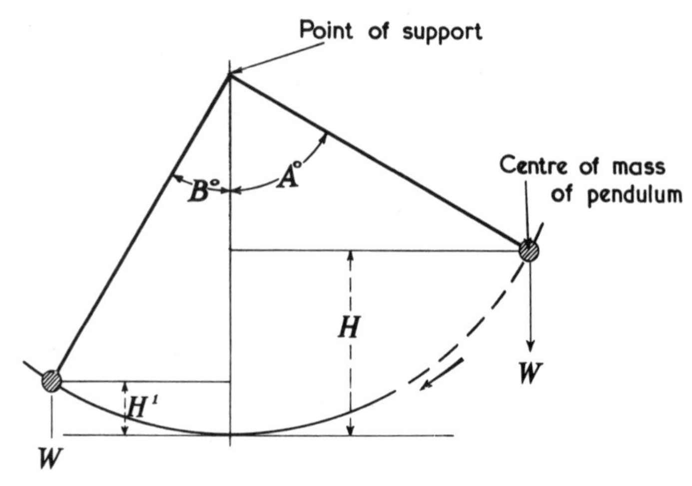

Figure 5.19 *Calculating the "energy to fracture" from an impact test.*

In general, the higher the value of "energy to fracture" the greater the notch toughness of the material. Temper brittleness, which occurs quite often in slowly-cooled *Ni-Cr* medium carbon steels, is quickly revealed by an impact test, but is not revealed by the normal tensile test. Table 5.2 relates to two samples of such a steel, one having been water-quenched after tempering, and the other air-cooled after similar heat treatment.

TABLE 5.2

TEMPER BRITTLENESS REVEALED BY THE IZOD TEST

Condition of Ni-Cr Steel	*Tensile Test Data*			*Izod*
	UTS	*Yield*	*% Elongation*	
Quenched after tempering	50.1 tsi	30.1 tsi	12%	60 ft lbf
Air-cooled after tempering	50.2 tsi	29.8 tsi	11.6%	15 ft lbf

In the second sample the temper brittleness is due to the coarse grain size developed as a result of air cooling.

FATIGUE IN METALS

Fatigue failures may occur in components subjected to many applications of a load that is well below that which would cause failure in a static test. Thus, stresses within the component must be cyclic if failure by fatigue is to occur, and may be caused by axial, shearing, flexural, or torsional loads, or by combinations of any of these types of loads. Fatigue failures are very common and account for about 80% of all failures of machine components.

A distinction must be made between those structures such as bridges and buildings in which stresses are not great enough and are not applied often enough to cause fatigue failure, and those components such as a connecting rod in a high-speed aero engine in which 25 million complete stress reversals could occur in about 200 hours of flying time. Axles, bolts, springs, gear teeth, turbine blades, engine parts, railway wheels, and bearings are subjected to fatigue-inducing stresses and must be designed and inspected with this in mind.

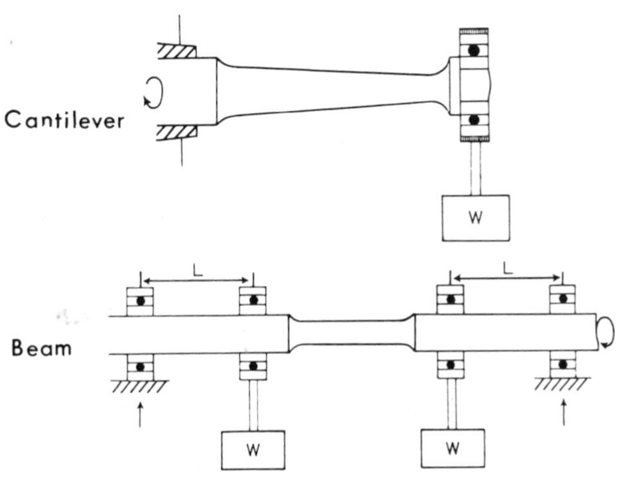

Figure 5.20 *Methods of fatigue testing.*

The only really reliable fatigue test is to take the actual structure or component and subject it to stress conditions approximating those encountered in service. This has been done with airframe components but is time consuming and expensive. Basic information is usually obtained from standardised tests on specially-shaped components.

One of the most common types of testing machine is the Wöhler machine in which the standard specimen is tested as a cantilever. A known "dead" load is applied by means of a ball-race fitted to the free end of the specimen, this producing a sinusoidal stress variation as the specimen revolves. A variation of this type of test is when a two-point loading system is used and the specimen is tested as a beam, this having the advantage that stresses are uniform over the gauge length of the test specimen. The results of such tests are plotted as graphs of applied stress against cycles to failure (S-N curves), the usual procedure being to make a number of specimens of the same material and test each one under different stress conditions.

Two different types of S-N curves occur, one showing a definite levelling off of the curve, which indicates that the material has a definite fatigue limit. Commonly, ferrous metals show distinct fatigue limits, while non-ferrous metals do not.

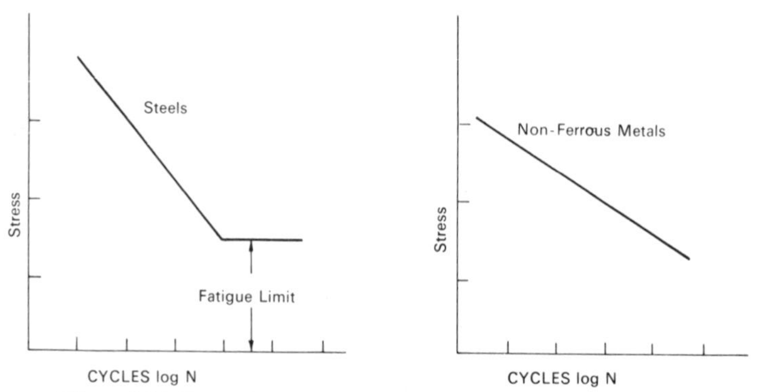

Figure 5.21 *Typical S-N curves for steels and non-ferrous metals.*

The fatigue limit of a material is defined as that value of (maximum) stress below which fatigue failure will not occur irrespective of the number of cyclic stress applications. This is to be contrasted with the endurance limit of a material, which is defined as the stress which just causes failure after a certain number of stress applications has occurred. Non-ferrous metals have endurance limits, while ferrous metals show definite fatigue limits.

The Nature of Fatigue Failure

Failures by fatigue occur suddenly with none of the features of ductile failures; there is no localised plastic deformation and the fractured surface appears coarsely crystalline. The appearance of fatigue failures led some early investigators to propose the "cold crystallisation" theory of failure but this is now known to be false. Fatigue failures usually reveal two distinct zones; one portion is smooth and discoloured and shows more or less concentric "ripple marks", while the other has a "torn" crystalline or fibrous appearance. The first-mentioned zone is formed when the minute crack or cracks (indicating the beginning of fatigue) begin to propagate across the cross-section of the component due to repeated stress applications; the "fibrous" zone is formed when the component ruptures suddenly because of its inability to withstand further the stress conditions. Thus, fatigue failures can be termed "progressive fractures".

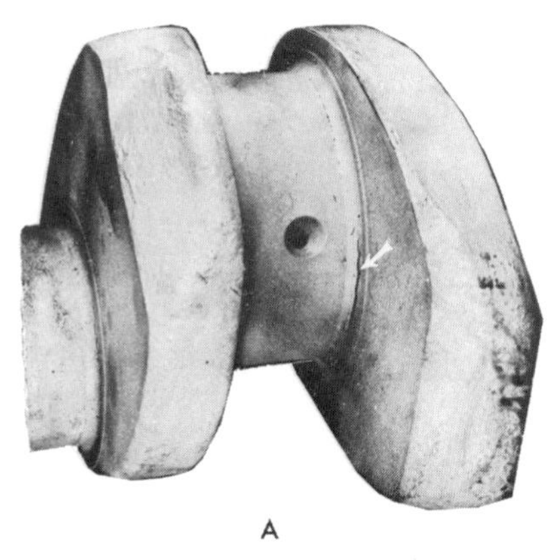

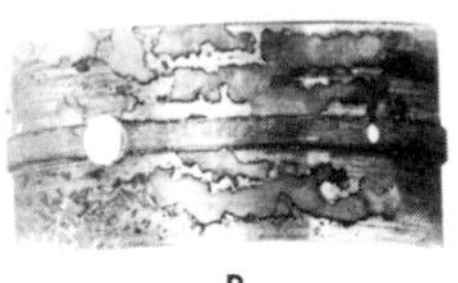

A B

Figure 5.22 *Types of fatigue failures: (A) shows a fatigue crack in a crankshaft (arrowed); (B) shows fatigue failure in white metal bearing slippers. (Photographs courtesy of the New South Wales Department of Railways.)*

The initial crack of a fatigue failure is formed by the concentration of stress at some point of weakness such as a sharp corner, file mark, keyway, or corrosion pit, such features being known as "stress raisers" for this reason. Surface finish becomes of great importance to parts subjected to fatigue-inducing conditions, since surface imperfections account for many instances of stress concentration. Connecting rods for high-speed engines are always highly polished to prolong the service life, while the design engineer avoids sharp corners, small fillet radii, and rapid changes of section since these are also common points of stress concentration in machine components. Non-metallic inclusions, particularly those near the surface, can act as stress raisers, particularly if they are brittle.

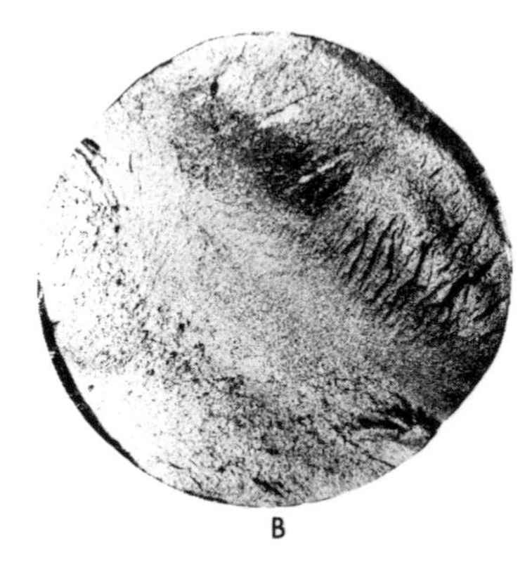

A B

Figure 5.23 *Types of failures in steel shafting: (A) shows the fatigue type of failure caused by crack propagation in a ductile metal; (B) shows a typical brittle failure. (Photographs courtesy of the New South Wales Department of Railways.)*

Shot peening, which hardens a metal surface by placing it in compression, and heat treatment processes which increase hardness and toughness, both increase fatigue strength since they strengthen the "skin" of the metal which is the area from which fatigue cracks begin.

HARDNESS TESTS

Hardness tests are the most commonly used non-destructive testing procedures in industry and research since they provide an easy and thoroughly reliable method of evaluating the effects of various hot and cold working techniques and heat treatments upon the basic properties of metals. Hardness can be evaluated in a qualitative way using Moh's scale, by scratching with a file, or by lightly touching the piece of metal on to a grinding wheel. However, more sophisticated and accurate procedures are necessary. Most modern hardness tests are of the indentation ("static") type, although several dynamic tests are available. The following are the six most common hardness tests used today.

(1) Rockwell Tests

In these tests a standard indenter is pressed into the surface of the material by a "dead" load of known magnitude, the index of hardness being obtained by measuring the depth of penetration. A number of different scales are used, each scale being suitable for certain classes of materials. It should be understood that each scale is entirely arbitrary, the hardness number obtained having relevance to that particular Rockwell scale only. Table 5.3 shows some of the more commonly used scales.

TABLE 5.3

ROCKWELL HARDNESS SCALES

Scale	*Indenter*	*Major Load, Kg*	*Dial Numerals*	*Typical Materials*
A	Diamond cone	60	Black	Cemented carbides, thin steel, case-hardened surfaces.
B	$\frac{1}{16}$" ball	100	Red	Copper, aluminium, brass, malleable iron, grey cast iron.
C	Diamond cone	150	Black	Hard cast iron, deep case-hardened surfaces, hardened steels.
D	Diamond Cone	100	Black	Thin steel specimens.
E	$\frac{1}{8}$" ball	100	Red	Soft aluminium and alloys, magnesium alloys, bearing metals.
F	$\frac{1}{16}$" ball	60	Red	Annealed copper and alloys, thin soft sheet metals, bearing alloys.

Table 5.3 shows that major loads and indenters vary from scale to scale. The diamond cone or brale is ground to an included angle of 120°, the tip being rounded off to a radius of 0.02 mm, while ball indenters are made of hardened tool steels or, less commonly, of tungsten carbide. Ball indenters tend to deform appreciably under load so that inaccuracies must be expected in readings near the top of scales employing such indenters; however, the use of carbide indenters minimises these errors.

The test procedure is as follows. The machine is set up with the appropriate indenter and proportional load and the test piece or component is placed on a suitably-shaped anvil. The anvil is then raised until the dial on the machine registers the application of a 10 Kg minor load; this forces the indenter just below the surface thus preventing surface irregularities from influencing test results. The dial indicator is then "set" to zero and the major load applied through an oil damper which controls the rate of load application. The load is applied and maintained for up to 15 seconds and then released, the *hardness number* then being read off the scale to the nearest whole number.

The test is easily applied and leaves only a small indentation on the specimen, and is thus very suitable for the routine testing of mass-produced articles as they move through various stages of manufacture. Its disadvantage lies in the fact that each scale is arbitrary and correlation with the other scales and indeed other tests is quite difficult.

(2) The Brinell Test

The Brinell test consists of pressing a hardened steel ball into the surface of the specimen using a known load, the surface area of the indentation so produced becoming the index of hardness. Various combinations of loads and indenters are used, the most common being a 3,000Kg or 500Kg load with a 10mm ball or a 750Kg load with a 5mm ball. Minor loads are not employed, so the surface finish on the specimen to be tested must be high. Considerable deformation can occur in the ball indenter when hard materials are tested, and, even though tungsten carbide balls are sometimes used, hardness numbers of 600 or more are rarely encountered. In stating test results it is most important to give details of load, indenter, and time of load application, the latter usually being about 15 seconds. The hardness number may be calculated from the following relationship:

$$\text{BHN} = \frac{\text{load on ball}}{\text{area of indentation}}$$

$$= \frac{P}{\frac{\pi D}{2}\left(D - \sqrt{D^2 - d^2}\right)} \qquad (15)$$

where: P = applied load in Kg
D = indenter diameter in mm
d = diameter of indentation in mm.

Indentation diameters must be read using a measuring microscope, the order of accuracy being ± 0.01mm. In practice, the hardness number corresponding to a particular indentation diameter is read off from a table in which load, indenter size, indentation sizes, and hardness numbers are correlated. It is often quite difficult to read indentation diameters accurately as "sinking" or "piling" often occur around the indentation.

The Brinell test suffers from some serious disadvantages:

(a) It produces a fairly large indentation, particularly on soft materials.
(b) It cannot be used near the edge of the specimen.
(c) It cannot be used on thin metals since reactions from the test anvil might influence test results.

However, irrespective of these limitations, it does give a linear scale of hardness and is particularly useful for research work. It is rarely employed for production testing since it is time consuming, requires expensive equipment, and leaves large indentations on the work.

(3) Vicker's Diamond Pyramid Tests

The test is similar to the Brinell in that the hardness number is derived from the relationship between the applied load and the surface area of the indentation. However, some of the difficulties of the Brinell are avoided by using a square-based, 136° included-angle diamond indenter and loads varying from 5 to 120Kg. The area of indentation is found using the diagonal measurement of the indentation which is measured to an accuracy of 0.001mm. As in the Brinell test, hardness numbers are read off tables, which in this instance correlate load and the length of the diagonal of the indentation.

The test is rapid, accurate, suitable for metals as thin as 0.006″, can be used up to values exceeding 800 Brinell, and is most suitable for determining the hardnesses of case-hardened or nitrided surfaces.

(4) Microhardness and Superficial Tests

The indentation tests described so far are, with the exception of the Vicker's test, unsuitable for thin materials and surface-hardened objects. For this reason special tests have been devised for such purposes, their main features being their smaller loads and special indenters. Such hardness tests have become known as superficial tests and microhardness tests, the Rockwell Superficial and the Tukon-Knoop Microhardness tests being typical. The Rockwell Superficial test employs a minor load of 3Kg and major loads of 15, 30 and 45Kg, while the Tukon-Knoop test uses a specially-shaped diamond indenter impelled by loads ranging from 25 to 3,600gm. These tests are exceptionally useful for small mass-produced articles such as watch springs and instrument gears, as well as for thin case-hardened or nitrided surfaces on larger components.

(5) Indentation Test for Wood

A very simple test is used to determine the hardness of wood. A hard steel ball of diameter 0.444″ is pressed to half its diameter into the wood, and the load necessary to do this is measured. This load, measured in pounds weight, becomes the hardness number, oregon having a hardness of about 900.

TEST	INDENTER	SHAPE OF INDENTATION SIDE VIEW	SHAPE OF INDENTATION TOP VIEW	LOAD	FORMULA FOR HARDNESS NUMBER
Brinell	10 mm sphere of steel or tungsten carbide	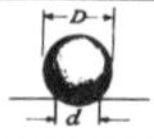		P	$BHN = \frac{2P}{\pi D(D - \sqrt{D^2 - d^2})}$
Vickers	Diamond pyramid	 136°		P	$VHN = 1{\cdot}72\ P/d_1^2$
Knoop Microhardness	Diamond pyramid	$l/b = 7.11$, $b/t = 4.00$	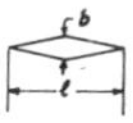	P	$KHN = 14{\cdot}2\ P/l^2$
Rockwell A, C, D	Diamond cone	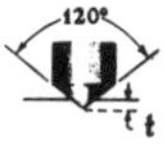 120°		60 kg, 150 kg, 100 kg	$R_A =$, $R_C =$, $R_D =$ } 100-500
B, F, G	1/16 in. diameter steel sphere			100 kg, 60 kg, 150 kg	$R_B =$, $R_F =$, $R_G =$ } 130-500
E	1/8 in. diameter steel sphere			100 kg	$R_E =$ } 130-500

Figure 5.24 *Indentation hardness tests in summary. (After Table 1.1 in Volume 3 of "The Structure and Properties of Materials" by Hayden, Moffatt, and Wulff. John Wiley and Sons Inc., N.Y. 1965.)*

(6) The Shore Scleroscope

In contrast to the previously discussed tests, the Shore Scleroscope test is dynamic in that the load is applied by means of a falling weight and the hardness is assessed from the height of rebound of this weight. The machine is portable and consists of a graduated tube and the standard weight or "hammer" which weighs about 1/12th of an ounce. The hardness scale is entirely arbitrary and runs from 140 to zero, the 140 corresponding to a rebound of 10″. In operation the machine is held firmly on the surface to be tested and the tube is held vertically. The hammer is dropped down the tube from a height of 10″, and the rebound noted. The surface should be flat, clean, and free from scale. A new spot should be used for each test as work hardening may influence test results. This is not really a valid test of hardness since the elasticity and resilience of the material will influence test results. It is almost impossible to correlate Scleroscope readings with those obtained from indentation tests.

Laboratory (25): *Determine the relative hardnesses of the following materials by using appropriate indentation tests: cast iron (grey), mild steel, hardened tool steel, annealed copper, work-hardened copper, aluminium, a clay-body ceramic, perspex, and oregon. List the results, together with details of the test procedure, loads, indenters, and any other important variables.*

TABLE 5.4

APPROXIMATE CONVERSIONS OF BRINELL, ROCKWELL AND SCLEROSCOPE NUMBERS AND ULTIMATE TENSILE STRENGTH

Brinell Number	*Rockwell Numbers*		*Scleroscope Number*	*Tensile Strength 1,000 psi*
	"C"	*"B"*		
780	70	—	106	384
745	68	—	100	368
682	64	—	91	337
653	62	—	87	324
627	60	—	84	311
601	58	—	81	298
555	55	120	75	276
534	53	119	72	266
514	52	119	70	256
495	50	117	67	247
461	47	116	63	229
429	45	115	59	212
415	44	114	57	204
401	42	113	55	196
388	41	112	54	189
341	36	109	48	165
321	34	108	45	155
302	32	107	43	146
293	31	106	42	142
277	29	104	39	134
262	26	103	37	128
248	24	102	36	122
241	23	100	35	119
229	21	98	33	113
223	20	97	32	110
217	18	96	31	107
207	16	95	30	101
197	13	93	29	97
192	12	92	28	95
183	9	90	27	91
174	7	88	26	87
166	4	86	25	83
159	2	84	24	80
146	—	80	22	74
140	—	78	21	71
134	—	76	21	68
131	—	74	20	66
126	—	72	—	64
121	—	70	—	62
116	—	68	—	60
107	—	64	—	55
101	—	60	—	52

INSPECTION OF MATERIALS

Techniques of inspection can be classed as non-destructive tests and may be divided into two main groups as follows.

(a) Tests to check the accuracy of the dimensions of a manufactured component.

(b) Tests to locate surface and internal flaws and defects both in stock material and in finished articles.

TABLE 5.5

INSPECTION PROCEDURES AND THEIR GENERAL APPLICATIONS

Testing Dimensional Accuracy	*Testing for Defects*	
	Surface	*Internal*
Accurate measuring instruments such as micrometers and vernier calipers; gauges of all types; magnifying comparators to check small parts and thread detail, etc.	Visual Examination	Hammer Tests
	Penetrant Tests	Radiation Tests
	Magnetic Particle Tests	Ultrasonic Methods

Materials are selected for particular applications largely because of the mechanical properties revealed by destructive tests. However, there is no guarantee that the actual piece used for a certain application has properties identical to those of the test piece. In fact, very dramatic reductions in strength properties occur due to surface and internal flaws, cracks, and voids which seem to act as points of stress concentration. For this reason the inspection of axles, connecting rods, crankshafts, and other parts where fatigue is likely to occur is most important. Similarly, welded joints are usually inspected in load-bearing structures to discover such defects as gas porosity, slag inclusions, and cracking, each of which severely weakens the weld and could cause premature failure.

Visual examination is always the first step in the inspection of a component, and may be done with the unaided eye or by using a hand lens or a microscope. Although a seemingly elementary procedure, it should be carried out carefully and systematically on every component.

Penetrant tests reveal discontinuities that are open to the surface and they may use dyes or fluorescent materials. They are especially useful for non-ferrous metals and non-metallics such as ceramics and plastics since these cannot be tested using magnetic methods. The simplest test involves dipping the component into kerosene, wiping it dry, and then coating it thinly with whiting. Cracks open to the surface will be revealed by a discoloured "line" appearing in the whiting, this being produced when kerosene trapped in the crack seeps out slowly. Dyes may be used instead of kerosene, but fluorescent materials are probably most common. The part to be examined is dipped into a tank of fluorescent substance, the excess is removed, and the part examined under ultraviolet light. Cracks appear as glowing lines, pores as glowing spots, and large discontinuities as fluorescent areas.

Magnetic particle tests are only suitable for ferrous metals capable of being magnetised. For example, 18–8 austenitic stainless steel cannot be examined by this technique. The component to be tested is magnetised and placed in a tank of kerosene containing a fine suspension of magnetic iron oxide. Cracks and voids are revealed since iron oxide particles collect around them due to their disturbing effects upon the distribution of magnetic flux. Either residual or continuous magnetism is employed, the latter being suitable for soft steels which cannot be permanently magnetised. The suspension containing the magnetic particles is sprayed or brushed on if continuous magnetisation is employed.

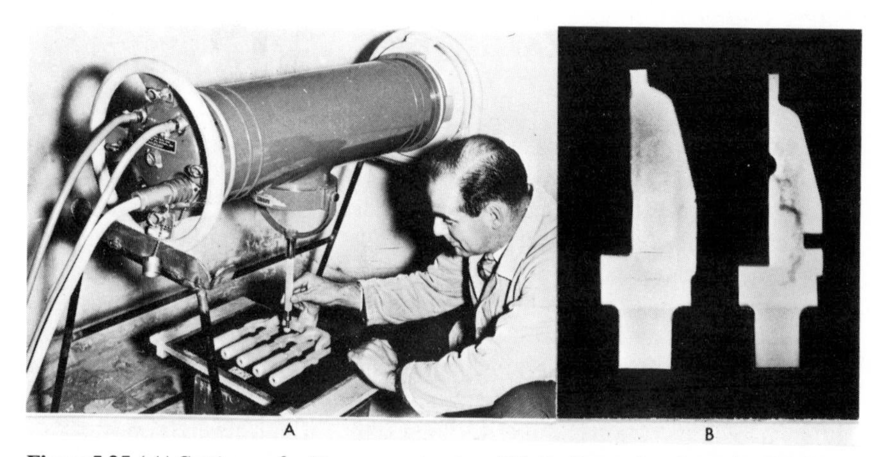

Figure 5.25 *(A) Setting up for X-ray examination. (B) Radiographs of sound and defective cast brake handles; the shadowed areas in the right-hand radiograph indicate extreme porosity in the defective casting. Note that the sound casting has not been machined. (Photographs courtesy of the New South Wales Department of Railways.)*

Radiation techniques are commonly employed to detect internal discontinuities, the method being more reliable than ultrasonic testing. Either X-rays or γ radiation is employed, the latter being more suitable for field applications since less complex equipment is required. In both methods the radiation is passed through the metal being examined and is then allowed to impinge upon sensitive film. The dark areas on this film are formed by the internal defects which absorb less radiation than the sound metal. In general, X-rays are preferred for laboratory testing since they offer greater control over intensity, and they produce sharper pictures.

Ultrasonic testing is a convenient method now in common use whereby discontinuities inside the specimen are detected by ultrasonic waves of frequencies between 100,000 and 20,000,000 cycles per second. The ultrasonic waves are usually produced by the piezoelectric effect within the crystal probe which is placed on the surface of the specimen. Discontinuities below the surface cause reflection of the ultrasonic waves which appear as peaks upon the cathode ray oscilloscope receiver. Of course, a normal reflection is obtained due to reflection from the opposite side of the specimen but this is very readily distinguished from other "abnormal" reflections. The size of the peak seen on the receiving tube is some indication of the size of the defect. The crystal probe thus becomes the receiver as well as the transmitter.

While radiation and ultrasonic tests are very sophisticated and yield excellent results under almost all circumstances, the older "hammer test" is still employed for the detection of internal defects. If a "sound" object, that is one free from large internal flaws, is struck sharply with a suitable hammer it emits a clear ringing note, whereas a defective object emits a flat note. This test, though completely qualitative, yields valuable information to a skilled operator about the quality of the object being tested.

In conclusion, it bears repeating that a manufactured component is only as good as the material from which it is made, and that materials testing before, during, and after manufacture is essential if consistent standards are to be maintained.

GLOSSARY OF TERMS

Elastic limit: the maximum stress that a material can endure without taking a permanent set.

Endurance limit: the measure of fatigue strength applied to metals not showing a definite fatigue limit; it is that stress which causes failure after a specified number of cyclic stress applications has occurred.

"Energy to fracture": the energy required to propagate a crack through a notched specimen subjected to an impact load; it is usually measured by an Izod or a Charpy test.

Engineering stress: the stress in a specimen calculated on the basis of the original cross-section area; it is always less than the true stress.

Fatigue limit: the maximum stress below which fatigue failure will not occur irrespective of the number of cyclic stress applications.

Modulus of elasticity: see "Young's Modulus".

Modulus of resilience: the amount of energy required to deform unit volume of a material up to its proportional limit.

Modulus of stiffness: see "Young's Modulus".

Percentage elongation: the total percentage strain occurring in a tensile test specimen tested to failure.

Percentage reduction in area: the difference, expressed as a percentage, between the original cross-sectional area of a tensile test specimen and the smallest area at the point of rupture.
Poisson's Ratio: the ratio existing between the lateral and longitudinal strains produced in a body subjected to a simple uniaxial stress within its elastic limit.
Proof stress: the stress that will cause a certain specified amount of permanent deformation in a test specimen; usually measured as 0.1 % or 0.2 % Proof Stress in the tension test.
Proportional limit: the maximum stress that a material can withstand without deviating from straight-line proportionality between stress and strain.
Shear stress: components of stresses that act along or parallel to a given plane, known as the plane of shear.
Strain: the ratio between total deformation in one direction and the length of the specimen as measured in that direction. Note that engineering strain is calculated on the basis of the original dimensions of the specimen.
Stress: the intensity of the (reaction) force at any point in a body subjected to a load. Stress is measured as the force per unit area of any given plane within the body.
Ultimate tensile strength: the maximum value of engineering stress that a tensile test specimen can withstand during the duration of a tensile test.
Yield strength: the stress at which a material exhibits some definite deviation from straight-line proportionality between stress and strain. Measures of yield strength include proportional limit stress, stress at the yield point, and proof stress.
Young's Modulus: the ratio between uniaxial tensile stress and elastic strain in a material obeying Hooke's Law. Also known as the "modulus of elasticity" and the "modulus of stiffness".

REVIEW QUESTIONS

1. Why is it important to conduct routine tests according to recognised standards?
2. Explain the essential differences between routine tests, experimental tests, proving tests, and inspection.
3. Define the terms "stress" and "strain", and explain how engineering stress differs from true stress.
4. In what units would strain be measured?
5. What is the significance of Poisson's Ratio?
6. List and describe the functions of the four main parts of a universal mechanical testing machine.

7. What data are usually calculated from routine tensile tests on (i) ductile materials (ii) brittle materials?
8. Distinguish between the elastic limit, the proportional limit, and the yield point of a metal.
9. The following data were obtained from a standard tensile test on a mild steel specimen:

Stress psi	*Strain*	*Stress psi*	*Strain*
10,000	0.00036	50,000	0.00184
20,000	0.00068	60,000	0.00272
30,000	0.00102	70,000	0.00418
40,000	0.00130	80,000	0.00710

Graph these results and determine the proportional limit stress, Young's Modulus, and 0.1% proof stress for the steel.
10. What is the significance of the Modulus of Resilience of a material?
11. Discuss four ways in which the ductility of a material may be assessed.
12. List and discuss four common difficulties encountered during compression testing.
13. For what materials would the compression test be regarded as a routine acceptance test? Why?
14. Show how the formula $\frac{3PL}{2bd^2}$ is derived for the transverse strength of a beam supported at both ends and loaded centrally.
15. Why are shear tests of rivets and bolts important?
16. A specimen from an Erichsen cupping test was found to develop circumferential cracks after only a shallow cup was formed. Explain why this material would be unsuitable for deep drawing applications.
17. What is meant by the term "notch toughness" and how is it connected with the energy to fracture of a material?
18. Compare and contrast the Izod and the Charpy tests.
19. Why are notched-bar impact tests suitable only for routine tests of quality control and not for investigations into the fundamental properties of materials?
20. What particular conditions of service induce fatigue failures in metal components?
21. What are stress raisers and how do they originate in metal specimens?
22. How does polishing a connecting rod in a high-speed internal combustion engine help to reduce the possibility of failure by fatigue?
23. Outline the significant differences between Rockwell and Brinell hardness tests.
24. Why is the indentation hardness test the most commonly employed test of quality control during the manufacturing of components such as axles, gears, and valves?

25. A material is to be subjected to a series of routine tests to determine its suitability as a structural material. Describe two tests that could be used and justify your selection.

26. It is suspected that the heat treatment of valves made from medium carbon steel is producing unreliable results. Describe two tests which, if taken together, would be an accurate check upon the suitability of the heat treating procedure.

27. What are the purposes of the systematic inspection of engineering materials?

28. Describe the principles of the following methods of crack detection: (i) ultrasonic testing; (ii) magnetic particle testing; (iii) radiation testing. What are the limitations of each method?

6

Casting and As-cast Structures

CASTING involves the pouring of molten metal into a prepared mould cavity which has the shape of the article to be made. Looked at in this way ingots are castings, which differ, however, from most other castings in that the latter are usually made as closely to their final shapes as possible, whereas ingots are subsequently changed in shape by rolling, extrusion, or some other forming process. In general, casting produces articles more cheaply than forging or fabrication, so casting is economically desirable. Metallurgical considerations also make casting a desirable procedure since, in some circumstances, castings possess no directional properties. Tank armour, for instance, is always cast, as are cylinder liners, engine blocks, and gun tubes.

While a large number of different and highly specialised casting processes are used in modern industry, the following four basic considerations apply to all forms of casting.

(a) The metal to be cast must be able to be melted cleanly and economically. Melting may be done in a variety of furnaces with coke, pulverised coal, fuel oil, gas, or electricity as fuel.

(b) A suitable mould cavity must be produced, enlarged sufficiently to compensate for the cooling shrinkage of the solidifying metal. This mould cavity must have some suitable means of metal access; must allow for the escape of gases either trapped in the mould during pouring or formed by the action of heat on the mould itself; and must not restrain the casting as it cools, since this type of restraint develops internal stresses within the casting which may cause cracking or premature failure in service.

(c) It must be possible to remove the solidified casting from the mould cavity. This is no problem when the mould itself is destroyed in order to remove the casting (e.g. sand moulding, shell moulding), but may become a problem when some form of permanent mould is being used (e.g. die-casting).

(d) Unless some form of permanent mould or die is being used, a pattern must be made. This is the responsibility of the patternmaker who shapes the original pattern in wood, plaster of Paris, or a suitable plastic, according to drawings supplied by the designer. If only a small number of castings are to

be made, this original pattern may be used directly, but more commonly it is used to make some form of pattern plate or die, these lending themselves to mass-production techniques. A pattern plate simply consists of a plate of suitable thickness having one half of the pattern on each side; it is usually cast in aluminium. If a die is made, patterns could be made in wax or frozen mercury, both of these materials being commonly employed for investment casting processes. Shrinkage allowances are added on to actual sizes as the pattern is being made, common allowances being—

Cast iron	$\frac{1}{8}''$ per foot
Aluminium	$\frac{5}{32}''$ per foot
Magnesium	$\frac{5}{32}''$ per foot
Brass	$\frac{3}{16}''$ per foot

Figure 6.1 *A wood pattern for an automobile roof rack bracket. (Courtesy D. V. Williams.)*

SAND CASTING

Sand, basically silica SiO_2 together with small amounts of clay, is used to form the mould, and each sand mould produces only one casting. The sand is reclaimable, but needs suitable reconditioning before it can be used again. Sand is fine enough to be packed into thin sections and refractory enough even for steel castings, which require very pure sand.

The moulder is responsible for the production of the sand moulds using suitable patterns and moulding boxes, this process being mechanised in modern foundries. A vital aspect of mould manufacture is concerned with the packing or ramming of the sand; if the sand is loosely rammed the mould may not be strong enough to withstand the metal-pouring process, and certainly a poor surface finish will result, and if it is too tightly rammed gases may not be able to escape and porosity or "blows" may result. The most successful mechanical method of filling sand moulds is the sand slinger, in which the conditioned sand is "flung" from a mechanical slinger into the mould in such a manner that the desired degree of compaction is obtained from this action alone.

Sand moulds are always gravity-fed; thus a good system of runners and risers is necessary if sound castings are to be produced. Risers are usually located so that they will "feed" the heavier sections of the casting as it is cooling and contracting. Holes and other cavities in the casting are cored out

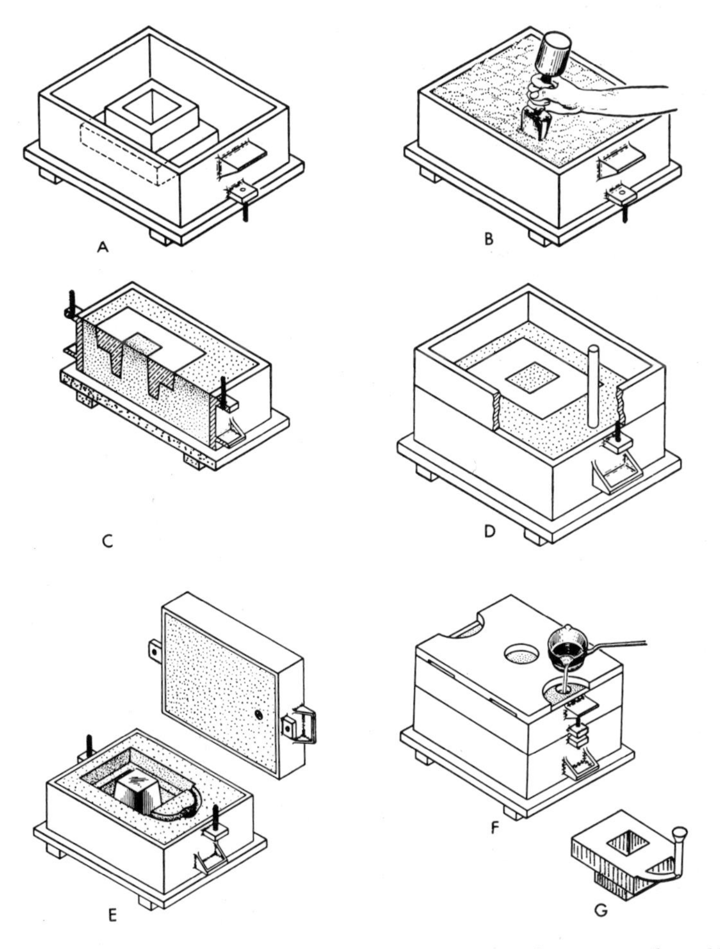

Figure 6.2 *Various stages in the ramming up and pouring of a green sand mould having a straight parting line; (G) shows the casting with the runner still attached. (Reproduced with permission from "Fundamentals in The Production and Design of Castings" by C. T. Marek. John Wiley and Sons Inc., N.Y. 1950.)*

using suitably shaped sand cores. These are made of sand containing about 2% linseed oil and are baked at about 230°C to give them strength. The use of cores conserves metal, cuts down on expensive machining, and prevents excessive shrinkage in thicker areas of the casting.

Sand moulds fall into the following four groups.

(a) Green Sand Moulds in which no attempt is made to dry out the slight amount of moisture added to the sand to bind it together.

(b) Dry Sand Moulds are those which are completely dried by heating to about 250°C, this giving them increased strength and at the same time effectively reducing the amount of steam generated when the hot metal enters the mould.

(c) CO_2 Sand Moulds use a mixture of sand and about 4% sodium silicate. After ramming the sand is "gassed" with CO_2 and becomes quite hard but not impermeable. This process is used for core-making and also for accurate moulds.

(d) Moulds Employing Synthetic Binders require no water. The most successful binders set at room temperature and must readily break down after casting so that the sand can be reclaimed. These binders are particularly useful for large castings where ramming is done by the sand slinger method.

SHELL MOULDING

Shell moulding is a recent development in the general field of sand casting. Very high-dimensional accuracy can be attained, tolerances of several thousandths of an inch being common. The four basic steps in the manufacture of a two-part shell mould are as follows.

(a) Fine clean sand, mixed with a small proportion of a thermosetting resin binder, is slung onto two metal pattern plates each of which has a matching half of the full pattern.

(b) The pattern plates are maintained at a temperature between 200°C and 250°C so that the layer of moulding sand-plastic mixture in contact with the mould is partially "cured"; this layer is usually kept between 1/8" and 1/4" thick.

(c) After the pattern plates have been inverted to remove the excess sand they are placed in ovens for a few minutes to complete the curing of the adhering sand.

(d) The shells of hardened sand are stripped from the half-moulds and are clamped or glued together to form the complete shell mould.

The completed shell mould may be placed in a moulding box and coarse sand packed around it to provide increased strength, usually only necessary for large castings.

Crankshafts for automobile engines are cast by this method and it has been found that the only finishing required is the grinding of all bearing surfaces. The process is very suitable for mass production since the equipment is expensive but this cost is offset by the high production rate.

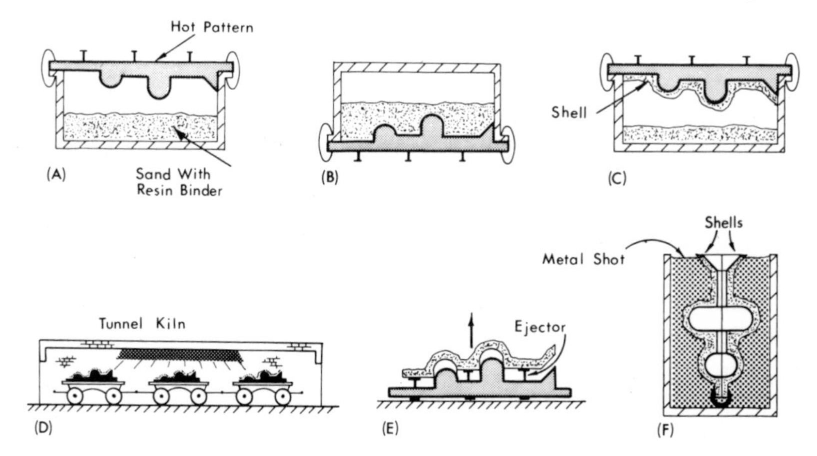

Figure 6.3 *The steps in making a shell mould: (A) the heated metal pattern is clamped to the dump box containing the resin impregnated sand; (B) the box is inverted and the hot pattern melts the resin in the sand layer nearest the pattern; (C) the box is again inverted, and a shell of sand remains attached to the pattern; (D) the curing of the resin binder is completed in a kiln; (E) the half shell mould is stripped from the metal pattern; (F) two halves are clamped together and packed into a moulding box ready for pouring.*

PERMANENT MOULD CASTING

The major cost factor in sand castings is the production of the moulds; hence the use of expensive but permanent moulds or dies from which several thousand castings can be made significantly reduces costs for long production runs.

Permanent moulds are usually made of cast iron, and metals such as cast iron, copper, aluminium, magnesium, lead, and tin are cast in commercial quantities in these moulds. The metal mould must be maintained at an elevated temperature in order to prevent chilling of the casting, and some sort of refractory wash is coated on to the mould surface just prior to pouring to prolong mould life. Cores, usually of the sand type, may be used, but the shapes that can be cast are somewhat limited by the problems associated with the removal of the solid casting from the mould. The whole process can be automated and little skilled labour is required.

One rather interesting variation, known as slush casting, involves letting the metal in contact with the mould surface solidify, and then inverting the mould so that the rest of the metal runs back into the ladle. This method produces hollow castings, and is commonly used to make toy lead soldiers, candlesticks, lampbases and the like.

DIE CASTING

Die casting, though similar in many respects, differs from permanent mould casting in that the metal is forced into the mould cavity under pressures which may be as high as 20,000 psi. The process is only economically feasible for large production runs since the cost of manufacturing the steel dies is very high. Furthermore, there are shape limitations, some components not lending themselves to die casting because they would require four, five or more piece dies.

Two different types of die-casting machines are in common use; the cold chamber type which operates on pressures up to 20,000 psi has molten metal introduced into it from a ladle, while the gooseneck machine, in which pressures rarely exceed 600 psi, melts its own metal and is thus self-contained.

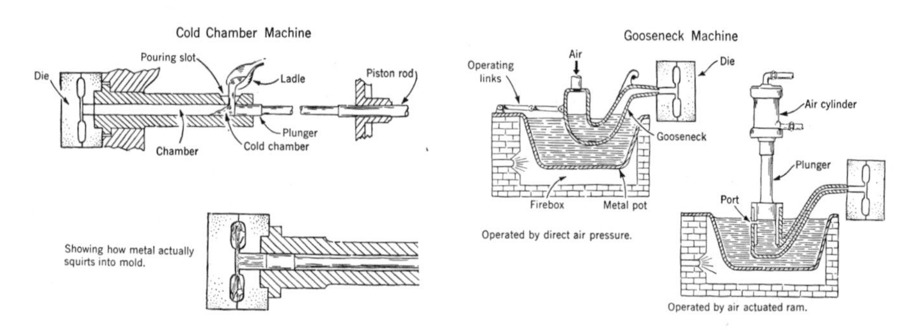

Figure 6.4 *The principle of two types of die casting machines. (Reproduced with permission from "Metallurgy for Engineers" by Wulff, Taylor, and Shaler. John Wiley and Sons Inc. N.Y. 1952.)*

The whole process is usually completely automated. The castings typically require little or no finishing; their surface finish is excellent, and hence they can be electro-plated after suitable cleaning. The dies themselves are water cooled to prolong their working life. Die-casting metals are commonly either zinc or aluminium-based, but the brasses can be satisfactorily die cast in the cold chamber machine.

INVESTMENT CASTING

Investment casting, although used extensively by artists over the centuries, has only gained industrial importance since World War II when it was extensively used to precision cast materials not readily machinable. In this process the pattern is expendable and is usually made of wax, although frozen mercury, plastic, tin, or other low melting-point metals may be used. A master pattern is made, usually from wood, and from this a set of master dies are prepared using some low melting-point metal. Alternatively, the master dies can be cut directly into steel, a more expensive method, but one which gives dies of exceptionally long service life. The master dies are then used to make the expendable patterns; if wax is used (this being the most common material) it is injected under pressure while in the molten state and allowed to solidify in the dies. The wax patterns are then given a thin refractory coating, placed into suitable moulding boxes, and a refractory plaster mixture is poured in and allowed to take an initial set. The hardened plaster cast is stripped from the moulding box, inverted, and passed through an oven where the wax is melted out. (Since the pattern is removed in this manner very complex shapes can be cast using this technique). The molten metal is then poured into the hot plaster mould which is later broken open to recover the solidified casting.

Excellent dimensional accuracy is possible, and a high degree of surface finish results from the very smooth mould surfaces. Sections as thin as 0.015″ can be cast with little difficulty. The process can be completely automated and indeed is only economically feasible when large numbers of identical, small, intricate, castings are to be made.

CENTRIFUGAL CASTING

Centrifugal force has been utilised in a number of interesting ways in the foundry, but generally speaking it offers two main advantages: (1) hollow objects such as pipes and gun tubes can be cast without using cores, and (2) the metal can be made to flow more rapidly into mould cavities, resulting in denser castings.

Three general types of techniques are used.

(a) True centrifugal casting, where the axis is either horizontal or vertical and the shapes produced are either circular or almost circular in cross-section.

(b) Semi-centrifugal casting, in which a casting such as a spoked wheel is spun about its own axis as the metal is being poured through a central runner.

(c) Centrifuged casting, in which several to many moulds are attached to a central runner which acts as the axis of rotation. This is simply a method of obtaining denser castings by forcing the metal into the moulds under centrifugal force.

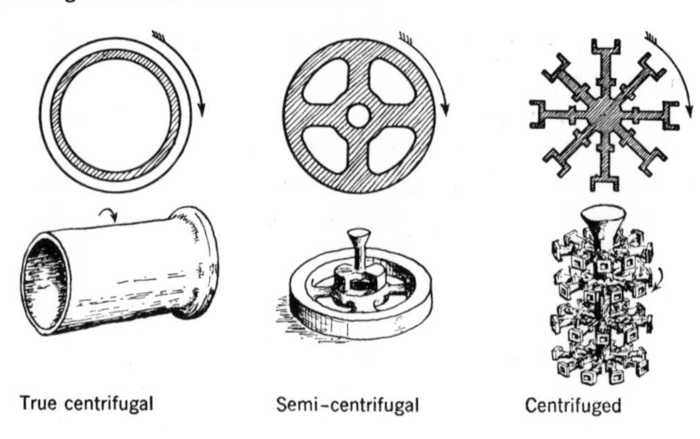

Figure 6.5 *The three principal types of centrifugal castings: (Reproduced with permission from "Metallurgy for Engineers" by Wulff, Taylor, and Shaler. John Wiley and Sons Inc., N.Y. 1952.)*

THE COOLING AND SOLIDIFICATION OF MOLTEN METAL

The solidification of molten metal is a process of crystallisation and results in the formation of a polycrystalline mass having as its principal unit the *metallic grain.* Metallic grains are true crystals in that they possess normal crystal lattices; however, they are bounded by irregular surfaces known as grain boundaries rather than by the plane faces so characteristic of minerals. The crystallisation of a pure metal has four stages.

(1) As the metal reaches its freezing point the latent heat of fusion begins to be given out and small groups of atoms form stable configurations known as nuclei, these being unit cells of the appropriate crystal lattice; copper, for instance, forms face-centred cubic unit cells as nucleation begins.

(2) Growth now begins from these nuclei as more and more atoms attach themselves to the ever-growing lattice. Growth is directional and crystal skeletons known as *dendrites* begin to form.

(3) The dendrites continue to grow until their arms almost meet, this being the initial stage in the formation of grain boundaries.

(4) The remaining liquid metal solidifies between the main arms of the dendrites by a process known as secondary growth, and the formation of the grains and grain boundaries is now complete.

Laboratory (26): *Cast a small quantity of bismuth into a metal mould, allow solidification to begin, and then pour out the molten metal from the central section before it solidifies. Examine your bled casting for cuboids of bismuth.*

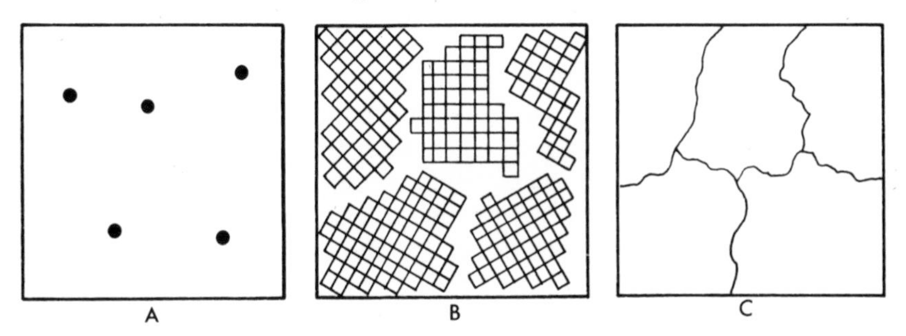

Figure 6.6 *Stages of nucleation and grain growth in a melt: (A) nuclei form as the freezing point is reached; (B) the lattices of the individual grains begin to grow; (C) grain boundaries form as crystallisation is complete.*

Grain Boundaries

The grain boundaries are disordered regions since they are the meeting places of crystal lattices that have different orientations; they therefore possess properties very different from the grains themselves. Because of the directional nature of solidification, impurities such as oxides and slag will tend to segregate around the grain boundaries. Segregation affects properties in one way or another; for instance, strength at elevated temperatures will be decreased if these impurities possess low melting points, while corrosion resistance will almost certainly be reduced irrespective of the nature of the impurity.

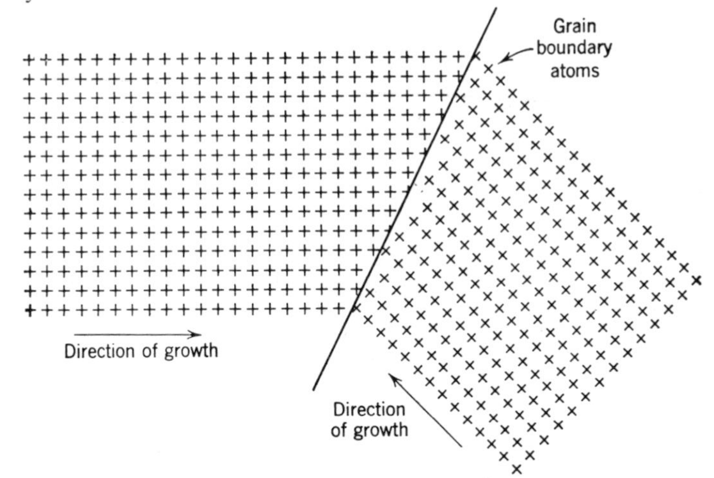

Figure 6.7 *The formation of a grain boundary. Since grain boundaries are disordered regions they are areas of high energy and are thus more susceptible to chemical attack. (Reproduced with permission from "The Nature and Properties of Engineering Materials" by Z. D. Jastrzebski. John Wiley and Sons Inc., N.Y. 1959.)*

Cooling Rates and Grainsize

The grainsize of the solidified metal is governed by the rate of nucleus formation which is, in turn, controlled by the cooling rate. Slow cooling produces few nuclei about which large dendrites grow, while fast cooling produces many nuclei and consequently small dendrites. If the metal is supercooled, that is, cooled so rapidly that its true freezing point is overshot, then the rate of nucleation is increased enormously, this rate being roughly proportional to the degree of supercooling employed.

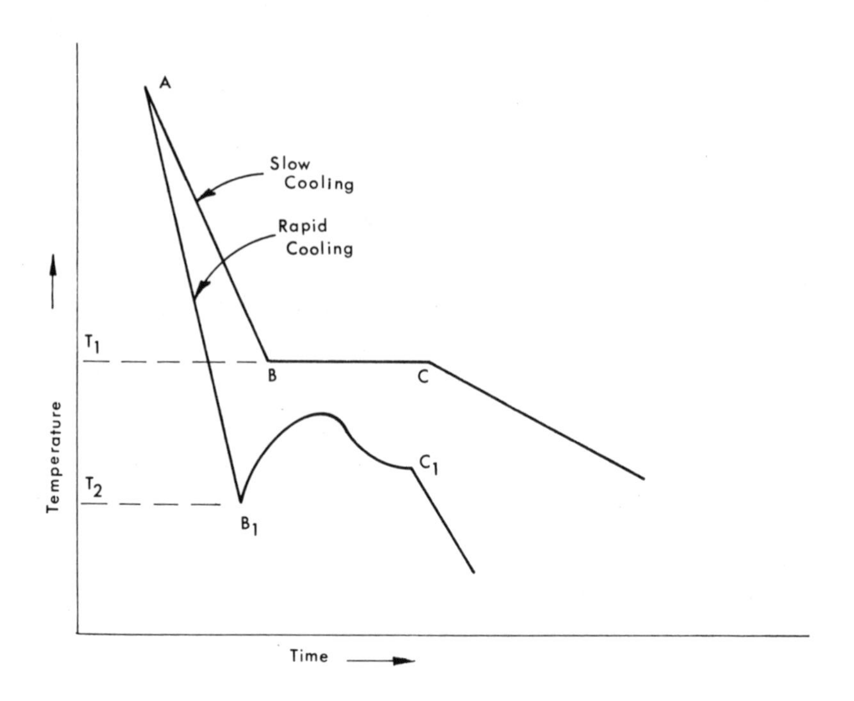

Figure 6.8 *Cooling curves for a pure metal obtained by slow and rapid cooling (undercooling). The line BC indicates the true freezing temperature of the metal; the line B_1C_1 on the undercooling curve indicates the temperature range of solidification for conditions of fast cooling.*

Agitation of the melt also increases the rate of nucleation and, if continued while the metal is in the "mushy" state (i.e. partly solidified), tends to break up the larger dendrites. Impurities present as small insoluble particles also act as nuclei, and melts may be "seeded" with such impurities so as to produce finer grainsize.

As-cast Structures

The typical as-cast metal is a fine-to-coarse crystalline mass which possesses no directional properties because of the random orientations of the grains. Many defects may be present, the most common being crystalline segregations, gas porosity, slag inclusions, and piping.

Crystalline segregations are normally found only in fairly complex alloys and may be considered to be compounds that are insoluble in the solid alloy. Such compounds can only be present in small percentages but frequently exert strong influences on the mechanical properties of the alloy.

Gas porosity occurs when gases trapped in the mould try to escape, or when steam is generated in green sand moulds by the action of the hot metal on the damp sand, or when gases dissolved within the metal try to escape as falling temperature decreases their solubility within the metal. Blowholes are formed by dissolved gases as they leave the surface of the solidifying metal; such defects are usually avoided by "degassing" the metal prior to pouring by adding some material that has a high affinity for the dissolved gas. For example, manganese is added to molten steel to remove dissolved oxygen, this process being known as deoxidising the steel.

Inclusions are usually slags of one kind or another, although fragments of furnace linings and ladles may sometimes be present. Slags are lighter than the molten metal and can thus be held back during the pouring process or alternatively skimmed off the surface just prior to pouring.

Various different types of grains may be present; this can best be seen if sections of an ingot cast into a cold metal mould are examined macroscopically.

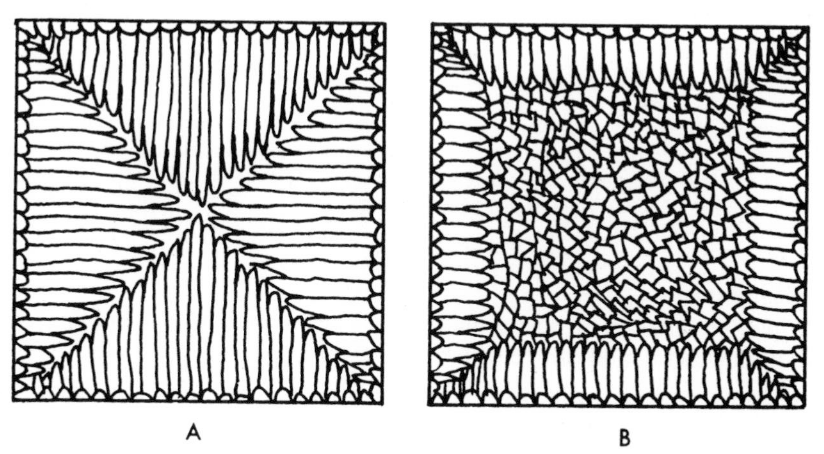

Figure 6.9 *As-cast structures in ingots. (A) shows the coarse columnar grains formed when the casting temperature is too high; (B) shows the centre of the ingot filled with the large polyhedral grains formed when the correct casting conditions are observed.*

A very thin layer of small polyhedral (or equi-axed) grains will form when the metal in contact with the mould surface cools rapidly, these grains being known as "chill crystals". Long columnar grains will then grow inwards towards the hot, molten centre of the ingot and they may meet to form planes of weakness if the casting temperature is high enough. However, correct casting temperatures result in the formation of fairly large polyhedral grains in the centre section of the ingot. Columnar grains (crystals) are weak and brittle and may cause cracking during subsequent rolling operations. Preheated metal moulds, refractory-lined moulds and sand moulds do not cause the development of columnar grains.

Figure 6.10 *As-cast deoxidised copper; the grain structure is polygonal or polyhedral. Magnification x100. (Original slide courtesy of the Copper and Brass Information Centre, Sydney.)*

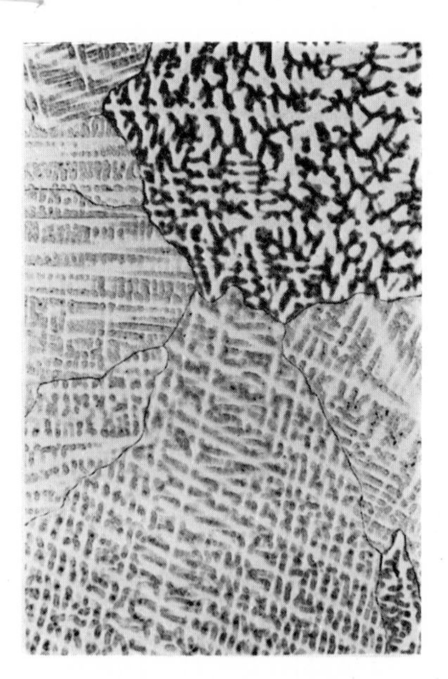

Figure 6.11 *As-cast cartridge brass showing well-defined dendrites and cored grains. Magnification x100. (Original slide courtesy of the Copper and Brass Information Centre, Sydney.)*

Piping in Ingots

Pipes are formed in ingots due to the contractions of the cooling and solidifying metal, and can be serious defects. Piping is of two types.

(a) Primary piping, which is the formation of a central, conical-shaped cavity in the upper end of an ingot.

(b) Secondary piping, which occurs inside the ingot due to lateral contractions once the top surface has solidified. In practice this occurs when moulds are used "little end up".

Primary pipes become heavily oxidised during cooling and must be cropped off before rolling; secondary pipes, however, will weld together during the rolling operation because their surfaces are clean.

Both primary and secondary piping can be avoided if a "hot top" is used on the mould; this consists of a refractory lid which is placed on the top of the molten metal after pouring and which ensures that the top of the ingot remains more or less flat. Some cropping is still necessary in order to remove the oxidised layer.

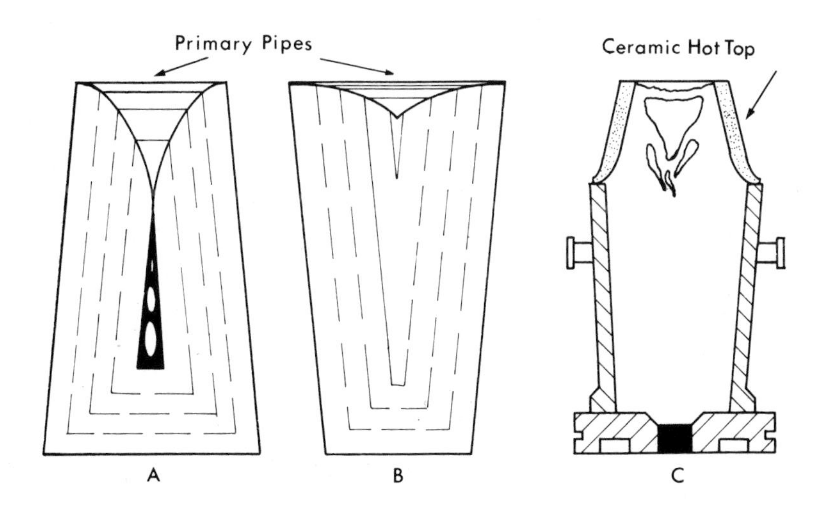

Figure 6.12 *Piping in ingots: (A) shows a primary and secondary pipe in a "little end up" ingot; (B) shows a primary pipe in a "big end up" ingot; (C) illustrates the use of a ceramic hot top to prevent excessive piping in the ingot (the pipe forms in the top and is trimmed away).*

Cored Grains

If an alloy such as brass is cast, there is a variation in composition across each grain and the dendritic mode of growth is clearly visible. Such grains are said to be cored, and occur because the original dendrites formed are richer in one of the constituents of the alloy. Brass, for example, forms dendrites rich in copper while the outer areas of each grain are rich in zinc. This type of grain structure reduces the formability of the alloy but can be removed by annealing.*

Cored grains do not occur in pure metals since the dendrites have the same composition as the rest of the grains.

*A more complete treatment of coring will be given in Chapter 9 and of annealing in Chapter 12.

CONTINUOUS CASTING

Continuous casting is a means whereby round ingots, rods, square and rectangular billets, and sometimes sheets can be obtained directly from molten metal using highly-specialised moulds and casting equipment. The method is not new, one of the earliest patents being taken out by that inventive Englishman, Sir Henry Bessemer, in 1858. Although copper,

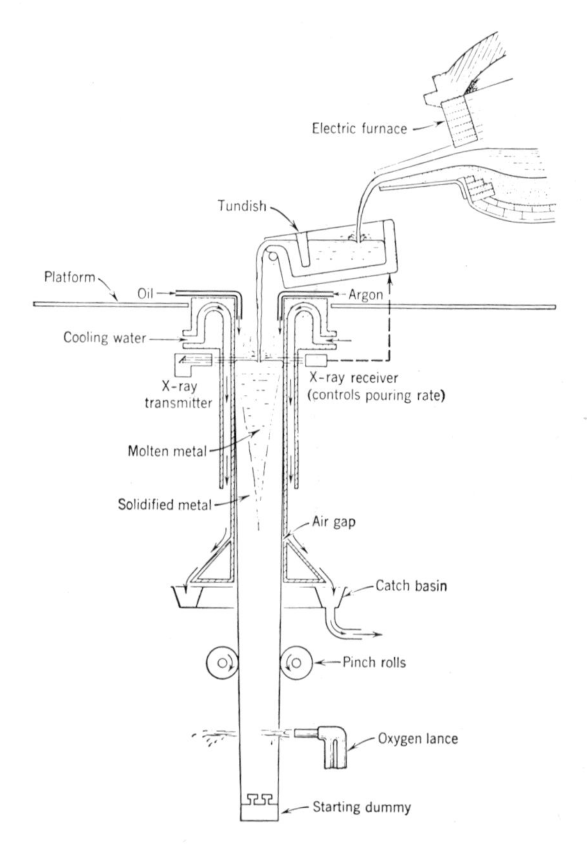

Figure 6.13 *The continuous casting of ingots. (Reproduced with permission from "Metallurgy for Engineers" by Wulff, Taylor, and Shaler. John Wiley and Sons Inc., N.Y. 1952.)*

aluminium, and other similar non-ferrous metals have been cast in vast quantities by this process, the ferrous metals have proven difficult to cast on a continuous basis. However, many modern steelworks have installed equipment in the last few years that continuously casts plain and alloy steels successfully.

The continuous casting process is normally confined to the production of ingots and the advantages offered by continuous casting are (1) far less wastage of metal since cropping is kept to a minimum; (2) ingot quality is higher, since segregation is avoided, piping does not occur, and grainsize can be closely controlled; (3) rough forming and breakdown rolling operations are not needed since ingots are produced in the sizes most suitable for final extrusion, drawing, or rolling processes; and (4) capital outlay is less than that required for older ingot casting equipment.

Although many types of continuous casting machines are available, the type illustrated in Figure 6.13 is typical. Slag-free, deoxidised molten metal is poured with a minimum of turbulence from the special ladle known as the tundish into the bottomless mould which has its lower end temporarily plugged with the starting dummy. Since the mould is usually water-cooled, solidification occurs around the mould-metal interface and the metal is allowed to move slowly downwards. The rate of travel of the metal is regulated by pinch-rolls so that solidification is occurring continuously within the mould. The oxy-acetylene cutter cuts the protruding billet into convenient lengths, and these cut billets are then rolled or otherwise hot worked without reheating if suitable plant is located nearby. The mould must be as frictionless as possible, and graphite, brass, and copper moulds are quite common. Since a pool of molten metal always exists at the upper end of the ingot no primary piping is possible, and mould shapes are regulated to eliminate the tendency for secondary piping to occur.

GLOSSARY OF TERMS

Casting: the process of pouring molten metal into a prepared mould cavity where it is allowed to solidify.

Centrifugal casting: a casting technique in which the mould is rotated during casting so that centrifugal force assists metal flow.

Columnar grains: elongated parallel grains formed by the directional solidification of a metal melt.

Coring: the variation in composition from the centre outwards within one grain; found only in solid solution-type alloys in the as-cast state.

Dendrite: the skeleton of an as-cast metal grain; usually has a tree-like structure composed of many branches.

Die casting: a casting technique in which molten metal is forced under pressure into a prepared (metal) die cavity.

Grain: individual crystals in polycrystalline aggregates bounded by irregular three-dimensional grain boundaries.

Grain boundaries: surface imperfections separating crystals of different orientations in a polycrystalline mass.

Inclusions: particles of impurities distributed throughout a solid metal.
Investment casting: a technique in which the pattern is made from a material such as wax or a low melting-point metal which is removed from the finished mould by melting it out.
Nucleation: the process by which nuclei form in a solidifying melt.
Nuclei: the first structurally determinate particles of a new phase or structure that is beginning to form as a result of the solidification of a melt, recrystallisation, or transformation.
Permanent mould casting: a casting technique in which the mould is made of metal and can be used many times; the molten material is fed into this mould under gravity only.
Pipe: a cavity formed by contractions in metal during solidification in a mould; most commonly found in ingots.
Polycrystalline: composed of an aggregate of crystals which may or may not be all of the same kind.
Polyhedral structure: a grain structure in which each grain is more or less equal in size in all directions; alternatively, a structure in which the grains are equi-axed.
Sand casting: a casting technique in which the mould is made from sand or loam, the mould being destroyed to recover the casting.
Segregation: the concentration of one constituent of a solidifying melt in one region within the casting.
Shell moulding: a casting technique in which the mould is made up of a relatively thin "shell" of sand held together by an organic binder.

REVIEW QUESTIONS

1. Outline the four basic considerations of any metal casting process.
2. Why are cores usually used when casting components have cavities or holes?
3. How does shell moulding differ from the more conventional green sand moulding?
4. What are the advantages and disadvantages of die-casting processes?
5. A foundry is about to mass produce, by casting, articles of a very complex shape which will require only a very small amount of finish machining. Describe a suitable casting procedure, making reference to the method of pattern production and the degree of automation that would be possible.
6. Explain the meanings of the following terms: grain; dendrite; nucleation; coring.
7. What is the nature of a grain boundary between two metal grains?
8. What general relationships exist between grainsize and strength properties of metals?

9. Discuss the four common types of defects that can occur in a metal casting.
10. What advantages does the continous casting of ingots offer over conventional procedures?
11. A cast iron engine block is to be manufactured by sand casting. Discuss each of the steps involved from pattern preparation to the pouring of the molten metal.

7

Mechanisms of Deformation in Crystalline Materials

ENGINEERING materials can be divided into the following three groups on the basis of the mechanisms that operate during the deformation brought about by an applied load.

(1) *Elastic materials*: all ionic and covalent (giant molecular) crystals belong to this group since they behave elastically until failure occurs. Such materials are brittle in the sense that plastic deformation under load is either nonexistent or negligible, and, although crystalline, must be contrasted with metals.

(2) *Elastoplastic materials*: the main structural metals are typical of this class since initial elastic deformation under load is followed by a considerable amount of plastic deformation which culminates in failure due to intracrystalline shearing. Such materials are characteristically ductile.

(3) *Viscoelastic materials*: polymers, vitreous materials, amorphous substances, and concrete belong to this group since they deform by a mechanism that can perhaps be best described as "viscous movement". Such materials are very variable, some being "elastic" like rubber, while others are brittle like glass.* Essentially, this "viscous flow" is the sliding of molecules, or groups of molecules, over one another as the stress increases.

This chapter is largely concerned with deformation within metals, but much of what will be said is applicable to other crystalline materials as well.

SINGLE CRYSTALS

A great deal of information concerning mechanisms of deformation has been gained from detailed studies of single crystals of metals. Such crystals may be produced by either of the following methods: (1) the carefully controlled solidification of a melt; and (2) the prolonged annealing of a piece of metal after suitable cold working. Using the latter method, pieces of wire of $\frac{1}{16}''$ diameter by several inches in length can readily be transformed into single crystals.

*More will be said about such materials in Chapters 15 and 16.

ELASTIC DEFORMATION

If a single crystal of a metal is stressed within its elastic range a temporary distortion or warping of the crystal lattice results. The atoms making up the lattice do not change their relative positions, but interatomic distances alter considerably, resulting in an increase of the energy within the lattice itself. When the deforming load is removed the atoms return to their original lattice positions and the excess energy is given out in the form of heat. The total amount of deformation as measured over unit length of the specimen is known as unit strain and is directly related to the degree of warping that occurs within the crystal lattice.

If a polycrystalline sample is similarly stressed, the lattices within each individual grain warp, the overall strain being related to the algebraic sum of all of the lattice displacements within each grain.

Laboratory (27): *Using the tensometer, determine the true elastic limits of annealed test-pieces of brass, copper, aluminium, and mild steel. This may be done by loading and unloading using progressively greater loads until the metal takes a very small permanent set. (Note: an extensometer is useful but not essential for this experiment.) Compare these stress figures to proportional limit stresses obtained by conducting normal tensile tests on similar specimens.*

PLASTIC DEFORMATION

Single crystals and individual grains in a polycrystalline aggregate undergo plastic deformation when either one of two mechanisms begin to operate, these mechanisms being known as *slip* and *twinning*.

Slip

Slip is by far the most common mechanism of plastic deformation, and occurs when one part of the crystal has become displaced with respect to another part along a particular crystallographic plane of the crystal. The particular crystallographic planes involved are known as slip or glide planes, and the direction in which movement occurs along such planes is known as the slip direction.

The cylindrical crystal shown in Figure 7.1 illustrates two aspects of crystalline slip: (1) the original circular cross-section of the crystal becomes elliptical; and (2) surface markings appear as the slip planes become exposed on the surface of the crystal and thus the outer surface beomes roughened.

An interesting analogy concerns the behaviour of a pack of playing cards which are neatly stacked and then gently pushed from one end. The top cards slide progressively over the lower cards so that the effect is similar to that shown in Figure 7.2

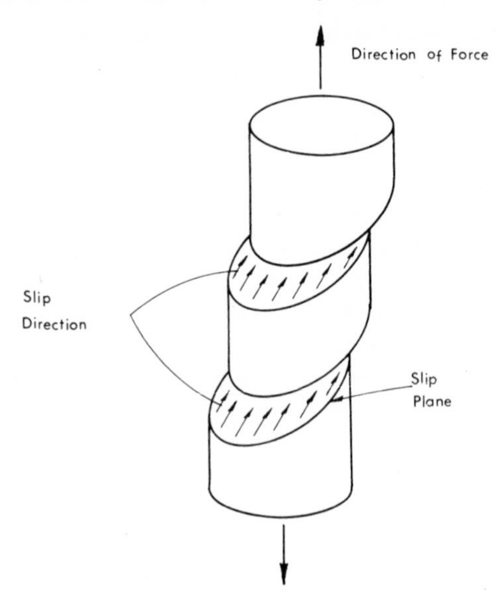

Figure 7.1 *Slip planes and slip directions in a single cylindrical metal crystal.*

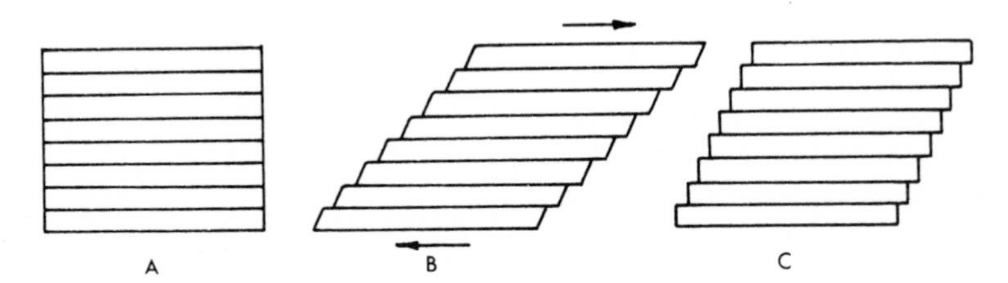

Figure 7.2 *Plastic deformation by slip: (A) unstressed simple cubic crystal with probable slip planes shown; (B) crystal plastically deformed with deforming load still applied; (C) residual elastic deformation released as the deforming load is removed from the crystal; slip has occurred along each of the planes indicated in (A).*

Three general observations can be made concerning slip movements in crystals.

(1) Slip planes are always planes of high atomic densities, while the direction of slip along these planes is always the direction of highest atomic density.

(2) Slip begins when the shearing stress acting along the slip plane in the direction of slip exceeds a certain value which is known as the critical shearing stress.

(3) Values of critical shearing stress depend upon the composition of the crystal, on the temperature, and on previous mechanical working; it is lowest for pure metals and decreases steadily with increasing temperature, while previous cold working increases it quite considerably.

Face-centred cubic crystals tend to slip along the (111) plane, while hexagonal closest-packed crystals slip along the basal (0001) plane. Copper and aluminium are examples of the former, while zinc is an example of the latter. If the crystal structure contains a number of planes of high atomic densities then slip will tend to begin on whichever of these planes are spaced furthest apart. In body-centred cubic crystals slip can occur on the (110), (112) and (123) planes, and frequently two of these planes will operate as slip planes more or less simultaneously. Iron is an example of this type of crystal.

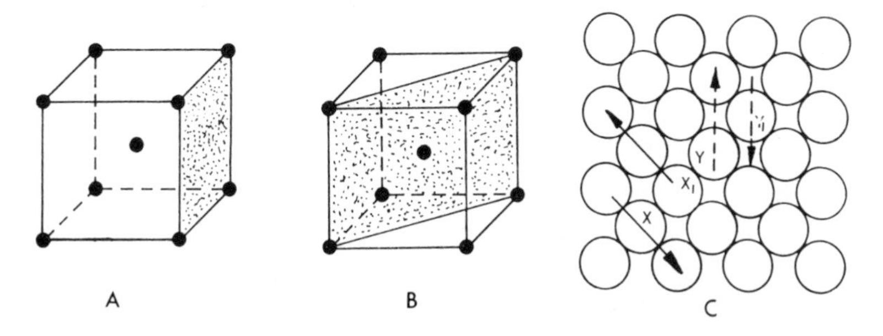

Figure 7.3 *(A) the (010) plane in a BCC crystal structure. (B) the (110) plane in a BCC crystal structure, a plane of higher atomic density and therefore more likely to undergo slip than the (010) plane. (C) slip occurs along the planes of high atomic density XX_1 rather than along the less dense planes YY_1 because there is no necessity for the sideways movement of atoms in the former situation and hence less energy is required.*

Twinning

Twinning is the second mechanism whereby plastic deformation can occur in crystals. While it does not occur as frequently as slip, it nevertheless is important since quite a few metals tend to deform under certain circumstances by twinning. A twin crystal is formed when one part of a crystal reorients itself so that it becomes, crystallographically, the mirror image of the remainder of the crystal. The crystallographic plane about which twinning occurs is known as the twin plane and is a plane of symmetry within the twinned crystal. Twinning differs from slip in that every plane of atoms suffers some movement, and the crystallographic orientations of many unit cells are altered.

TABLE 7.1
OBSERVED SLIP SYSTEMS IN SELECTED CRYSTALS

Structure	*Slip Plane*	*Slip Direction*	*Number of Slip Systems*
FCC Cu, Al, Ni, Pb, Au, Ag, γFe, . . .	$\{111\}$	$\langle 1\bar{1}0\rangle$	$4 \times 3 = 12$ 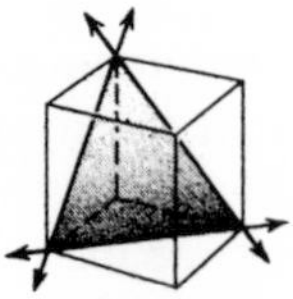
BCC αFe, W, Mo, β Brass	$\{110\}$	$\langle \bar{1}11\rangle$	$6 \times 2 = 12$ 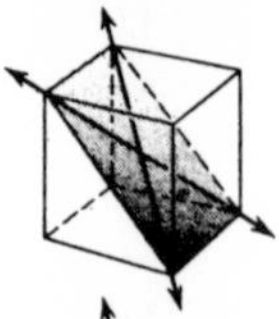
αFe, Mo, W, Na	$\{211\}$	$\langle \bar{1}11\rangle$	$12 \times 1 = 12$ 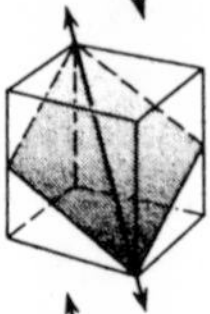
αFe, K	$\{321\}$	$\langle \bar{1}11\rangle$	$24 \times 1 = 24$ 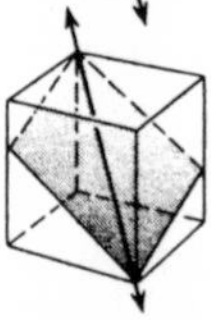
HCP Cd, Zn, Mg, Ti, Be, . . .	(0001)	$\langle 11\bar{2}0\rangle$	$1 \times 3 = 3$ 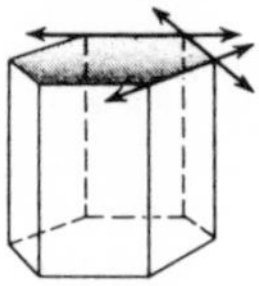
Ti	$\{1010\}$	$\langle 11\bar{2}0\rangle$	$3 \times 1 = 3$ 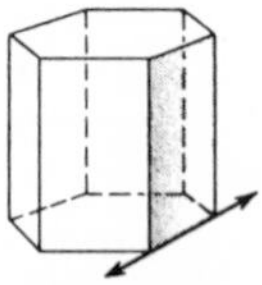
Ti, Mg	$\{10\bar{1}1\}$	$\langle 11\bar{2}0\rangle$	$6 \times 1 = 6$ 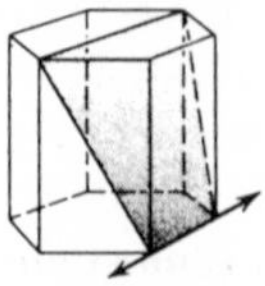
NaCl, AgCl	$\{110\}$	$\langle 110\rangle$	$6 \times 1 = 6$ 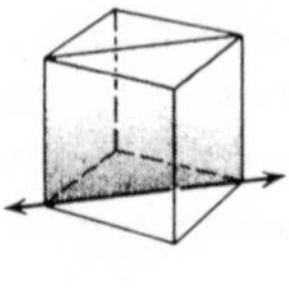

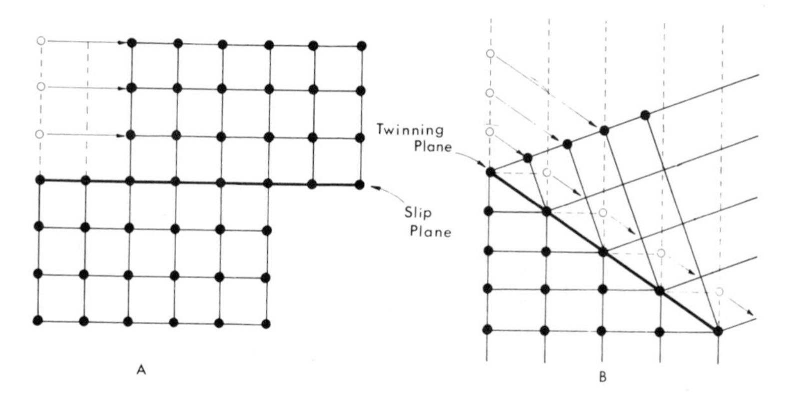

Figure 7.4 *Relative atom movements in (A) slip and (B) twinning.*

Twinned crystals are of two kinds: (1) *strain twins* which are produced as an immediate result of applied stress in the manner as outlined, and (2) *annealing twins* which only appear in annealed metals that have been subjected to prior mechanical working at ordinary temperatures. These are more common than strain twins.

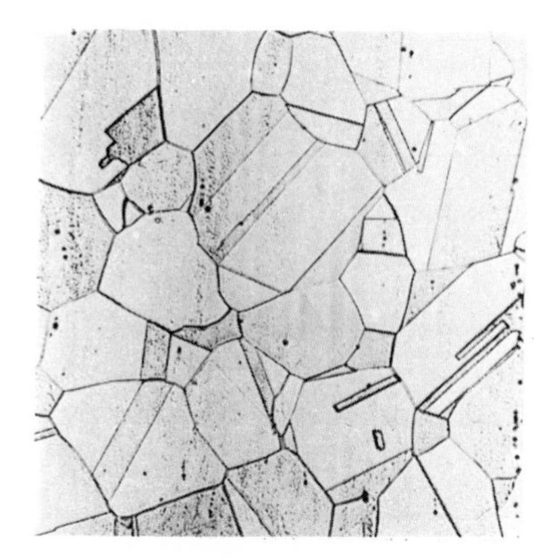

Figure 7.5 *Annealing twins in alpha-brass. Magnification x100. (Original slide courtesy of the Copper and Brass Information Centre, Sydney.)*

Figure 7.6 *Slip lines in deformed phosphor bronze. Magnification x200. (Original slide courtesy of the Copper and Brass Information Centre, Sydney.)*

It is most important to note that neither slip planes nor twin planes extend beyond the grain boundaries of the individual crystals in which they occur. Thus, in polycrystalline aggregates both slip and twinning are intracrystalline, this indicating that the grain boundaries are stronger than the grains themselves.

Laboratory (28): *Take 3/4" cubes of zinc and brass and suitably anneal each. Polish, etch, and examine any three adjacent surfaces microscopically. Place each cube in a parallel-jawed vice and deform each in turn by compression. Re-polish, re-etch, and microscopically re-examine the previously prepared surfaces. Now anneal each cube as before and re-examine microscopically. Tabulate your results.*

DISLOCATION THEORY

Early theories of crystalline slip envisaged all of the atoms on any one slip plane moving simultaneously from one set of lattice positions to the next. This type of movement has become known as bulk shearing. Calculations were carried out using known information about bond strengths and lattice arrangements, and it was found that the values of theoretical shear stress so obtained were often 1,000 times as high as the experimentally determined values. Obviously this simple theory was not adequate.

A breakthrough occurred in the 1930's when it was suggested that slip does not occur simultaneously but starts from structural defects within the crystal lattice and spreads from these over the entire slip plane in question. Such crystal defects have become known as *dislocations*, and calculations based upon the progressive movements of dislocations give results that correlate well with experimental values of critical shear stress.

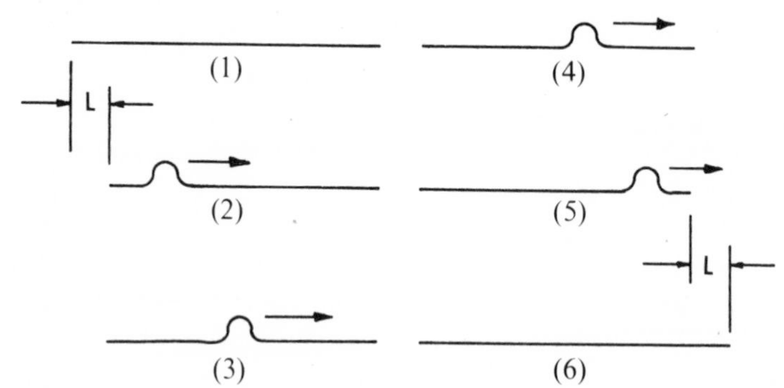

Figure 7.7 *The "carpet analogy" of a dislocation in a crystal. By moving the wrinkle from one end of the carpet to the other the carpet is shifted a distance of L units to the right. The energy required to progressively shift the wrinkle is considerably less than that required to bodily drag the carpet across the floor.*

Dislocations in crystalline materials can be regarded as *edge dislocations* or *screw dislocations*. All dislocations may be regarded as linear disturbances of the three-dimensional lattice of the crystal, with many dislocations possessing at one time or another both *edge* and *screw* components. Dislocations may occur during the growth of crystals from a melt, but they are more frequently produced by mechanical strain. Other crystal imperfections such as vacant lattice sites, interstitial atoms and voids also weaken the crystal and promote slip, and in a sense these too can be regarded as forms of dislocations.

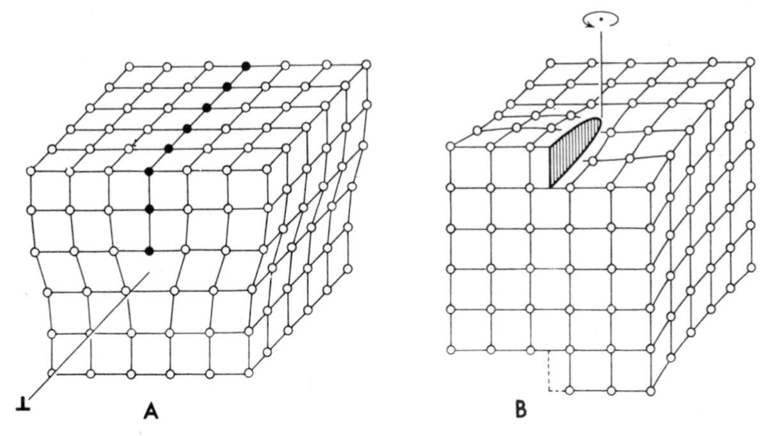

Figure 7.8 *The geometry of simple dislocations: (A) an edge dislocation; (B) a screw dislocation. The lines normally used to represent the dislocations are indicated, as are the symbols for them. (Reproduced with permission from Volume 3 of "The Structure and Properties of Materials" by Hayden, Moffatt, and Wulff. John Wiley and Sons Inc., N.Y. 1965.)*

WHISKERS

Whiskers are very small single crystals which are virtually free from imperfections and dislocations. They usually possess diameters of a few microns (10^{-4} cm) and rarely exceed one centimetre in length. They have exceptionally high critical shearing stresses and can usually be given about 2% strain before plastic deformation begins. Tests conducted on whiskers of various metals have shown that perfect crystals do possess the very high strengths predicted from theoretical calculations. Further proof comes from X-ray and electron microscope studies, both of which have yielded "photographs" of dislocations moving across a slip plane under the influence of applied stress.

WORK HARDENING

When a crystalline material undergoes a significant amount of plastic deformation at ordinary temperatures it becomes harder and stronger and its electrical properties are affected. The general effect on yield strength can be illustrated by the tensile curves shown in Figure 7.10(A). The medium carbon steel test piece is stressed to within the range of plastic deformation (B), unloaded, and re-stressed. The overall effect is a dramatic increase in yield strength from 57,400 to 85,000 psi. The general effects of work hardening on strength, ductility, hardness, and electrical resistivity are shown graphically in Figure 7.10(B).

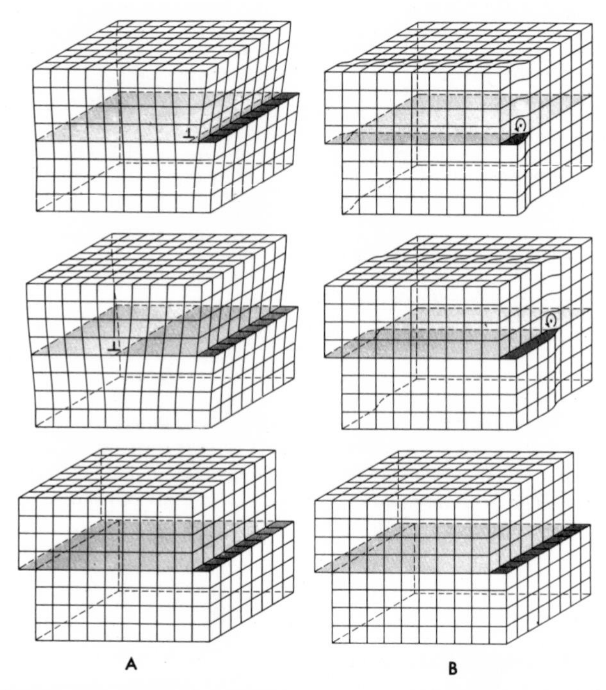

Figure 7.9 *(A) slip resulting from the movement (right to left) of a pure edge dislocation through a simple cubic lattice. (B) slip resulting from the movement (front to back) of a pure screw dislocation through a simple cubic lattice. (Reproduced with permission from Volume I of "The Structure and Properties of Materials" by Moffatt, Pearsall, and Wulff. John Wiley and Sons Inc., N.Y. 1964.)*

Laboratory (29): *Place a standard tensile test piece of brass or aluminium in the tensometer and load the specimen into the range between the yield and the ultimate strengths. Slowly unload the specimen and plot an unloading line. Reload the specimen in the normal manner and plot the new load-elongation curve to failure. How has the yield strength been affected, and why? How ductile is the specimen? Has the elastic modulus altered?*

The changes in properties are brought about by the increasing internal distortion caused by the continual "piling-up" of dislocations within the crystal lattice as the metal is cold worked. However, deformation carried out above a certain temperature does not cause work hardening effects. This temperature is known as the recrystallisation temperature for the particular metal, and varies with different amounts of prior cold working.

It is generally accepted that work hardening effects result from the *tangled forest of dislocations* that is present in a heavily cold worked piece of metal. This tangled structure makes further plastic deformation more difficult, and hence the magnitudes of stresses needed to cause further plastic deformation must increase. If the tangle of dislocations is such that further deformation is completely prevented, the application of higher stresses will simply cause the metal to fracture.

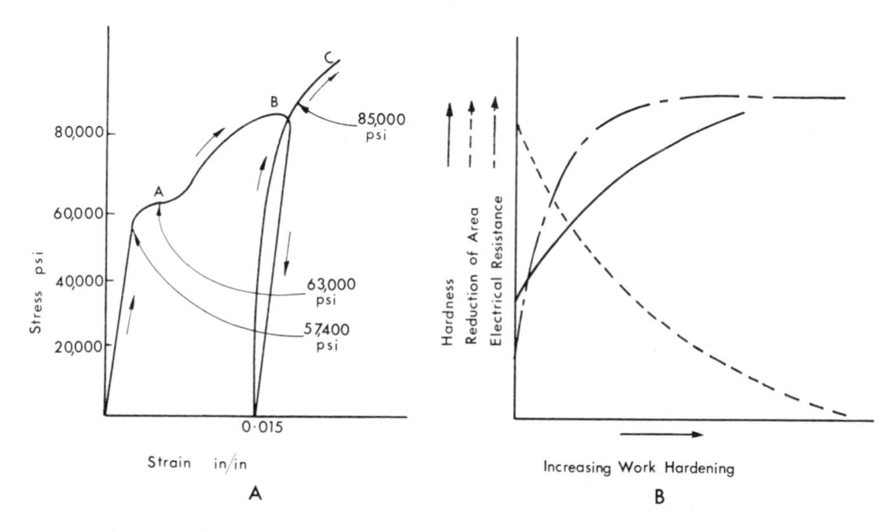

Figure 7.10 *(A) the effect of mechanical working on the yield strength of steel as shown by an interrupted tensile test. (B) effects of work hardening on the ductility, hardness, and electrical resistance of metals.*

FRACTURE

A material will fail by fracture when deformation becomes excessive and work hardening effects reach a maximum; the internal cohesion of the material is broken down, this resulting in the separation of one section from another.

Fractures are usually said to be either *brittle* or *ductile*, depending upon the amount of plastic deformation preceding failure. It follows that brittle failures occur suddenly with little or no prior deformation. Many ductile metals can be cooled to a temperature at which the mode of failure alters from ductile to brittle.

It must be realised that fracture at elevated temperatures is often quite different from fracture at ordinary temperatures. Below a certain temperature (known as the *equi-cohesive temperature*) which is different for every material, failure is intracrystalline with separation occurring within the grains themselves. Above this temperature the grain boundaries weaken and failure becomes inter-crystalline. It follows that while fine-grained metals are stronger at ordinary temperatures, coarser-grained metals are stronger at elevated temperatures. This fact is of extreme importance in metals subjected to conditions causing high-temperature creep.

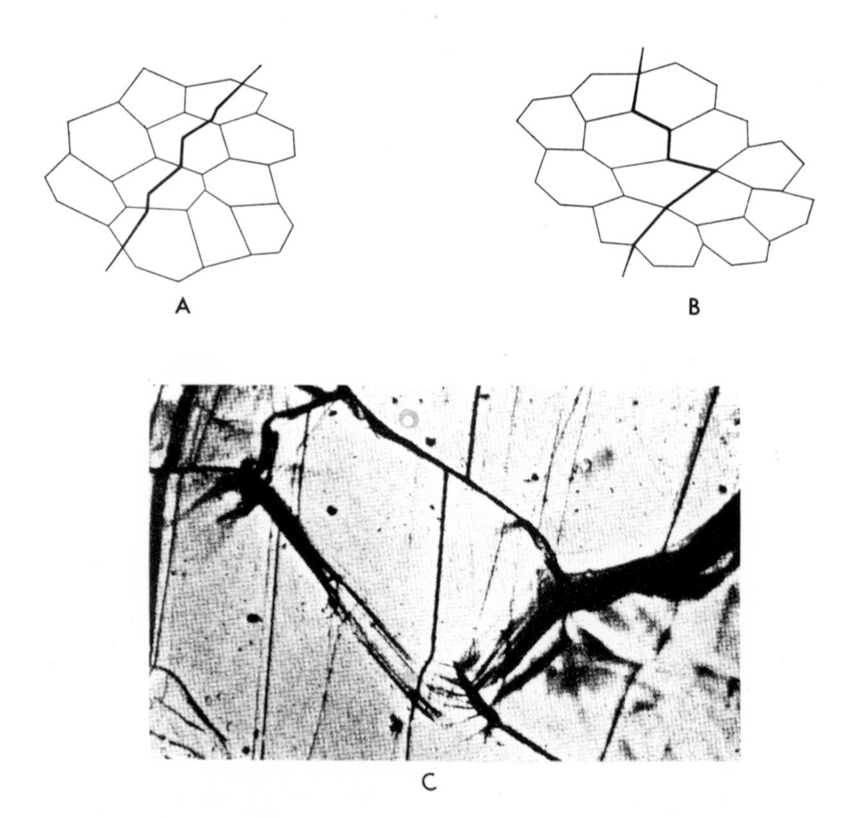

Figure 7.11 *(A) intracrystalline and (B) inter-crystalline failures. (C) grain separation in aluminium. The aluminium sample was subjected to a low stress for hundreds of hours at 300°C. The following points are important: at the right of the section two grains have separated sufficiently to produce a large gap at the boundary; in the upper centre, relative movement of two adjoining grains has produced a jog in the scratch line where it crosses the grain boundary; in the lower centre, the boundary of the central grain has moved several times, leaving traces of its original positions and a pronounced curve in the central scratch line. (Reproduced with permission of W.A. Rachinger, Department of Physics, Monash University.)*

DEFORMATION BY CREEP

Many different types of metal components are subjected to steady stresses while at elevated temperatures; boilers, fireboxes, steam pipes, gas turbine parts, and components of internal combustion and jet engines being common examples. Such metal components may—depending upon the type of metal from which they are made, the service temperature, and the stresses involved—exhibit a form of plastic deformation known as *high-temperature creep.* It is important to realise that creep strains may be induced in certain materials at ordinary temperatures; this form of plastic deformation is known as *low-temperature creep*, and can be readily demonstrated in lead and glass. Indeed, detailed examinations of sheet lead roofing materials on very old churches have revealed that the thickness of the sheet is greatest at the gutter and least along the ridging, indicating that, over the years, plastic flow has occurred due to the weight of the lead itself.

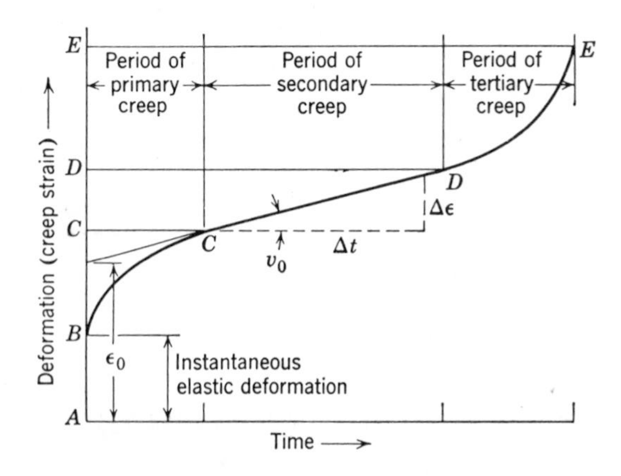

Figure 7.12 *Creep curve at constant temperature and stress. The minimum creep rate is determined by the slope of the curve in the stage of secondary creep. (Reproduced with permission from "Nature and Properties of Engineering Materials" by Z. D. Jastrzebski. John Wiley and Sons Inc., N.Y. 1959.)*

Time is one of the vital factors involved in deformation by creep. Standardised tests have been evolved in which creep strain is plotted against elapsed time, the tests being conducted at controlled temperatures under conditions of constant load. Tensile creep tests are most common. The loads involved are always much less than those likely to cause the elastic limit of the material to be exceeded at normal temperatures.

Graphs resulting from such tests are known as *creep curves* and may exhibit three well-defined stages:

(1) *Primary creep*—creep begins and proceeds at a decelerating rate until it becomes constant;

(2) *Secondary creep*—the rate of deformation by creep remains approximately constant;

(3) *Tertiary creep*—the creep rate rapidly increases and failure becomes inevitable.

Laboratory (30): *(a) Take a strip of lead 1″ wide by ⅛″ thick, and 8″ in length. Suspend it from a rigid support and hang a weight of 5 lbf from its bottom end. Record the overall length of the specimen daily for several months or until failure occurs. Plot creep strain against time in days.*
(b) Suspend a strip of soda-lime glass in a similar manner and gently warm the glass by running a batswing burner flame slowly up and down each side in turn. Measure overall length every two minutes and plot creep strain against time.

Creep appears to occur due to some sort of viscous movement within the metal, the normal slip mechanism usually being absent. It can also occur in glass, the mechanism again being one of slow viscous flow. Perhaps the most important design consideration with respect to creep is that creep can take place and lead to failure at static stress levels much lower than those needed to cause failure when the load is applied fairly rapidly. Creep tests enable the design engineer to predict the service lives of plant and components, the data contained in Table 7.2 being typical.

TABLE 7.2

SERVICE LIVES OF MACHINE COMPONENTS SUBJECTED TO CREEP

Components	*Creep Rates (in/in/hr)*	*Maximum Permissible Strain (in/in)*	*Service Life (hrs)*
Turbine rotors	10^{-9}	0.0001	100,000
Steam pipes, boiler tubes	10^{-7}	0.003	100,000
Superheater tubes	10^{-6}	0.02	20,000

The less-dense metals are particularly prone to creep at elevated temperatures; accordingly aluminium alloys used by the aircraft industry are fully creep-tested. The most useful group of creep-resisting light alloys are the so-called "Y-alloys" or duralumins, a typical member containing 4% copper, 1.5% magnesium, and the balance, aluminium. In general, alloying decreases creep, and coarse-grained metals are more creep-resistant than fine-grained metals of the same composition. It has been found that creep will not occur to any measurable extent below a certain temperature and this is different for each metal. Table 7.3 gives approximations for five common metals.

TABLE 7.3

TEMPERATURES BELOW WHICH CREEP DOES NOT OCCUR

Metal	*Temperature at which Creep begins °C*
Lead	–70°
Aluminium	40°
Copper	180°
Iron (pure)	330°
Tungsten	940°

GLOSSARY OF TERMS

Creep: plastic deformation within a material that occurs as a function of time when that material is subjected to constant load.
Dislocations: linear imperfections in a crystal which move during plastic deformation.
Elastic deformation: the change of shape of a body due to an applied load which is fully recoverable when the load is removed.
Equi-cohesive temperature: the temperature at which the mode of failure of a crystalline material alters from intracrystalline to inter-crystalline.
Plastic deformation: the permanent change of shape of a body as a result of an applied load.
Slip: the sliding displacement of one part of a crystal relative to another caused by the movement of dislocations along slip planes.
Slip planes: crystallographic planes along which dislocations move under the influence of applied stress. Slip planes are always planes of high atomic density.
Slip system: the combination of a slip plane and a slip direction on that plane which operates under a specific load condition.
Twin crystals: crystals in which two regions separated by a definite plane (the twin plane) are crystallographic mirror images of one another.
Whiskers: very small single crystals (usually of metals) which are virtually free from imperfections and dislocations.
Work hardening: (also **strain hardening**) the increase in hardness and strength of a metal occurring with increasing plastic deformation.

REVIEW QUESTIONS

1. Compare and contrast elastic and elastoplastic materials in terms of their behaviour under a tensile load. Give examples of each class of material.
2. What occurs to the lattice structure within individual grains when a piece of polycrystalline metal is deformed to within its elastic limit?
3. What is meant by the term "critical shearing stress"?

4. Which planes within a crystal are most likely to act as slip planes? Illustrate your answer by reference to the BCC unit cell.
5. How is twinning different from slip?
6. Under what circumstances would you expect (i) annealing twins or (ii) strain twins to occur in metal crystals?
7. What effects do dislocations have on the mechanical properties of crystalline materials?
8. What is meant by (i) an edge dislocation, (ii) a screw dislocation?
9. What are the general effects of work hardening on the yield strength, ductility, electrical conductivity, and hardness of metals?
10. What is the significance of the "equi-cohesive temperature" of a metal?
11. How does deformation by creep differ from deformation brought about by a steadily-increasing load?
12. Sketch a typical creep curve and label important areas. In which portion of this curve is the industrial designer interested? Why?
13. What relationship exists between grainsize and creep within a metal?
14. What are crystal whiskers?

8

Methods of Forming and Working Metals

MOST forming processes involve the controlled plastic deformation of a piece of metal such that a permanent change in shape occurs. Forming processes may be carried out at ordinary or elevated temperatures and, depending upon temperature, are classified as either hot working or cold working.

RECRYSTALLISATION

If ductile metals are worked (plastically deformed) at ordinary temperatures they undergo work hardening and may eventually fracture. Work hardened metal is in a state of high internal energy and atomic diffusion may begin either while the metal is at room temperature or as its temperature is raised. This results in the formation of nuclei which cluster mainly around the grain boundaries of the distorted grains, the grain boundaries being the areas of highest energy within the metal. Each nucleus then becomes the starting point for the development of a new crystal lattice, so that after a certain time the distorted strain-hardened grains are completely replaced by smaller, stress-free grains. This is the process of recrystallisation and it proceeds in two stages:

(1) nucleation by atomic diffusion, this occurring in high-energy areas under suitable temperature conditions;
(2) the growth of new unstressed, equi-axed grains.

When recrystallisation is complete some grains begin to grow at the expense of others if the temperature conditions are still favourable. This is the process of *grain growth* and is generally considered to be undesirable since it is accompanied by a decrease in normal strength properties. The temperature at which recrystallisation occurs is known as the *recrystallisation temperature* and is different for different metals; it also varies according to (1) the amount of plastic deformation to which the metal has been subjected, and (2) the time during which this temperature is maintained. Some metals, notably lead, recrystallise below normal atmospheric temperature, and thus cannot be work hardened. The approximate recrystallisation temperatures for some of the more common metals are given in Table 8.1.

TABLE 8.1

RECRYSTALLISATION TEMPERATURES AND MELTING POINTS OF SOME METALS

Metal	*Recrystallisation Temperature °C*	*Melting Point °C*
Aluminium	150	660
Magnesium	150	651
Lead	below 20	327
Tin	below 20	232
Silver	200	961
Copper	200	1083
Nickel	600	1452
Iron	450	1535
Molybdenum	900	2620
Tungsten	1200	3400

Figure 8.1 depicts the changes occurring during the period when a piece of metal is slowly raised to beyond its recrystallisation temperature.

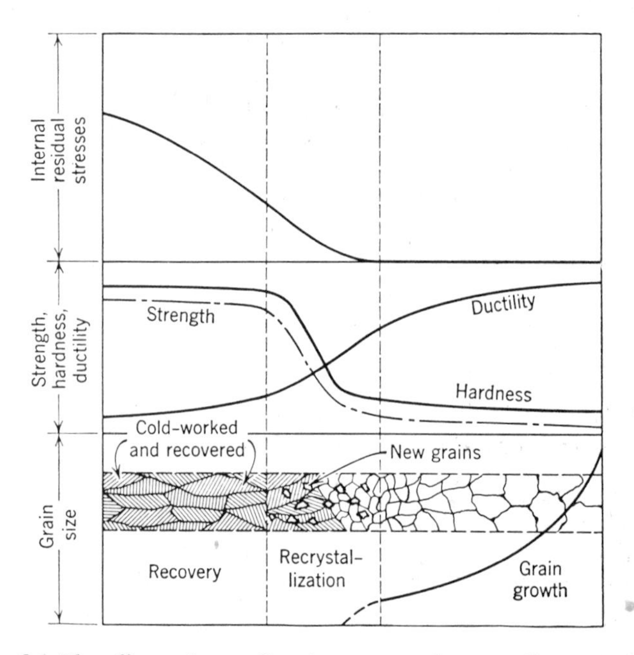

Figure 8.1 *The effects of annealing (recovery and recrystallisation) on cold deformed metals. (Reproduced with permission from "Nature and Properties of Engineering Materials" by Z. D. Jastrzebski. John Wiley and Sons Inc., N.Y. 1959.)*

It is apparent that a period of *recovery* precedes actual recrystallisation. During recovery the greater proportion of internal stresses are released at temperatures well below those necessary to cause recrystallisation. Microstructures are not visibly affected by recovery and mechanical properties remain much the same; however, electrical conductivity increases quite markedly. Recovery is often termed *stress-relieving*, this being a heat treatment process commonly applied to castings, forgings, welded, and fabricated components, cartridge cases, and boiler tubes in order to remove high internal stresses that could promote premature failure. Articles work hardened during manufacture can be stress-relieved by low-temperature heat treatment without appreciable loss of hardness or strength.

During the period of recrystallisation strength and hardness fall away while ductility and electrical conductivity increase significantly. Complete recrystallisation occurs when a metal is *fully annealed*,* and annealing times and temperatures are closely controlled in order to keep grain growth to a minimum.

Laboratory (31): *Take three pieces of ½″ diameter cold drawn 70/30 brass rod ½″ long and prepare by filing and polishing a transverse and longitudinal section on each. Etch one specimen in ammonium persulphate solution and examine microscopically. Place the three specimens into an electric muffle furnace and heat to 750°C. Remove one specimen after 10 minutes, the next after 20 minutes, and the third after 40 minutes of soaking at this temperature. Fine polish the prepared surfaces, etch in the ammonium persulphate solution, and examine microscopically. Sketch all microstructures and comment on their differences.*

HOT WORKING

Hot working involves the plastic deformation of a metal at a temperature above its recrystallisation temperature. This means that no work hardening occurs, no grain distortion remains, and the final structure of the metal is similar to that resulting from cold working plus annealing. In effect, the working promotes continuous recrystallisation, the final grainsize of the metal being determined by the temperature at which the deformation is carried out.

The following classification of hot working processes lists the processes important in modern manufacturing:

*Other special annealing processes, such as sub-critical annealing, do not cause complete recrystallisation.

1. Rolling
2. Forging
   (a) Hammer (or smith) forging
   (b) Drop forging
   (c) Press forging
   (d) Upsetting
   (e) Swaging
3. Extrusion
4. Drawing
5. Piercing
6. Spinning

Rolling

The basic principle involved in rolling is that the thickness and cross sectional shape of a piece of metal can be altered by passing the metal through a set of rolls of a certain shape set a certain distance apart. The rolls are revolved in opposite directions at controlled speeds so that the rolling action pulls the metal through the rolls at the correct rate. The maximum amount of reduction possible by a single pass through the rolls is determined by the "angle of bite" as shown in Figure 8.2, this being determined principally by roll diameter.

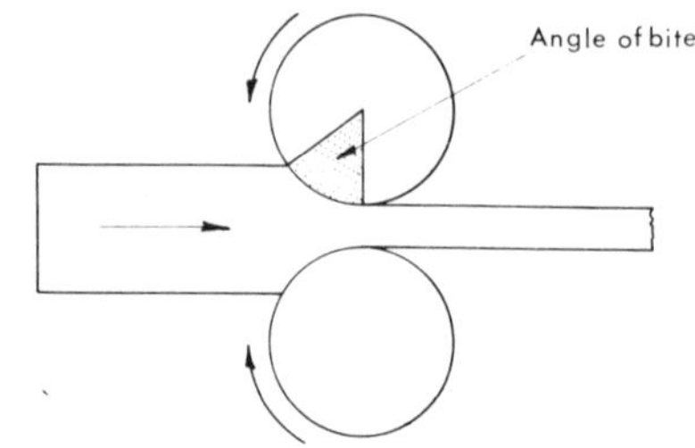

Figure 8.2 *The angle of bite of rolls.*

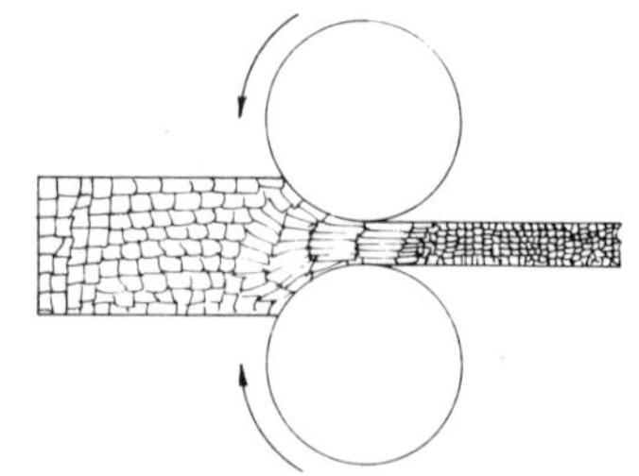

Figure 8.3 *The effects of hot rolling on the grain structure of a metal; as the metal passes through the rolls the grains are flattened and elongated, and then recrystallisation occurs.*

Hot rolling is usually carried out at temperatures 40° C to 95° C above the recrystallisation (or critical) temperature of the metal, this promoting complete grain refinement and the development of a uniform fine-grained structure in the metal. The typical cycle of events in hot rolling is shown in Figure 8.3.

The following shapes and sections are commonly produced by hot rolling.

(1) *Blooms*: the first stage in the breakdown of an ingot and are large bars, often square in section, having a minimum thickness of 6″.

(2) *Billets*: rolled bars formed by a further breakdown of the ingot; they vary from 1½″ to 6″ thick and are not of finished shape.

(3) *Slabs*: blooms are often broken down into slabs which are much wider than they are thick. Slabs provide the raw stock from which sheet material is made and are never less than 10″ wide.

(4) *Special Sections*: railway line sections, rolled steel joists (RSJ's), I beams, round, square, and rectangular sections are all hot rolled to shape. In contrast to billets, blooms, and slabs, these special sections are rolled in closed rolls that control the lateral spread of the metal during rolling.

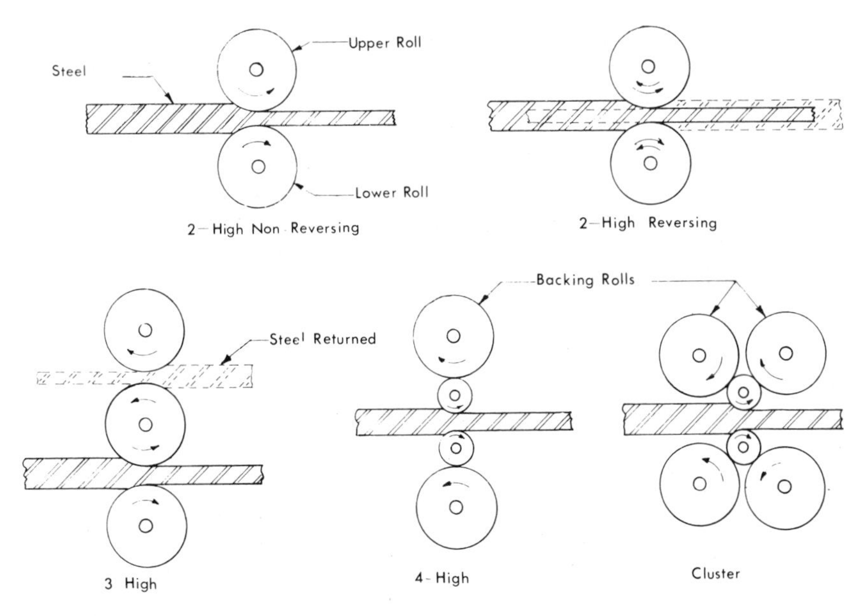

Figure 8.4 *Various roll configurations used in rolling stands.*

Continuous Rolling Mills: These are used when the volume of production justifies the high installation costs, and they consist of a series of two-high rolling stands through which the metal passes in one continuous piece. Each stand reduces the metal by a certain amount, and, because of the accompanying increase in length, each stand must run at a certain speed, with the finishing stand running much faster than the roughing stands. The metal sheet or strip leaves the final rolling stand with a linear speed of 60 mph or more and is usually coiled as it leaves this stand.

Quality of Hot Rolled Stock: Hot rolled stock usually leaves the rolling mill covered with a strong and hard mill scale which forms at the high temperatures used. Dimensional tolerances vary from about 2% to 5%, and uniform quality is obtained by the close control of rolling procedures and temperatures. Directional properties usually result from hot rolling, but these are not of great significance except in the case of special sections, such as I beams, where one part of the product undergoes considerably more deformation than the other.

TABLE 8.2

EFFECTS OF HOT ROLLING ON MILD STEEL

State of Material	*UTS (tsi)*	*% Elongation*	*% Reduction in Area*	*Izod (ft lbf)*
As Cast	27	15	10	5
Hot Rolled: tested across direction of rolling	28	19	25	12
Hot Rolled: tested along direction of rolling	30	35	52	30

Forging

If metal is worked by the application of localised compressive forces it is said to have been forged, and the following three basic processes are used in forging: (1) The metal may be *drawn out*, so that its length is increased while its cross-sectional area is reduced; (2) it may be *upset*, in which case the section is increased while length is reduced; and (3) it may be *squeezed* or *hammered* in between special dies so that the metal flows and takes up the shape of the die cavity.

Forging may be done by hand using the tools and techniques of the old "village blacksmith"; or steam, air, or mechanical hammers may be used. Presses and forging machines, however, are always used if special dies are to be employed. Perhaps the most important forging process is *drop forging* in which single and double-action drop forging machines of up to 1,000 tons capacity are used. Drop forging is an essential mass-production technique and a series of dies, or one die with several impressions, is nearly always needed. Figure 8.5 shows the steps in forging one half of a universal joint for an automobile, the machined and sectioned component being shown for comparison.

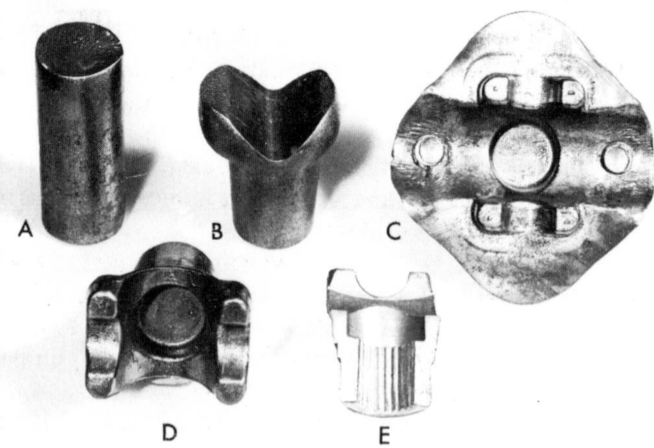

Figure 8.5 *The steps in the forging of a universal joint yoke from a cylindrical blank. A machined and sectioned yoke (E) is shown for comparison. (Original specimens courtesy of Duly and Hansford Ltd., Sydney.)*

Dimensional accuracy is not high with drop forged components and machining is always necessary. Die design can be a real problem, the following being some of the more important factors that must be considered:— (1) the parting line should lie in a single plane and near the centre of the forging; (2) generous fillet curves are necessary, and at least 7° draft or draw must be built into the dies or the forging will not be able to be withdrawn; (3) extreme differences in section thickness should be avoided if possible; and (4) dimensional tolerances should be kept as wide as possible.

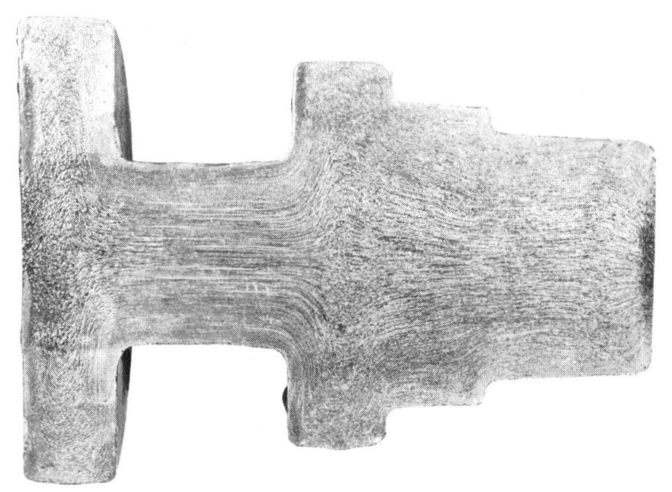

Figure 8.6 *Grainflow in a forged gear blank. The fibrous pattern follows the directions of deformation within the metal. (Original specimen courtesy of Duly and Hansford Ltd., Sydney.)*

"Grainflow" in Forgings: Forging at a temperature above the recrystallisation temperature results in the immediate plastic deformation of the as-cast grains of the cast billet followed by immediate recrystallisation. Segregations and slag inclusions become dispersed and cored grains disappear, the overall effect being a marked increase in the uniformity of grain structure. Forging usually produces a particular type of grain structure known as "grainflow" or "fibre" which results in the development of directional strength properties within the finished forging. Grainflow can be explained in terms of recrystallisation in preferred directions. If a forging is polished, etched, and examined microscopically it is seen to consist of small more or less equi-axed grains. Deep etching and macroscopic investigation, however, reveal that these grains are orientated in particular directions within the forging. Such "preferred orientations" occur because of the tendency for the grains formed during recrystallisation to assume the orientations of the larger grains from which they were formed.

Grainflow is related to directional properties within the forging in that the elastic limit, ultimate strength, ductility, and toughness are all greater when measured along the direction of the grainflow than when measured across it. Thus, the teeth cut into a forged gear are strengthened by grainflow. Also, other machine components which are subjected to high levels of fluctuating stress (for example, crankshafts, connecting rods, bell-crank levers) are usually forged rather than cast or machined from stock.

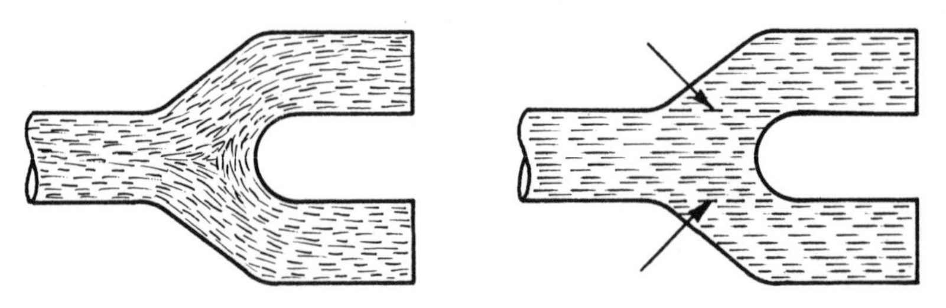

Figure 8.7 *Sections through typical forged and machined yokes showing fibre patterns. Arrows indicate planes of weakness in the yoke machined from bar stock.*

Laboratory (32): *Section a small forged component such as a steel connecting rod by hacksawing, and file and polish the sectioned surface. Etch for thirty minutes using a suitable etchant, wash, and examine macroscopically for grainflow.*

Figure 8.8 *A range of extruded shapes in aluminium. (Original specimens courtesy of Alcan Australia Ltd.)*

Extrusion

Extrusion can be likened to the squeezing of toothpaste from a tube. The extrusion of most metals involves many complex problems which are caused by the strength of the metals themselves. While some metals can be extruded cold, most metals are heated to temperatures at which plasticity is considerably increased; for example, brass is normally extruded between 700–800°C and aluminium alloys are extruded between 400–500°C. The cross-section of the extruded metal is determined by the shape of the die, Figure 8.8 showing some common shapes produced in aluminium.

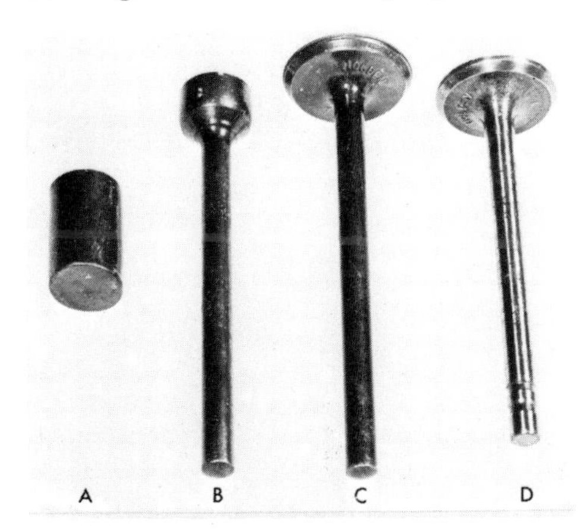

Figure 8.9 *The steps in the forming of a valve for an auto engine: (A) the steel blank; (B) the valve stem is formed by extrusion; (C) the valve head is formed by upsetting (heading); (D) the finished machined valve. (Original specimens courtesy of Duly and Hansford Ltd., Sydney.)*

Metals with low yield strengths and low extrusion temperatures, for example, lead, brass, bronze, copper, aluminium, and magnesium alloys, are most commonly extruded. Steel, however, can be extruded if suitable lubricants are used. The dies used for extrusion forming are usually relatively cheap, particularly for non-ferrous alloys. Dimensional tolerances can be held to $\pm 0.003''$ per inch. Directional properties are usually quite marked, and suitably-prepared samples show grainflow similar to forgings.

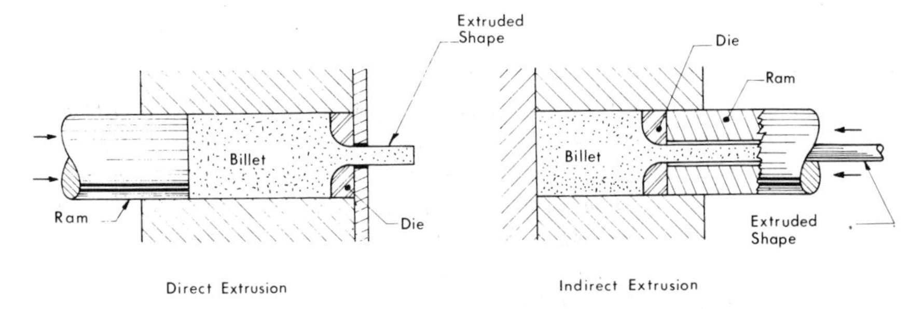

Figure 8.10 *Methods of direct and indirect extrusion of metal.*

Basically three methods of extrusion are used in modern manufacturing: (1) *direct extrusion*, in which the plunger forces the metal through the die; (2) *indirect extrusion*, in which the die itself becomes the ram and is forced into and through the metal billet; and (3) *impact extrusion*, which is a cold forming process and which will be discussed later.

Indirect extrusion requires less power and is more suitable for the less ductile metals; however, machines become more complex and therefore direct extrusion is used wherever possible.

Hollow sections, including tubes and pipes, are readily extruded; however, dies are more costly due to the need for the mandrel which forms the hole. Several different methods of tube extrusion are shown in Figure 8.11.

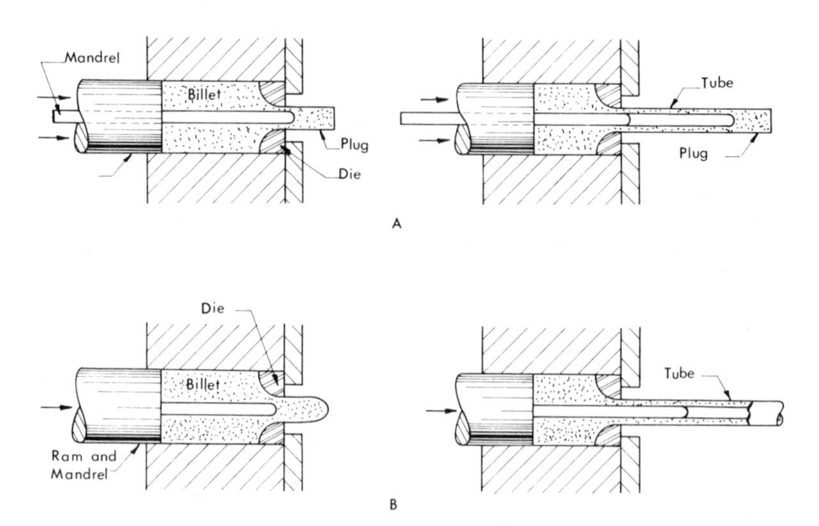

Figure 8.11 *Two methods of extruding tube or other hollow shapes.*

Drawing

Hot drawing is generally restricted to the production of relatively thick-walled cylindrical sections such as oxygen tanks, thick-walled pipes, and large artillery-shell cases. The drawing process starts with a large round blank which is forced through a suitably-shaped die by a punch impelled either hydraulically or mechanically. Further reductions in diameter with consequent extensions in length are possible if smaller die sets are used, and some machines allow for multiple reductions to be made by employing several dies in a line and a longer ram or plunger.

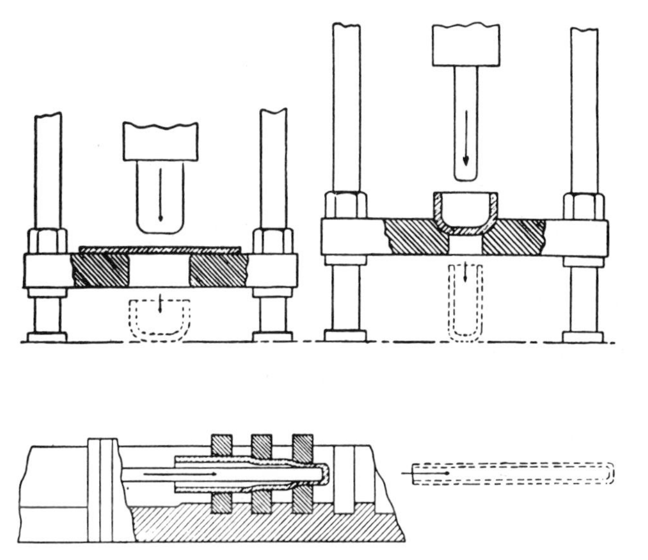

Figure 8.12 *Methods of cupping or hot drawing by the use of single or multiple dies. (Photograph courtesy of The United States Steel Corporation.)*

Piercing

The Mannesmann process is a hot forming process that is both fascinating to observe and of great industrial significance since nearly all seamless tube is produced by this method. In this process a solid rod of metal is rotated between two mutually-inclined heavy rolls rotating in the same direction.

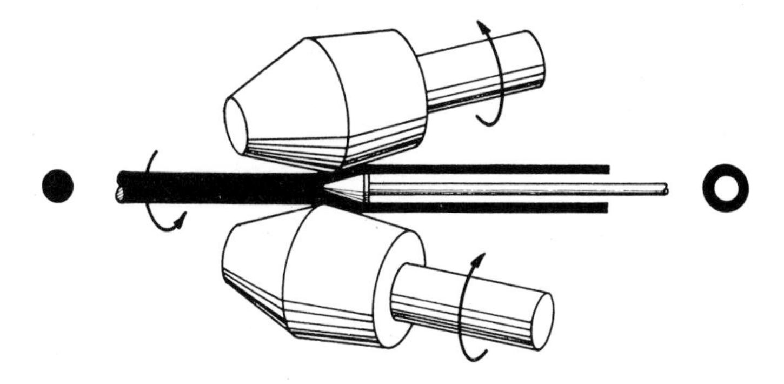

Figure 8.13 *Tube piercing by the Mannesmann process.*

The rolling action causes the rod to assume an elliptical section and at the same time draws the metal *away from* the centre of the rod. The cavity so formed is further opened out by the rotating mandrel, the diameter of which determines the bore of the seamless tube so produced. The tube may be passed through sizing and straightening rolls after it leaves the Mannesmann machine.

Spinning

Hot spinning is restricted to the production of fairly simple "dished" shapes in thick plate and is quite unlike cold spinning in that it is machine-controlled using automatic equipment. A typical application is the manufacture of curved ends for pressure vessels.

COLD WORKING

Cold working involves the controlled plastic deformation of a metal at a temperature well below its recrystallisation temperature. Although usually carried out at ordinary temperatures, elevated temperatures are used in cases where this results in increased ductility; for example, magnesium alloys are cold worked between 150°–200°C. Cold working processes are becoming increasingly important in manufacturing and their advantages and disadvantages compared with hot working are set out in Table 8.3.

TABLE 8.3

COLD WORKING COMPARED WITH HOT WORKING

Advantages of Cold Working	*Disadvantages of Cold Working*
No necessity for heating	Heavier, more powerful and more rigid equipment required
Improved dimensional control	Higher deformation forces required
Improved surface finish	Strain hardening occurs
Improved strength properties	Metals are usually less ductile at ordinary temperatures
Development of directional properties	Undesirable directional properties may develop
Less metal loss and tool wear from scaling	Metal surfaces must be free from oxides and scale prior to working

Thus, the major limitations of cold working or forming processes are those of high force and power requirements and reduced ductility at low temperatures. The former limitations are largely economic; thus if a certain component is to be mass-produced in large quantities this limitation may cease to exist since the high production offsets the initial capital outlay. The problem of reduced ductility is being overcome by the development of better materials and improved forming processes, an example of the latter being high-rate forming. A typical example of a cold formed component is the piece of a tie-rod end shown in Figure 8.14; this component was formed in a two-stage die using a 250 ton press and required no machining, being accurate to 0.0002″ on all diameters.

Cold forming processes may be classified in the following manner:

Squeezing Operations

Rolling	Coining
Cold Forging	Peening
Swaging	Staking
Extrusion	Riveting
Sizing	Thread Rolling

Bending Operations

Angle bending	Flanging
Roll bending	Seaming
Roll Forming	Straightening

Shearing Operations

Shearing, Slitting	Notching, Nibbling
Blanking	Trimming
Piercing, Lancing	Shaving

Drawing Operations

Bar Drawing	Embossing
Tube Drawing	Stretch Forming
Spinning	(deep drawing)

Figure 8.14 *A tie-rod end produced by cold pressing.*

Figure 8.15 *A defect (arrowed) in a piston pin caused by rolling a piped ingot into bar stock. (Original specimen courtesy of Duly and Hansford Ltd., Sydney.)*

Preparation of Metals for Cold Working

Cold forming operations offer the advantages of very high-dimensional accuracy combined with excellent surface finish but in order to achieve these advantages, special preparation of metal blanks is necessary. Three common preparatory steps are— (1) Scale removal by pickling in a suitable acid solution or by sand or shot-blasting. (2) The dressing or careful forming of the blank to size. Sheet materials are often lightly cold rolled to obtain accurate and uniform thickness prior to heavy cold forming. Sheets, blanks, etc., may be straightened prior to cold forming. (3) Annealing is carried out so that maximum ductility is available; however, annealing is also often carried out at certain stages during cold forming to remove the effects of work hardening.

Typical Effects of Cold Forming

Cold rolling can be used to illustrate the important effects of cold forming. Strip, sheet, bar, and rod sections are cold rolled in order to produce materials having accurate dimensions, improved surface finish, and increased strength. The general effect of cold rolling is to modify the existing grain structure by elongating the grains in the direction of rolling, producing either flattened or "fibrous" grains. The tensile strength may be greatly increased in the direction of rolling; for instance, while as-cast copper has a UTS of about 10 tons/sq in, heavily cold rolled copper sheet may have a UTS as high as 31 tons/sq in.

Cold rolled sheet and strip is classified as skin-passed, quarter-hard, half-hard and hard-rolled, depending upon the amount of working carried out during manufacture. In general skin-passed metal is given only about a 1% reduction by cold rolling, while the harder grades may have undergone reductions of up to 50% or more.

TABLE 8.4

RANGES OF ULTIMATE STRENGTH AND HARDNESS FOR VARIOUS GRADES OF COLD ROLLED SHEET MILD STEEL

Classification of sheet in terms of amount of cold rolling	*Ultimate Tensile Strength (tsi)*	*Vickers Pyramid Hardness*
Dead Soft*	18—22	90—100
Skin Passed	22—24	95—105
Quarter Hard	25—30	105—130
Half Hard	30—35	130—160
Hard Rolled	37—45	170—220

*Sheet fully annealed

In general, it is possible to bend a piece of quarter-hard steel strip back across itself without causing fracture, while half-hard and hard-rolled sheets will only withstand bends of 90° and 45° respectively, if the radius of bending is equal to the metal thickness.

Work hardening effects are not concentrated in the outer portion of a fairly heavily-reduced section.The following Vickers hardness determinations were taken at regular intervals from the outside to the centre of a cold rolled bar of 2″ diameter.

(skin) 172, 172, 173, 165, 158, 152, 152, (core)

Thus while some skin hardening effects do occur, they are not very pronounced even on a 2″ diameter section.

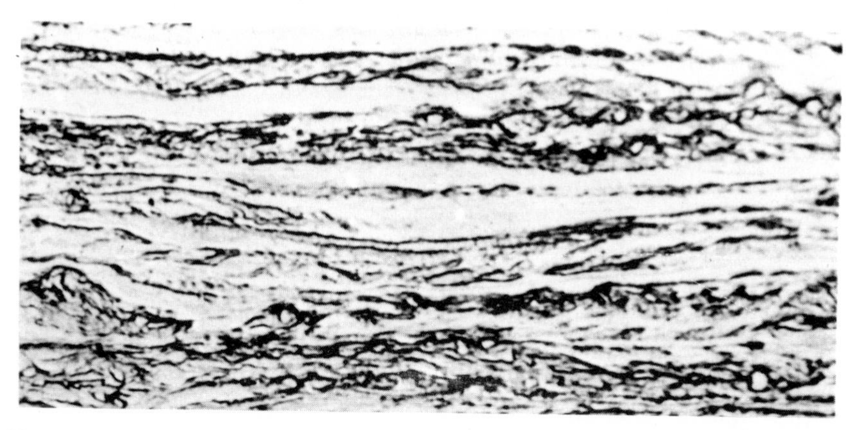

Figure 8.16 *A photomicrograph showing the effects of cold rolling on the grain structure of steel. Magnification x500 (Photograph courtesy of Australian Iron and Steel Pty. Ltd.)*

Springback

Springback is associated with many cold forming processes, particularly those involving bending, deep drawing, and spinning. If a metal is cold deformed to some point between its yield and ultimate strengths, the deformation will be partly plastic and partly elastic. When the deforming load is removed, the residual elastic deformation is released, and springback occurs. Thus cold deformation must always be carried out beyond the desired point by an amount that will compensate for springback effects. This is of particular importance in most angle bending operations.

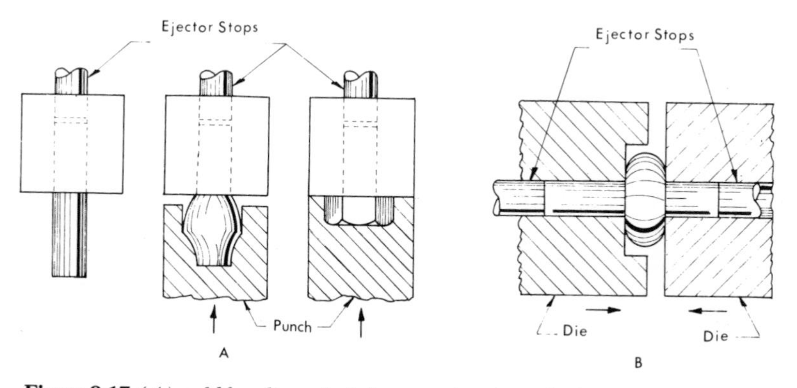

Figure 8.17 *(A) cold heading a bolt by upsetting the end of a rod (B) upsetting the centre section of a rod.*

SOME SPECIAL COLD FORMING PROCESSES

Cold Forging and Pressing

Many small components such as bolts, nails, and rivets have their heads formed by cold heading. By using closed dies made to close tolerances extremely accurate components can be made by other similar cold squeezing or pressing operations. Such cold forming processes are very economical in that no prior heating is required, little or no machining is necessary, and metal waste is kept to an absolute minimum. Surface finish is good and no scaling occurs during forming.

Impact Extrusion

The extrusion punch operates at a fairly high energy, the article being formed by a single blow of the punch into the extrusion die. Only metals having low strengths can be formed; lead, tin, and aluminium alloys being most commonly used. Articles produced include toothpaste tubes, cans, and special tubes. Two variations of impact extrusion are used, these being shown in Figure 8.18.

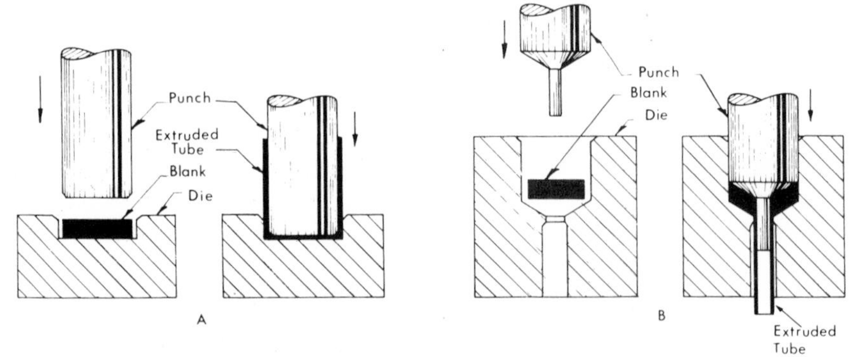

Figure 8.18 *(A) direct impact extrusion. (B) indirect impact extrusion.*

Hydrostatic Extrusion

In this process a fluid such as glycerine, castor oil, or a dispersion of molybdenum disulphide in oil is used to transmit the pressure from the ram to the billet. Since direct contact between the ram and the billet has been eliminated, lower pressures are required and die lubrication is effected by the fluid being used.

An important feature of this process is billet preparation; the billet must be free from scale and its nose portion must be accurately machined to fit the entrant angle of the die being used. It is this latter feature that results in the often dramatic reduction in extrusion pressure required. Aluminium is ideally suited to this type of cold extrusion; however, various steels including annealed tool steel have been shaped by this process.

Cold Drawing

Articles such as small cups, rod, wire, auto turrets and fenders, and brass shell casings are formed accurately and economically by cold drawing processes. The cold drawing of rod and wire is a very simple process; essentially the stock which is always round in cross section and slightly larger in diameter than the die through which it is to be drawn is tapered at one end, pushed through the die, gripped by mechanical tongs, and slowly drawn through the die. A schematic illustration of the process is shown in Figure 8.19, the carbide dies being used to reduce die wear to a minimum.

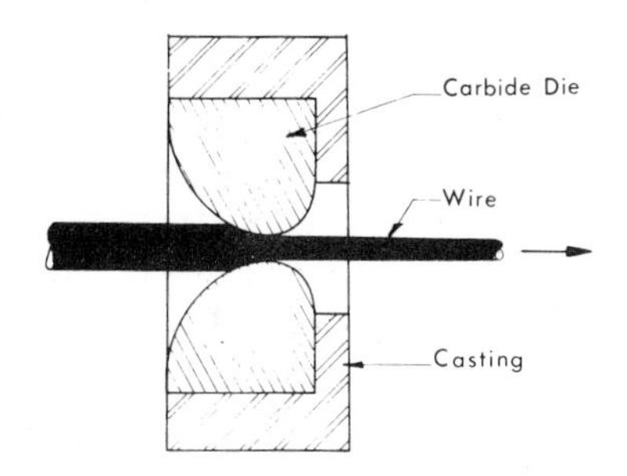

Figure 8.19 *The principle of a wire drawing die.*

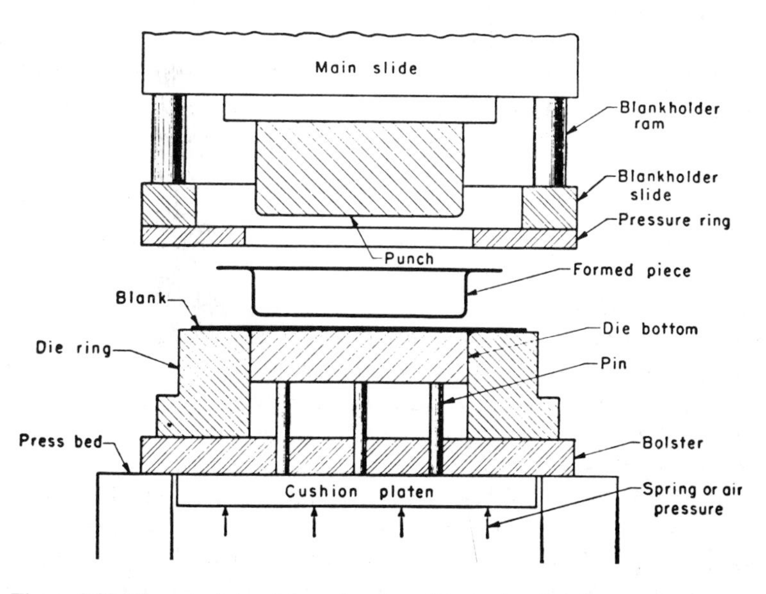

Figure 8.20 *The principle of deep drawing. (Reproduced with permission from "Materials and Processes in Manufacturing", 2nd Edition, by C. De Garmo. Macmillan, N.Y., 1957.)*

Many passes through successively smaller dies are usually necessary in order to achieve the required amount of size reduction, and intermediate annealings may be necessary to remove work hardening effects.

Deep drawing is really a pressing operation but in certain circumstances can be carried out on a drop forging machine. The steel dies used must be extremely accurate and must be highly polished, and some type of lubrication is usually used to reduce die wear during drawing. Materials to be deep drawn must possess low yield strengths and high ductility, the Erichsen cupping test being an excellent test of the overall deep drawing qualities of a sheet metal.

High-Rate Forming

Double-acting presses have recently been developed which are capable of driving dies together at very high speeds. The dies used are always closed impression dies of extremely high precision and the metal billet must always be of the exact size required. The extremely high rate of deformation that results from this type of forming process leads to very high dimensional accuracy combined with excellent surface finish, so that machining is negligible or unnecessary. This forming process is still largely experimental and is only economically feasible for long production runs where the output offsets the high tooling costs.

Explosive forming, however, can be regarded as a special type of high-rate, high-energy forming process in which tooling costs can be kept to a minimum. Only a female die is required since the forming pressure from the explosive charge is usually transmitted to the sheet metal blank through a fairly deep layer of water. If a one-off piece is required an easily-made plaster of Paris

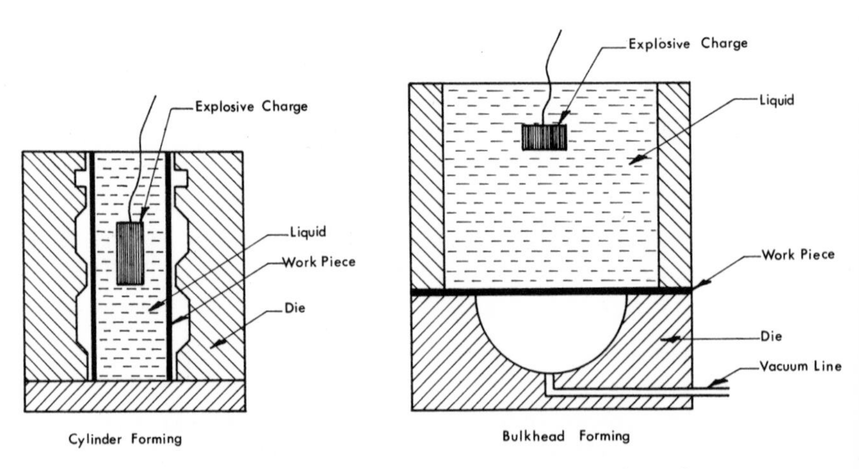

Figure 8.21 *Principles of the explosive forming of metals.*

or wooden die can be used since the blank is formed to shape before the force of the explosion destroys the mechanically weak die.

Extremely hard high-strength metals such as stainless steel and titanium can be formed precisely to shape by explosive forming, and springback effects are negligible due to the high energies used. There appears to be no real limit to the size of the piece that can be shaped by this process; for instance, the dish-shaped ends of huge boilers have been explosively formed in lakes where the female die has been suspended from a floating crane. The metal appears to suffer no work hardening effects, thus remaining soft and pliable; this is related to the extremely high deformation rate which creates enough internal heat within the work-piece to anneal it during forming.

POWDER METALLURGY

If metals are to be formed by the techniques of powder metallurgy, they must first of all be powdered. Metals and alloys have such diverse properties that various different techniques must be used to prepare them in powdered form; the following are the most important methods.

Mechanical Disintegration is usually used to powder the more brittle metals, and includes such techniques as grinding, machining, milling, and disintegration in hammer mills, eddy mills, and in stamp mills. The individual particles are work hardened and jagged in shape.

Atomised Powders are produced when a stream of molten metal meets a powerful jet of air, gas, liquid or steam. Atomised powders are usually spherical or pear-shaped and coated with a thin oxide layer; the more ductile or soft metals like aluminium, cadmium, copper, lead, tin, zinc, brasses, bronzes, and solders are prepared by this technique.

Chemical Methods of powder preparation include reduction, condensation and precipitation. Metals like iron, cobalt, nickel, and tungsten are commonly powdered by reduction from oxides; zinc is commonly powdered by a condensation reaction; while precipitation is commonly used to prepare very fine powders of such metals as silver and copper.

Electrolytic Deposition is commonly used to prepare very pure powders of metals like copper and iron, porous copper bearings commonly being made from electrolytic powders.

After the powders are prepared they must be stored under dry conditions prior to use. Just before use, they must be carefully blended in correct proportions, pressed at the optimum pressure in a suitable die, and then sintered to promote adhesion between all of the particles in the component being manufactured.

Pressing

Pressing, usually carried out in steel or carbide dies at room temperature, has the following effects upon the powder.

(1) It reduces the porosity of the mass as particles are forced closer together.

(2) It causes particle deformation, some particles "keying" into others, causing an increase in the mechanical strength of the pressed articles.

(3) Cold welding will occur between adjacent particles, the extent depending upon the hardness of the metal powder and the pressure used.

Presses may be mechanical or hydraulic, the latter type being used for pressures in excess of 30 tons per square inch. It has been found that there is an optimum pressure range for most powders, this range giving maximum results for a given expenditure of energy. Table 8.5 shows the ranges of pressure used for the common metals.

TABLE 8.5

PRESSURES FOR METAL POWDERS

Metal Powder	*Pressure (Tons per Sq In)*
Soft lead-based bearing alloys	5—10
Soft brasses	15—25
Copper and Bronzes	20—30
Iron and Steels	Up to 50

In general, a fine powder requires a higher pressure to give it the necessary strength and density than a coarser powder, and, other things being equal, powders of hard metals give weaker and less dense compacts than do those of softer metals if both types are subjected to the same pressure in the press.

Sintering

After the pressed compact has been produced it is sintered by heating to a temperature below its melting point, usually in a continuous furnace of the "tunnel kiln" type. If a mixture of powders is being sintered, the temperature may be regulated so that the lowest melting-point constituent melts, forming a liquid phase that assists in holding the article together. This is common practice with cemented carbides, when the "cement" consists of a tough metal like cobalt or nickel.

Sintering proceeds most rapidly at the recrystallisation temperature of the metal or alloy. If the compact is sintered for too long, grain growth may increase to the point where the coarseness of the grains will weaken the metal, causing premature failure in service. Porous bronze bearings are sintered at about 750–850°C and iron powders often above 1100°C. In order to prevent oxidation of the metal powders during sintering, an inert or controlled atmosphere is maintained in the sintering furnace.

If machining is to be carried out on a sintered component it is often done after a *pre-sintering operation* in which the powder compact is heated for a short length of time at a temperature considerably lower than its final sintering temperature. This imparts sufficient strength to the compact to allow machining to be carried out. Pre-sintering is most useful for those sintered articles that would be too hard and brittle for machining after final sintering.

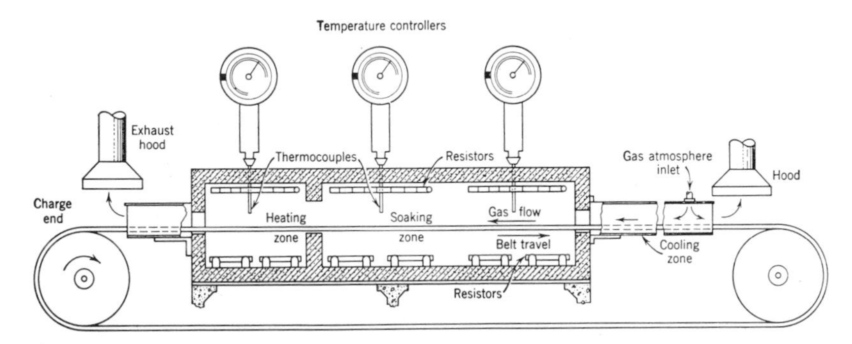

Figure 8.22 *Cross-section of a continuous sintering furnace. (Reproduced with permission from "Metallurgy for Engineers" by Wulff, Taylor and Shaler. John Wiley & Sons Inc., N.Y. 1952.)*

Sintered compacts are often finished to high-dimensional accuracy and surface finish by a pressing or coining operation in which the pressures used are often greater than those used in the initial pressing operation. Such finishing "sizing" operations increase the strength of the sintered component by up to 50%.

Products of Powder Metallurgy

The following four groups illustrate the range of components produced by pressing and sintering metal powders.

(1) *Porous Products*: porous bronze bearings, impregnated with up to 40% of oil, constitute the largest volume of powder metallurgy production; while filters, with pores as small as 0.0001″, are produced easily and economically.

(2) *Products of Complex Shapes*: small gears, pawls, cams and actuating levers that would require considerable machining if produced by conventional methods are often produced by powder metallurgy, with many such components being made self-lubricating as well.

(3) *Products Difficult to Machine*: tungsten carbide and other hard ceramics are readily formed into cutting tips by this process, and thus machining is avoided. Cemented carbides are obtained if a tough metal like cobalt or

nickel is powdered and mixed to the extent of 5–10% into the ceramic powder prior to pressing.

(4) *Composites* may be formed between materials normally difficult to alloy or mix. For example, copper and graphite are formed into brushes for electric motors and generators, while copper or silver are combined with tungsten, nickel or molybdenum for electrical contacts. In the latter example, the copper or silver provides electrical conductivity while the tungsten, nickel, or molybdenum provides resistance to fusion and oxidation.

Composites may also be formed by infiltration. In this process the porous sintered component is immersed in a bath of molten metal of lower melting point, which then fills all pores in the component. Infiltration may be assisted by pressure or vacuum techniques, and increases the strength of the component by 50–100%.

Table 8.6 lists the advantages and disadvantages of sintered metals.

TABLE 8.6

ADVANTAGES AND LIMITATIONS OF FORMING BY POWDER METALLURGY

Advantages	*Limitations*
Elimination of, or great reduction in, machining.	The strength properties of the finished article are usually lower than those of a similar article produced by conventional means.
Elimination of scrap.	
High production rates possible since process is automated.	The metal powders are expensive to produce.
Production of many complex shapes easily achieved.	Die costs are high and therefore production runs must be long.
Wide variations in material compositions can be achieved.	Die design limitations limit the types of shapes that can be produced.

GLOSSARY OF TERMS

Atomising: the dispersion of molten metal into small particles by a high-velocity stream of air, steam, or other gas.

Cold working: controlled plastic deformation carried out at such temperatures and rates that recrystallisation does not occur.

Drawing: the process whereby material such as rod or wire is simultaneously lengthened and reduced in section by pulling it through hard dies.

Extrusion: shaping a billet of metal to a given cross-section by pushing it through a die of suitable shape.

Forging: the shaping of metal by the application of localised compressive forces.

Grainflow: a structure found in hot worked components in which both the grains and impurities are arranged in a fibrous pattern which follows the pattern of deformation; grainflow causes directional properties in hot worked metals.
Hot working: controlled plastic deformation carried out at temperatures above the recrystallisation temperature of the metal.
Infiltration: the process of filling the pores in a sintered compact by immersing it in a bath of oil or a bath of low melting-point metal.
Recovery: the partial release of internal stresses in metal caused by heating the metal to well below its recrystallisation temperature; the process of recovery increases ductility and electrical conductivity quite considerably.
Recrystallisation: the nucleation and growth of stress-free grains in a deformed crystalline matrix.
Recrystallisation temperature: the lowest temperature at which recrystallisation will proceed at a reasonable rate; recrystallisation temperatures vary considerably with the amount of strain energy within the metal.
Sintering: the process whereby cohesion is promoted in a compacted powdered material by heating it in a controlled atmosphere.
Sizing: a final pressing of a blank or sintered compact used to reduce it to its final size and shape.
Springback: the movement in a cold deformed metal component caused by the release of residual elastic strain.
Stress relieving: a heat treatment process in which recovery but not recrystallisation occurs.

REVIEW QUESTIONS

1. What alterations to mechanical and electrical properties occur when a piece of work hardened metal undergoes complete recrystallisation?
2. What two factors influence the temperature at which a particular metal will recrystallise?
3. Why does nucleation during recrystallisation begin at the grain boundaries?
4. What effect does excessive grain growth have on metals?
5. Compare and contrast the processes of (i) recovery and (ii) stress-relieving.
6. Distinguish between hot and cold working in terms of working temperatures and general effects on mechanical properties.
7. Explain the meaning of "fibre" (or "grainflow") in hot deformed metals, and outline its effects on mechanical properties.
8. What properties are desirable in metals that are to be shaped by (i) extrusion; (ii) drawing; (iii) spinning?
9. List and discuss some of the important advantages and limitations of cold forming processes.

10. What is "springback" and how does it influence tooling design in cold forming processes?

11. How does high-rate forming differ from the conventional forging processes?

12. Auto hubcaps are to be mass produced. Select a suitable combination of a material and a manufacturing process; describe the process, and justify your selection.

13. Briefly describe three methods by which powders suitable for powder metallurgy can be produced.

14. What are the effects of sintering on the powder compact produced by pressing?

15. Outline the advantages of (i) pre-sintering and (ii) coining or sizing on metal compacts.

16. List and discuss the advantages and limitations of the powder metallurgy process.

17. The following two products have been mass produced by powder metallurgy: (i) porous, self-lubricating bronze bearings, and (ii) small steel pawls of considerable strength. Compare and contrast the manufacture of these two products, referring particularly to powder preparation, pressing pressures, sintering cycles, and any treatment that may be necessary after sintering.

9

Single and Multi-phase Materials

DETAILED microscopic examinations of metals, rocks, minerals, and other solid materials reveal that while vast differences exist between some materials at this level, others possess certain similarities of structure. For instance, while pure copper and a thin section of quartz each consist of only one distinct type of grain, materials like cast iron, 60/40 brass, mild steel, and granite each consist of several distinctly different types of grains or micro-constituents. In effect, then, a broad distinction must be drawn between those materials that are single-phase and those that are multi-phase.

It must be borne clearly in mind that this distinction is quite different from that existing between a single crystal and a polycrystalline material. If a piece of material is a single crystal then it will possess a crystal lattice of the same orientation throughout, the outer surfaces of the piece of material being the crystal (or grain) boundary. A polycrystalline material consists of a number of crystals or grains each of which possesses its own peculiarly-orientated crystal lattice.

A *phase* as seen under the microscope is a physically distinct, chemically homogeneous and (theoretically, at least) mechanically separable portion of a material. It thus becomes apparent that a single crystal of a material can consist of one phase only, while polycrystalline materials may be single- or multi-phase, depending upon the natures of the individual crystals or grains present.

PURE METALS AND ALLOYS

All industrially-produced metals are polycrystalline materials. However, while all pure metals are single-phase, alloys may be single-phase or multi-phase depending upon the solid solubilities of the metals melted together to form a particular alloy.

An alloy may be composed of two or more elements, one of which must be a metal, and the alloy itself must exhibit metallic characteristics. The base metal of an alloy is the metal present in the greatest proportion, while other metallic or non-metallic elements present are known as alloying elements.

However, if certain elements are present in a metal or alloy as a result of imperfect refining or smelting operations, these elements are known as impurities. The properties of an alloy sometimes differ in significant respects from the properties of the base metal even when only small amounts of the alloying elements are present. Hence, it is important that the mechanisms of alloying be fully understood.

TABLE 9.1

SOLUBILITY POSSIBILITIES* FOR METALS MELTED TOGETHER TO FORM AN ALLOY

In the Molten (Liquid) State	*In the Solid State*
Insoluble	Insoluble
Soluble (Miscible)	Soluble
	Partly Soluble

*The proviso must be added that the metals involved may react together chemically to form a definite chemical compound. These substances, known as intermetallic compounds, will be discussed later in this chapter.

SOLID SOLUTIONS AND COMPOUNDS IN ALLOYS

All solutions are homogeneous mixtures in which the atoms or molecules of the solute material are more or less uniformly dispersed throughout those of the solvent material. Whereas the relative sizes of the atomic particles of the solute and solvent materials are of little importance in a liquid solution, they are of extreme importance when a solid solution is formed, such size differences accounting for the fact that while most metals are soluble in each other in the liquid state, solid state solubility is more restricted.

Solid Solutions

If one metal dissolves in another to form a solid solution, then either one of two different kinds of solutions will be formed.

Interstitial solid solutions are formed when the solute atoms are very small in comparison with the solvent atoms, the former thus being able to occupy some of the interstices present in the crystal lattice of the latter. Perhaps the most important interstitial solid solution present in ferrous metals is ferrite, which consists of pure iron containing up to a maximum of 0.008% carbon at room temperature. In general, hydrogen, carbon, nitrogen and boron can form interstitial solid solutions with the transition metals.

Substitutional solid solutions, on the other hand, form by the operation of a completely different mechanism, since atoms of the solute material come to take up positions normally occupied by some of the atoms of the solvent material. In this case, solid solutions form readily if solute and solvent atoms are of about the same size. Copper and nickel form a continous series of substitutional solid solutions for all percentages of copper and nickel, while copper and zinc form a similar solid solution for all compositions up to about 36% zinc.

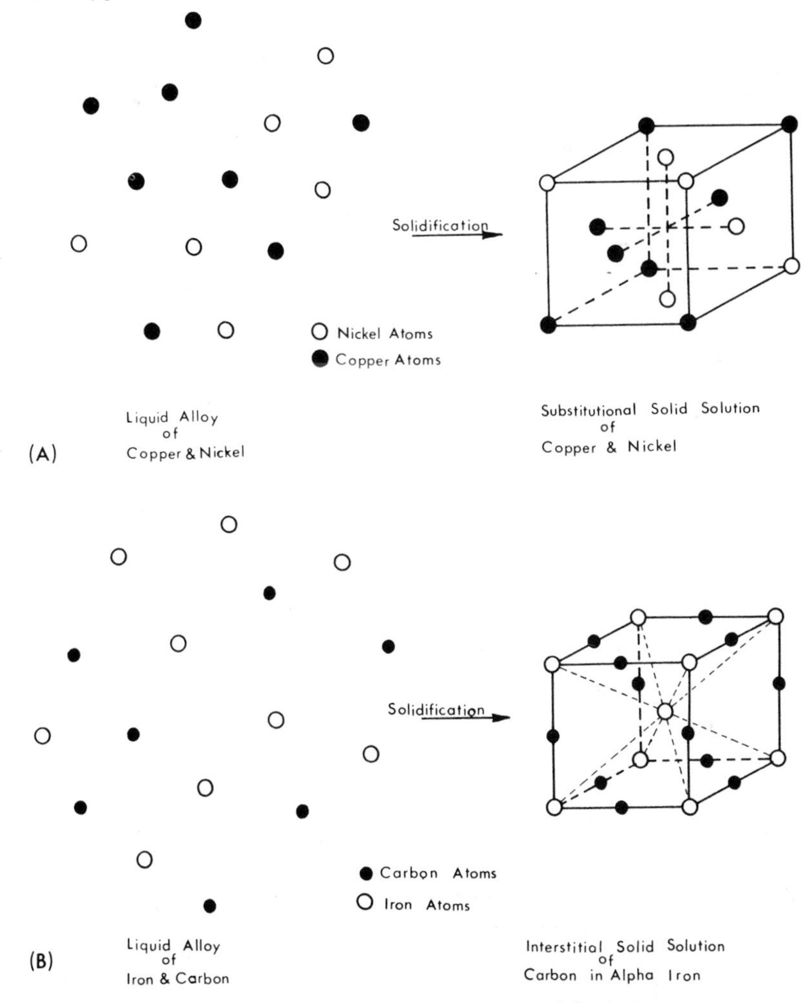

Figure 9.1 *Interstitial and substitutional solid solutions.*

Substitutional solid solutions may be *ordered*—if the atoms of the solute material occupy similar lattice points within the crystal structure of the solvent material, or, *disordered*—if substitution is random. Very often a substitutional solid solution will be ordered or disordered depending upon temperature. An example is the β phase in the copper-zinc alloy system, which is ordered only below about 450°C.

Compounds in Alloy Systems

Under certain well-defined conditions definite compounds will be formed between the constituent elements of an alloy. Such compounds differ from solid solutions principally in that, while a range of compositions is possible for any given solid solution, a compound of the particular type being discussed here is similar to more normal compounds in that the constituent elements usually combine in fixed proportions by weight. Compounds formed in alloy systems may be classified into one of two broad groups:

Intermetallic compounds—being formed between two or more metals which combine together usually without regard to the normal rules of valency. Both ionic and covalent bonding can occur in such compounds, which are characteristically hard and brittle. Magnesium silicide Mg_2Si and copper aluminide $CuAl_2$ are examples of intermetallic compounds.

Interstitial compounds, on the other hand, are formed between a metal and a non-metal, and occur particularly between elements where extreme differences in atomic sizes exist. A most important example is Fe_3C, iron carbide or cementite, which is responsible for the hardness of normalised high carbon steels. Its structure is shown in Figure 9.2 below.

Figure 9.2 *The orthorhombic crystal structure of iron carbide (Fe_3C). The unit cell contains four carbon atoms and twelve iron atoms. (Reproduced with permission from Volume 2 of "Crystal Structures", 2nd Edition, by R. W. G. Wyckoff. John Wiley and Sons Inc., N.Y. 1964.)*

It is important to realise that iron carbide, even though written Fe_3C, does not possess molecules Fe_3C; the formula simply indicates that the unit cell of the crystal lattice of iron carbide contains iron and carbon atoms in a three to one ratio.

Like intermetallic compounds, interstitial compounds do not obey normal valency rules and are hard and brittle.

THERMAL EQUILIBRIUM DIAGRAMS

An equilibrium diagram is a graphical method of showing the phases present in an alloy system at different temperatures and for different compositions. All equilibrium diagrams have temperature as the vertical scale (ordinate) and percentage composition by weight as the horizontal scale (abscissa). It is essential to realise that time cannot be represented on an equilibrium diagram since the diagram is constructed on the basis that equilibrium always exists between the phases present. In other words, if an equilibrium diagram is being used to explain the cooling and solidification of a certain alloy, this alloy must be cooled at a rate such that all reactions occurring within it both while it is solid and while it is liquid are allowed to reach equilibrium. This may involve very slow cooling or even soaking at certain temperatures. In general, industrially-produced alloys are not cooled under equilibrium conditions and therefore their patterns of solidification deviate from those indicated in the appropriate equilibrium diagrams to a greater or lesser extent.

The Construction of Thermal Equilibrium Diagrams

Equilibrium diagrams, also known as constitution diagrams or phase diagrams, are usually constructed on the basis of data obtained from cooling curves which are obtained by plotting falling temperature against time for small samples of selected alloys allowed to cool under equilibrium conditions. Figure 9.3 illustrates the three main types of cooling curves obtained using this technique.

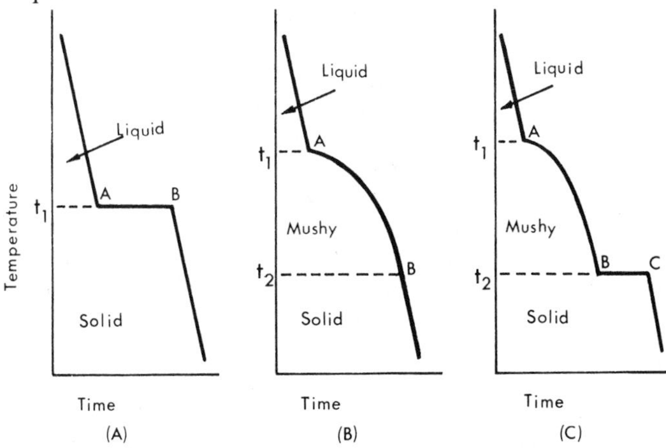

Figure 9.3 *Cooling curves for pure metals and alloys. (A) for a pure metal, solidification occurs at constant temperature as indicated by the horizontal line AB; (B) for a solid solution, solidification occurs over a temperature range as indicated by the interval AB; (C) for a multi-phase alloy containing a eutectic, the solidification range is indicated by the interval ABC, the horizontal section line BC indicating the eutectic reaction.*

The arrestments in cooling curves provide the data used to construct equilibrium diagrams, each arrestment indicating the start or finish of some sort of phase transformation. If a pure metal is cooled, the arrestment zone of the cooling curve is horizontal, thus the transformation from liquid to solid occurs at constant temperature. However, all alloys except eutectic mixtures continue to fall in temperature while phase changes are occurring. Thus, an arrestment point occurs at the beginning of transformation and again at its completion.

Laboratory (33): *Plot cooling curves for pure lead, 50/50 solder, and 62/38 solder by melting samples of each in suitable crucibles and using a 0-360°C thermometer to measure temperature. The portion of the thermometer immersed in the molten metal should be placed in a brass sleeve made by drilling out a piece of 3/8" diameter rod, and the top of the crucible covered with a piece of asbestos to prevent excessive heat loss. Explain the differences between these cooling curves.*

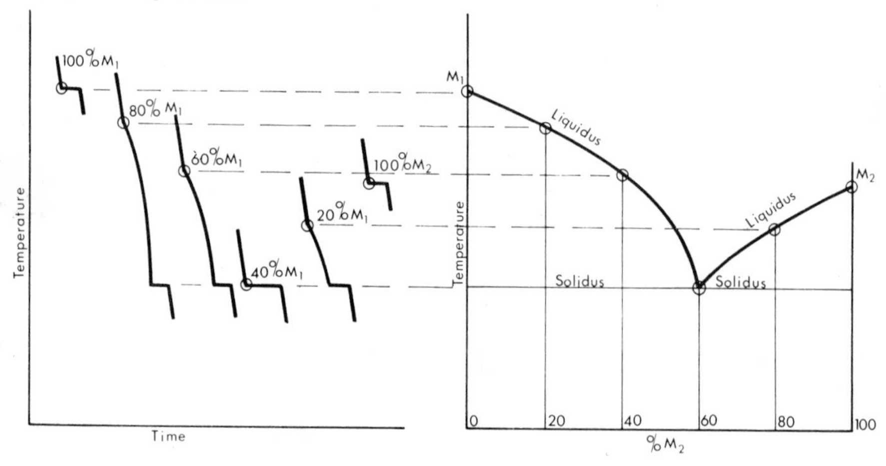

Figure 9.4 *A series of cooling curves for various alloys of the two metals M_1 and M_2 which are insoluble in the solid state.*

Figure 9.5 *Constructing the equilibrium diagram for the alloys of M_1 and M_2 using the data from the cooling curves shown in Figure 9.4. Note that the equilibrium diagram, unlike the cooling curves, does not show time as a variable.*

Figure 9.4 shows a series of cooling curves plotted for various mixtures of the two metals M_1 and M_2, with the cooling curves for each pure metal also being included. The arrestment points have been joined by dashed lines. If this data is used to construct an equilibrium diagram, time must be removed as a variable, and the appropriate arrestment points plotted on vertical composition lines as shown in Figure 9.4. If these arrestment points are joined by lines as shown in Figure 9.5 the equilibrium diagram for all alloys of the metals M_1 and M_2 can be constructed. The line ABC is known as the *liquidus* since all alloys are liquid above this line, while the line DBF is called the *solidus* since all alloys are solid below this line. The point B is a eutectic point and represents the point of solidification of the lowest melting-point mixture of the two pure metals M_1 and M_2 (40 %M_1, 60 % M_2 by weight). The area between the solidus and liquidus represents "mushy" metal; that is, both liquid and solid phases are present at temperatures and compositions within this zone.

Equilibrium Diagrams for Binary Alloys Forming a Continuous Series of Solid Solutions

This type of binary alloy system is not common. It is formed when the two metals have the same crystal structure and their atoms exhibit similar chemical characteristics and are roughly the same size. The alloy systems formed between copper and nickel, copper and platinum, copper and gold, gold and silver, iron and vanadium, and bismuth and tin are of this type.

All alloy systems of this type exhibit the following characteristics. (1) No eutectic exists within the system since alloys of every composition form solid solutions; (2) the solid solutions formed by the solidification of melts of varying percentage compositions reveal little variation with respect to their mechanical properties, while electrical properties and corrosion resistance may alter drastically with only minor compositional variations; and (3) alloys of this type cooled under industrial (non-equilibrium) conditions exhibit coring, but this is readily removed by appropriate annealing.

The *copper-nickel system* (see Figure 9.6A) is typical and will serve as a useful example since these alloys are of industrial importance. Consider the cooling and solidification of an alloy of composition 40 % copper 60 % nickel (see Figure 9.6B).

(1) Above the liquidus line a uniform liquid solution of the two metals exists, nickel being considered the solvent since it is present in greater proportion.

(2) When the temperature falls to that indicated as t_1, solidification begins with nucleation and the growth of dendrites, the composition of which may be determined by drawing a horizontal line across to point A on the solidus and then vertically downwards to the composition scale. Using this technique it is found that the first few dendrites to form contain about 79 % nickel; in other words, they are richer in nickel than the melt itself.

(3) During the temperature interval from t_1 to t_3 grain growth continues as more and more metal solidifies around the original dendrites. If equilibrium cooling conditions are present the composition of the solidified metal progressively alters as atoms diffuse from one area to another. However, under non-equilibrium cooling conditions the solidifying metal is laid down in layers around the dendrites and each layer varies in composition, becoming

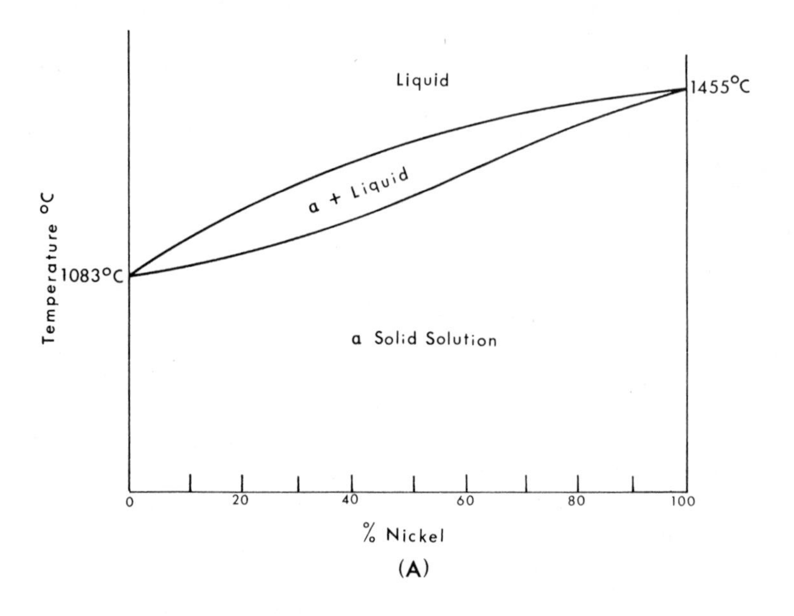

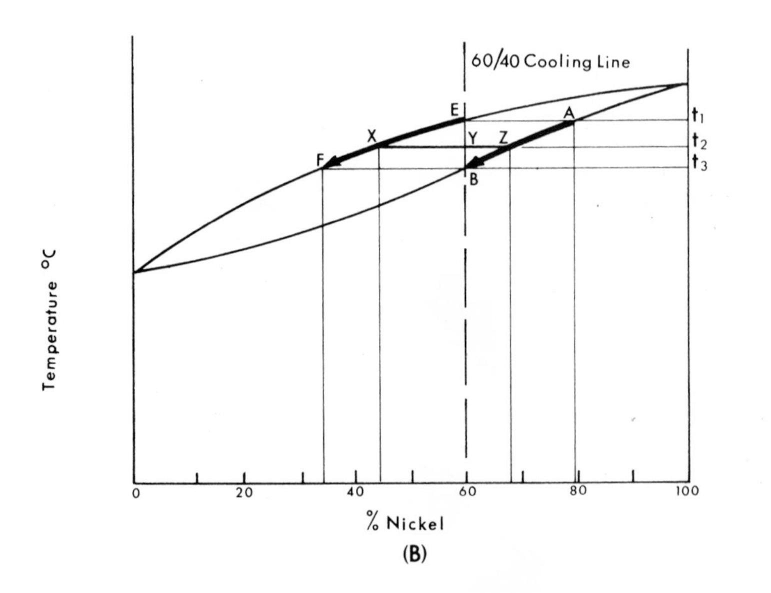

Figure 9.6 *(A) The copper-nickel equilibrium diagram. (B) The cooling and solidification of a 60% nickel, 40% copper alloy.*

progressively richer in copper as the temperature falls. The composition of the solid metal forming at any particular temperature in between t_1 and t_3 can be found by again taking a horizontal line across to the solidus, while the composition of the remaining liquid phase can be found by continuing this line in the other direction until it intersects the liquidus. For example, at the temperature, t_2, the solid phase separating has the composition of about 68% nickel and 32% copper while the residual liquid phase contains only 44% nickel.

(4) Solidification will be complete when the temperature t_3 is reached, and the remaining liquid solidifying at this temperature will have the composition of about 34% nickel, 66% copper, this being determined by drawing a horizontal line across to the liquidus from the intersection of the 60/40 composition line and the solidus.

(5) Thus, it becomes apparent that during the solidification of this alloy the composition of the solid phase separating during the interval from t_1 to t_3 varies along the solidus from A to B while that of the residual liquid varies along the liquidus from E to F.

(6) During the interval of cooling from t_3 to room temperature solid state diffusion occurs with nickel atoms slowly moving from the nickel-rich centres towards the outer portions of each grain while copper atoms move from the copper-rich outer zones in towards the centre. Under non-equilibrium cooling conditions this diffusion process could not go to completion and so coring would be present within the alloy.

The Lever Rule

The lever rule is a convenient method of calculating the relative proportions of solid and liquid material present at any given temperature. Consider the alloy containing 60% nickel at the temperature, t_2: then—

$$\frac{\text{weight of solid solution}}{\text{weight of liquid solution}} = \frac{XY}{YZ}$$

Therefore, (1) the pereentage of solid material present in this alloy at this temperature equals

$$\frac{XY}{XZ} \times \frac{100}{1} \text{ (where XZ = total lever arm)}$$

$$= \frac{(60-44)}{68-44} \times \frac{100}{1}$$

$$= \frac{16}{24} \times \frac{100}{1}$$

$$= 66.7\%$$

and (2) the percentage of liquid material present equals

$$
\begin{aligned}
& \frac{YZ}{XZ} \times \frac{100}{1} \\
&= \frac{68-60}{68-44} \times \frac{100}{1} \\
&= \frac{8}{24} \quad \frac{100}{1} \\
&= 33.3\%
\end{aligned}
$$

It is convenient to calculate the lengths of the lever arms using the percentages obtained by dropping vertical lines down to the composition scale of the equilibrium diagram. In this example, the relative percentages of nickel have been used to determine the lengths of the levers XY, YZ, and XZ.

Eutectic Mixtures

A eutectic mixture is formed in an alloy system when two distinct solid phases separate simultaneously and at constant temperature from a single liquid phase. The phases present in a eutectic are pure metals or solid solutions depending upon the type of solubility exhibited by the metals within that system. In general the eutectic within a simple binary alloy system is the lowest melting-point mixture within that alloy system. However, alloy systems made up of three or more metals may exhibit several eutectic points and, of course, several eutectic mixtures.

Equilibrium Diagrams for Binary Alloys Forming Eutectics

Binary alloy systems containing eutectics are most conveniently divided into two groups: *(a)* those systems in which the two pure metals are mutually insoluble in each other in the solid state; and *(b)* those systems in which partial solid solubility exists. This latter group is by far the most important, containing as it does many important industrial alloy systems; however, the former group is a little less complex and will be considered first.

Group (a): Binary Alloy Systems in which there is Mutual Liquid Solubility but Solid Insolubility

All such alloy systems have the same general type of equilibrium diagram, this being illustrated in Figure 9.7.

Note the following points concerning the equilibrium diagram shown in Figure 9.7: (1) The melting point of metal "A" is shown at F, that of metal "B" by point G; (2) the eutectic point is indicated by the letter E, while the line CED is the eutectic isotherm and also the solidus, indicating the temperature at which separation of the eutectic will begin for any given alloy composition; and (3) The line FEG is the liquidus, and indicates the start of solidification of any given alloy composition.

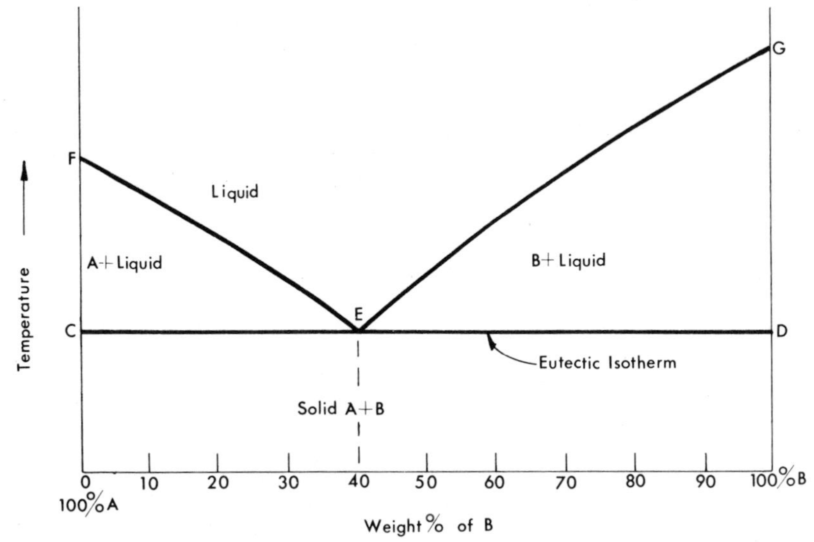

Figure 9.7 *A typical binary equilibrium diagram for an alloy system possessing complete solid insolubility (simple eutectic system).*

The *cadmium-bismuth alloy system* is a good example of complete solid insolubility, and the equilibrium diagram for this system is given in Figure 9.8. The eutectic formed is a finely-divided intimate mixture of cadmium and bismuth.

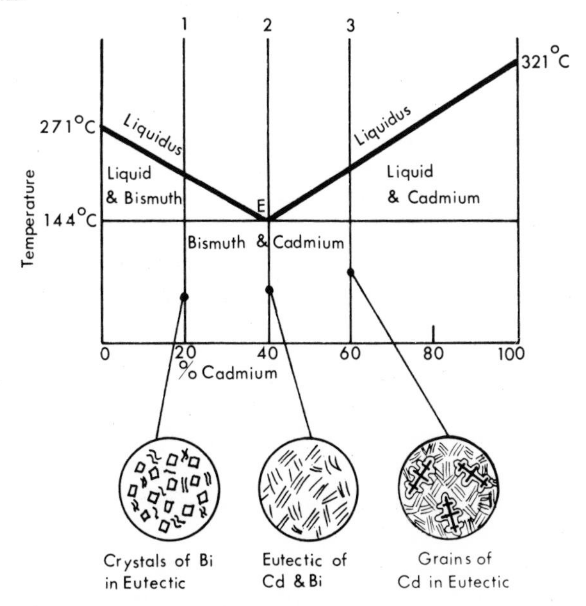

Figure 9.8 *The cadmium-bismuth equilibrium diagram.*

Consider the cooling and solidification of an alloy of composition 20% cadmium and 80% bismuth (cooling line 1 in Figure 9.8).

(a) As the cooling line reaches the liquidus, the liquid solution becomes saturated with bismuth, and dendrites of pure bismuth begin to form.
(b) As cooling continues more bismuth separates as solid metal, and the liquid phase becomes progressively richer in cadmium. The composition of the liquid phase at any given temperature can be read off the appropriate point on the liquidus.
(c) When the eutectic point E is reached (at a temperature of 144°C) the remaining liquid solidifies as the cadmium-bismuth eutectic, and the final microstructure consists of grains of pure bismuth surrounded by eutectic. (The eutectic itself is a finely divided mixture of 40% cadmium and 60% bismuth intimately mixed together.)

Now consider the cooling of a melt of composition 60% cadmium and 40% bismuth (cooling line 3 in Figure 9.8).

(a) As the cooling line reaches the liquidus, dendrites of pure cadmium are formed.
(b) The remaining liquid becomes progressively richer in bismuth, until the eutectic point is again reached, when the remainder of the liquid solidifies.
(c) In this case the solid metal consists of grains of cadmium embedded in the eutectic mixture.

Finally, consider the behaviour of a melt of eutectic proportions (cooling line 2 in Figure 9.8).

(a) As the molten alloy cools no change occurs until the eutectic temperature of 144°C is reached.
(b) The temperature then remains constant at 144°C until the whole of the liquid solidifies as the eutectic mixture.

Coring cannot occur in alloys of this type, since the precipitating metal is always a pure metal.

Group (b): Binary Alloy Systems Exhibiting Partial Solid Solubility Together with a Eutectic

If the solvent metal of a binary alloy can accommodate only a certain percentage of the solute metal, then a limited solid solution is formed, and the two metals involved are said to exhibit only partial solid solubility. This type of solid solubility varies tremendously with temperature, decreasing in almost every instance with falling temperature. In some such alloy systems a eutectic is formed, the eutectic being a mixture of several solid solutions. Figure 9.9 is typical of such an equilibrium system.

The following points are relevant to the type of equilibrium diagram shown in Figure 9.9.

(1) No pure metals exist in a solid alloy of any composition.

(2) Two solid solutions are formed

(a) α, which is mostly metal "A" with some metal "B"

(b) β, which is mostly "B" with some metal "A". The points H and J indicate the maximum solid solubilities of B in A and A in B respectively.

(3) The line CED is the liquidus and point E is the eutectic point, the eutectic having a composition as indicated by the point K.

(4) The solvus lines indicate the falling solid solubility of one metal in the other.

(5) Since the eutectic itself is a mixture of the α and β phases any alloy made up of either α or β plus eutectic can be considered to be merely composed of the α and β phases.

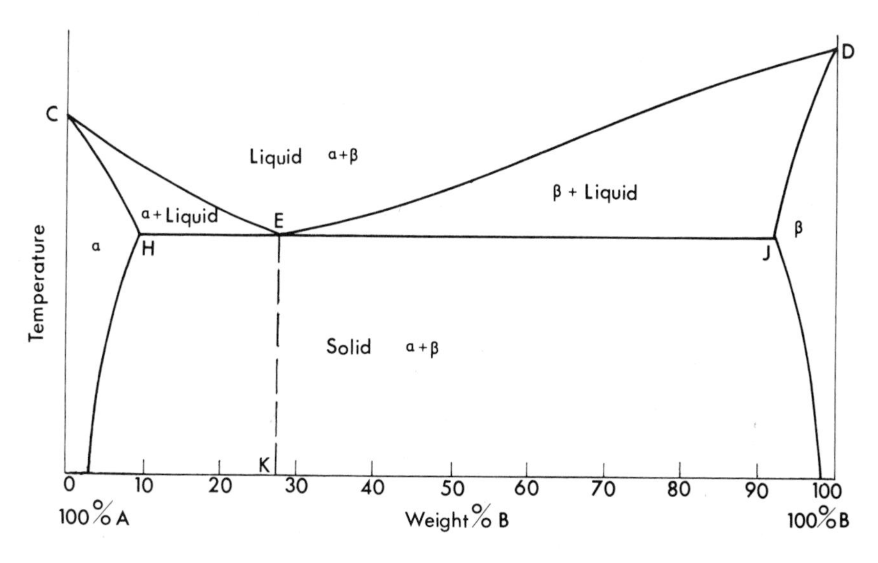

Figure 9.9 *A typical equilibrium diagram for two metals exhibiting partial solid solubility and which form a eutectic.*

The lead-tin equilibrium system is typical of this class of alloys, the diagram appearing in Figure 9.10. In this system the α phase is lead-rich, while the β phase is tin-rich.

Consider the cooling and solidification of a melt of 40 % tin and 60 % lead:

(a) When the liquidus is reached, dendrites of the α solid solution begin to form, the remaining liquid becoming richer in tin.

(b) This precipitation of the α phase continues as the temperature falls, until at a temperature of 183°C the remaining liquid solidifies, forming the eutectic of α and β phases. This eutectic has an overall composition of 62% tin and 38% lead.

The final microstructure reveals dendrites of α solid solution in a matrix of the eutectic.

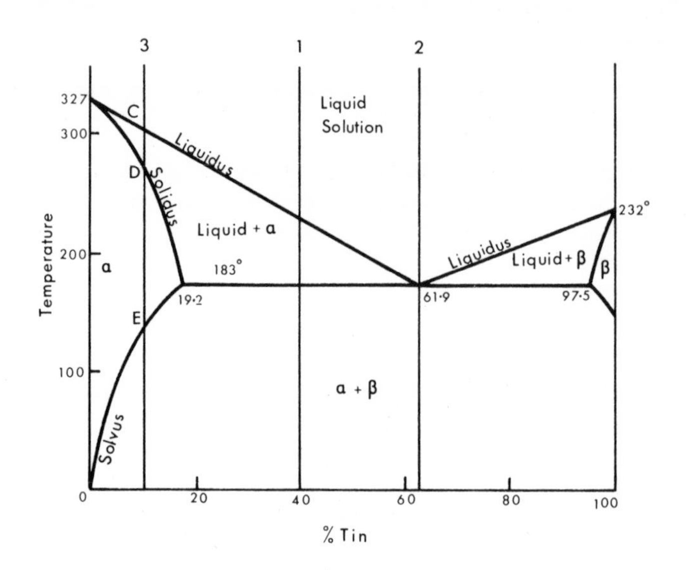

Figure 9.10 *The lead-tin equilibrium diagram.*

Now consider the cooling of an alloy of 62% tin and 38% lead:

(a) No change occurs until the eutectic temperature is reached.
(b) The temperature remains constant at 183°C until the whole of the melt has solidified as the eutectic, a finely divided and intimately mixed mixture of α and β phases.

Finally, consider the cooling of a melt of 10% tin with 90% lead:

(a) At the liquidus (point C), dendrites of the α solid solution begin to form. These continue to form and grow until the solidus is reached (point D).
(b) The solidified alloy now consists wholly of grains of α solid solution. Further fall in temperature causes no change until the solvus line is reached (point E), when, due to the decreasing solid solubility of the tin in the lead, some tin is thrown out of solution together with a little lead, forming material of β composition. This β phase will occur mainly along the grain boundaries and along preferred planes within the α crystals. This kind of structure is termed *Widmanstätten*.

Equilibrium Diagrams for Binary Alloy Systems Exhibiting Partial Solid Solubility but no eutectic

Quite a number of binary alloy systems are of this type, notable examples being the copper-zinc, copper-tin and aluminium-copper systems. Generally speaking, such equilibrium systems are more complex than those previously discussed due to the often complex phase changes that occur after an alloy has solidified. The portion of the copper-zinc equilibrium diagram up to about 60% zinc is shown in Figure 9.11, this portion containing all commercial brass alloys.

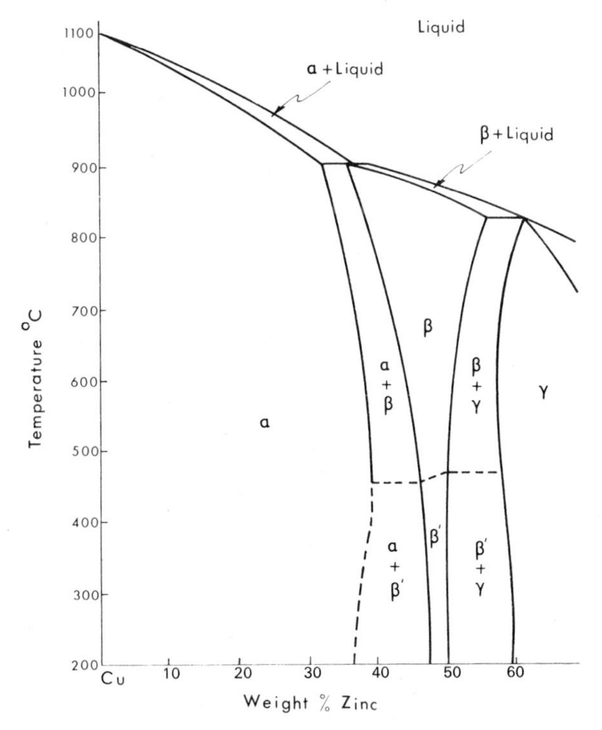

Figure 9.11 *Portion of the copper-zinc equilibrium diagram.*

Three phases are present in the portion of the copper-zinc equilibrium diagram shown in Figure 9.11.

(a) *The alpha phase*: a solid solution of up to 36% zinc in copper, which is malleable, ductile, and hence readily cold worked.

(b) *The beta phase*: another solid solution, characteristically hard and brittle; it can, however, be hot worked very readily.

(c) *The gamma phase*: also a hard and brittle solid solution, differing from the beta phase in that it cannot be hot worked. Thus, very few commercial brasses contain more than about 52% zinc, this being the limit above which the gamma phase appears.

Laboratory (34): *Examine the following alloys under the metallurgical microscope: 70/30 cupro-nickel; 62% tin — 38% lead; 30% tin — 70% lead; annealed alpha brass; 70% cadmium — 30% bismuth.*
Relate their structures to the appropriate equilibrium diagrams.

Peritectic Changes: During the solidification of a cooling alloy the solid already formed may react with the residual liquid to form another solid solution or compound intermediate in composition between the first solid present and the liquid. Such a reaction is termed peritectic, and always occurs at a constant temperature. Consider the portion of the equilibrium diagram shown in Figure 9.12. The line BCD indicates the temperature at which the peritectic change occurs.

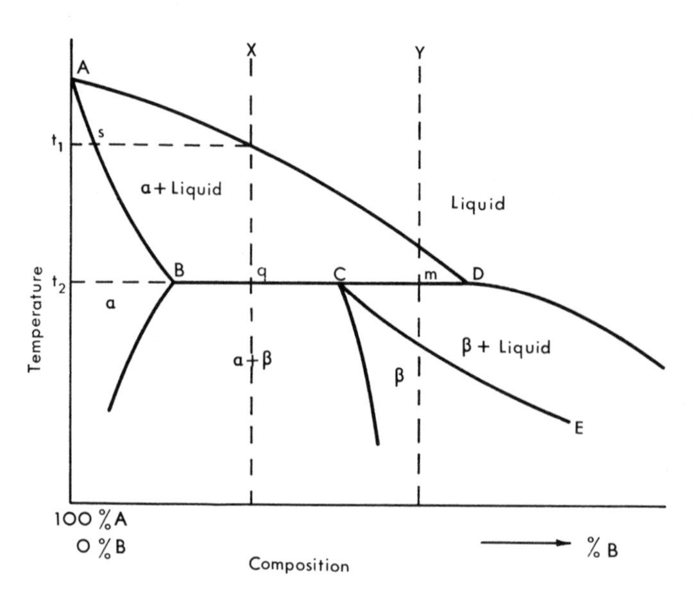

Figure 9.12 *Portion of an equilibrium diagram indicating a peritectic reaction.*

Consider the cooling of an alloy of composition X. Solidification begins with the formation of grains of the α solid solution and α continues to form until the temperature t_2 is reached, when the solid α reacts with the residual liquid to form a new solid solution termed β. Since the original melt did not contain enough of metal "B" to form all β, some residual α remains in the alloy. The alloy of composition "Y" beings to solidify in a similar manner, but transforms entirely to β after undergoing its peritectic change.

Equilibrium Diagrams for Binary Alloy Systems Exhibiting Partial Solid Solubility together with the Formation of an Intermetallic Compound.

The tin-magnesium equilibrium diagram is typical. It is important to note that the compound Mg_2Sn behaves as if it were a new substance and two eutectics are formed, one at A, and the other at B.

Thus, this type of diagram allows for greater complexities within alloys than do those of the types discussed earlier.

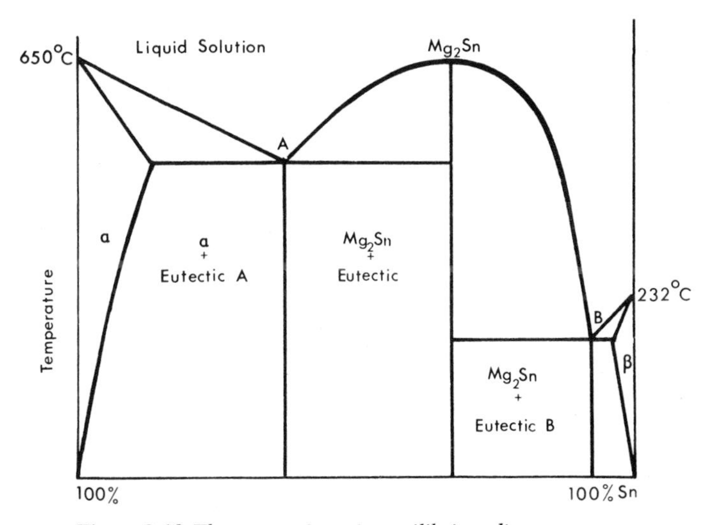

Figure 9.13 *The magnesium-tin equilibrium diagram.*

Layer-Type Equilibrium Diagrams

If the two metals are insoluble in both the liquid and the solid states then a layer-type diagram is formed; a good example is lead-aluminium.
The progress of a cooling alloy of any composition is very simply described. When the melting point of aluminium is reached, crystals of aluminium begin to separate out of the melt, and the temperature remains constant until all the aluminium has solidified. The temperature then drops to the melting point of lead, when it again remains constant until all of the lead solidifies.

Eutectoid Transformations

A eutectoid transformation occurs when an already-existing single solid phase breaks down at constant temperature into two distinct and separate phases, which are finely divided and intimately mixed in the transformed alloy. The most common example is pearlite, which is formed in the iron-carbon system when austenite breaks down to ferrite and cementite at a temperature of 723°C.* Eutectoids are similar in appearance to eutectics.

*See Chapter 11 for details.

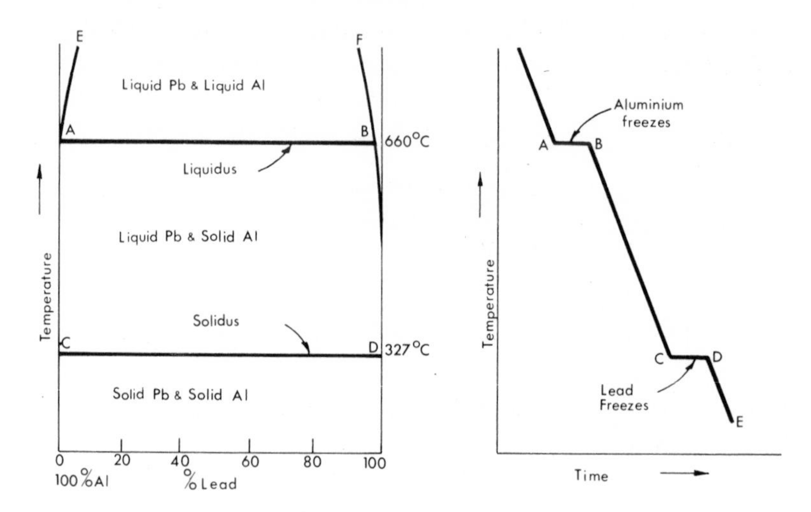

Figure 9.14 *The aluminium-lead equilibrium diagram. The cooling curve shown at right is typical for aluminium-lead alloys.*

GENERAL PROPERTIES OF ALLOYS

In general, the physical, chemical, and mechanical properties of an alloy are not found by averaging the properties of its component pure metals. However, the following generalisations are of interest.

Solid Solutions

(a) The hardness generally rises to a maximum at a composition of about 50% of each metal for complete solid solubility, while the tensile strength increases gradually towards that of the stronger component. If two metals form a limited solid solution, then, in general, the greater the amount of solute (i.e. dissolved metal) the higher the hardness and strength. For example, a brass containing 10% zinc is softer and weaker than a brass containing 30% zinc.

(b) Electrical conductivity is always lower in solid solutions than in the constituent pure metals, and even very small amounts of solute metal are sufficient to drastically lower electrical conductivity.

Two-component Alloy Systems

(a) No general rules can be stated concerning the strength and hardness of such alloys, since these properties depend upon the nature of the phases present and the way in which they are dispersed with respect to one another.

(b) Conductivity depends upon the proportion of each constituent present and their conductivities, so again generalisations are difficult to make.

SOME IMPORTANT NON-METALLIC MULTI-PHASE MATERIALS

It is generally held that structure-property relationships in all multi-phase materials depend upon the following four factors:

(1) the physical and chemical natures of the phases present
(2) the percentage amount of each phase present
(3) the distributions of each phase in relation to one another
(4) the size of the "domains" occupied by each phase.

The three non-metallic multi-phase materials discussed below verify these generalisations.

Rocks are multi-phase materials in which the phases are naturally-occurring minerals. The sizes of the domains occupied by the mineral phases are dependent upon the rate of cooling; in general, the slower the cooling in or on the earth's crust, the coarser the grainsize of the rock. Igneous rocks —rocks solidifying from a melt (magma or lava)—are generally crystalline. However, not all rocks contain distinct crystals or grains, examples being some types of volcanic rocks which because of very fast cooling contain glassy phases. Obsidian (volcanic glass) is a volcanic rock with a completely "glassy" structure.

Ceramic Materials such as fired building bricks, fired clay products in general, and cement are also multi-phase solids, differing from naturally-occurring rocks in both their origins and phases present. While most rocks nucleated and grew during the cooling from a complex melt, the matrix phase in a fired ceramic based upon a clayey material is formed during the firing of the mixture in a kiln and is usually a complex silicate, glassy in nature. This "glassy" matrix in the fired ceramic bonds the residual crystalline material into a solid mass, giving the article strength and brittleness.

Concrete, a mixture of sized aggregate (gravel) and sand, is bonded together by a silicate gel which is formed by hydration of the Portland cement itself. This gel behaves as a noncrystalline single phase, even though it is really a sub-microscopic mixture of two separate phases.

GLOSSARY OF TERMS

Arrestment point: the point or zone on a cooling curve indicating the temperature at which some phase change occurs, begins to occur, or goes to completion.
Equilibrium: a state in which no changes take place with changing time.
(thermal) Equilibrium diagram: a graphical representation of the tempera-

tures and compositions for which various phases within an alloy system are stable at equilibrium.

Eutectic: the micro constituent formed by the simultaneous precipitation, at constant temperature, of two solid phases from a single uniform liquid phase.

Eutectoid: the micro constituent formed by the simultaneous precipitation of two phases, at constant temperature, from an already-existing uniform solid phase.

Intermetallic compound: a compound of two or more metals that has a characteristic crystal structure and a definite chemical composition.

Interstitial compound: a compound of a metal plus a non-metal that has a characteristic crystal structure.

Liquidus: the line on an equilibrium diagram indicating the commencement of solidification of melts of all possible compositions.

Peritectic change: a phase change in which an already-formed solid phase reacts with the residual liquid phase at constant temperature to form a new solid phase and a new liquid phase.

Phase: a chemically homogeneous and physically distinct portion of a material.

Phase transformation: the process, usually temperature controlled, whereby the nature of a phase is altered.

Solid solution: the phase formed when the atoms of one solid substance "dissolve" in another solid substance; the solute atoms either take up some of the normal lattice sites of the solvent metal (substitutional solid solution), or they fit into the interstices of the lattice of the solvent metal (interstitial solid solution).

Solidus: the line on an equilibrium diagram indicating the completion of solidification of metals of all possible compositions.

REVIEW QUESTIONS

1. Define the term "phase". How is a phase different from a microconstituent?

2. List the solid solubility possibilities for two metals melted together to form an alloy, and give an example of each.

3. Distinguish between an interstitial solid solution and a substitutional solid solution.

4. How does an ordered substitutional solid solution differ from a disordered solution? Give an example to illustrate your answer.

5. What are the characteristics of (i) interstitial compounds and (ii) intermetallic compounds?

6. Sketch typical cooling curves for (i) a pure metal, (ii) a eutectic, and (iii) an alloy of a single solid-solution type. Indicate the arrestment points on each type of curve.
7. The copper-nickel equilibrium system is shown in Figure 9.6. Use this diagram to explain the cooling and solidification of a melt of 50/50 composition cooled under equilibrium conditions. How would the final structure be altered if a faster rate of cooling occurred?
8. How does a eutectic differ from a eutectoid? Give examples of both types of microconstituents.
9. Explain the stages of solidification of a melt of 85% lead and 15% tin cooled under equilibrium conditions. The equilibrium diagram is given in Figure 9.10.
10. Calculate the percentage of solid metal present in the alloy referred to in Question 9 at a temperature of 250°C.
11. Describe a typical Widmanstätten structure.
12. Explain, in terms of the phases present in each alloy, why a 70/30 brass is readily cold worked while a 60/40 brass must be hot worked.
13. Explain how a peritectic change occurs within an alloy.
14. List and discuss the four factors that influence structure-property relationships in multi-phase materials.
15. Two metals A and B are mutually soluble while liquid and exhibit partial solid solubility. Given the following data, draw an approximate equilibrium diagram for A and B: A melts at 623°C, B melts at 770°C; the eutectic mixture occurs at 50% A; the eutectic temperature is 380°C; the maximum solid solubility of B in A is 10% and of A in B is 5%, both occurring at the eutectic temperature; the solid solubilities fall away to almost zero at room temperature.
16. Sketch and describe the microstructures of the following types of metals: (i) polycrystalline single phase; (ii) polycrystalline showing coring; (iii) having a matrix and a dispersed phase. What typical mechanical properties would you expect in each type of metal?

10

Non-ferrous Metals and Alloys

ALTHOUGH non-ferrous metals and alloys are not produced in as great tonnages as ferrous metals they are vital to industrial growth since they usually offer combinations of properties not found in irons and steels. Table 10.1 lists some of the more important properties of non-ferrous metals and alloys.

TABLE 10.1

SOME PROPERTIES OF NON-FERROUS METALS AND ALLOYS

Good formability	Good corrosion resistance
Low density	Strength and stiffness usually lower than for ferrous metals
High thermal and electrical conductivities	Poor weldability
Attractive colour	

It must be realised that no one non-ferrous metal or alloy possesses all of these properties; however, some will possess nearly all of them while others possess only a few. It is also important to realise that some ferrous metals possess some of these properties; stainless steel, for example, has excellent corrosion resistance. However, unless the high strength of stainless steel is required it may be more economical to manufacture the particular article in question from a corrosion resistant non-ferrous alloy, such as bronze.

COPPER AND ITS ALLOYS

The most important ores of copper are the sulphides, chalcocite ($Cu_2 S$) and chalcopyrite ($CuFeS_2$). However, copper also occurs as the carbonates, azurite ($CuCO_3.Cu(OH)_2$) and malachite ($CuCO_3.2Cu(OH)_2$), as the oxide, cuprite (Cu_2O), or as the "native metal". The ores are concentrated by flotation and then smelted in a reverberatory furnace where a *matte* of copper and iron sulphides is formed. This matte is then further refined in a converter, very similar to the Bessemer converter. The refined copper, often about 99 % pure, is then cast into rectangular slabs known as *"blister" copper ingots* because of their rough blister-like surfaces. These slabs become the electrodes in the electrolytic refining tanks where copper of up to 99.997 % purity can be produced. Alternatively, the blister copper may be fire-refined in another type of reverberatory furnace to a purity of 99.96%. Fire-refined copper is suitable for wire, general casting applications, and sheet production; electrolytic copper, being purer, is used where maximum corrosion resistant and electrical conductivity are required.

Copper has widespread industrial applications, mainly because of its high electrical conductivity (second only to silver), high thermal conductivity, excellent formability, and extremely high corrosion resistance. The following grades of copper are commercially available:

(a) Tough pitch copper, produced by electrolytic refining, contains a very small amount of dissolved oxygen in the form of very tiny particles of copper oxide scattered around the grain boundaries. If this grade of copper is heated to about 400°C in a reducing atmosphere, "gassing" occurs due to a reaction between the cuprous oxide impurity and the hydrogen or carbon monoxide which penetrates into the metal during heating. This reaction produces steam which forces the copper grains slightly apart, the intercrystalline cracks so produced severely reducing ductility. For this reason tough pitch copper is not used when welding or brazing is involved.

(b) Deoxidised copper is formed when electrolytic copper is remelted and deoxidised with about 0.5% phosphorus, the phosphorus having a much higher affinity for dissolved oxygen than the copper itself. This grade is most suitable for gas welding and brazing applications. However, the residual 0.05% or so of phosphorus in the copper severely reduces its electrical conductivity.

(c) Oxygen-free high-conductivity copper (OFHC Copper) is made by remelting and casting tough pitch copper either in vacuo or in a non-oxidising atmosphere. It therefore contains neither dissolved oxygen nor residual deoxidant, and is the preferred material for all current-carrying members such as wires and cables, busbars, and terminals.

(d) Free machining copper is produced by adding about 0.5% tellurium or selenium to copper. This improves its machinability without greatly reducing its electrical conductivity.

Copper Alloys

Copper may be alloyed with a wide range of other elements to produce many different alloy groups of industrial importance. The most important are:

(a) Copper-Zinc (the Brasses)
(b) Copper-Tin (the Tin Bronzes)
(c) Copper-Tin-Phosphorus (the Phosphor Bronzes)
(d) Copper-Aluminium (the Aluminium Bronzes)
(e) Copper-Nickel (the Cupro-Nickels)

These alloys, with the exception of the cupro-nickels, all exhibit the following sequence of structural similarities.

(1) the first phase to appear (α) is always soft and ductile and has a similar appearance under the microscope.

(2) the second phase (β), appearing after the limit of solid solubility has been exceeded for the α phase, always strengthens the alloy but reduces ductility.

(3) further additions of alloying elements cause the formation of hard and brittle phases (γ, δ, etc.) which generally induce lack of formability in the alloy.

The Brasses: The copper-zinc equilibrium diagram has already been briefly examined and the characteristics of the α and β phases noted (see Chapter 9). The following alloys are of commercial importance:

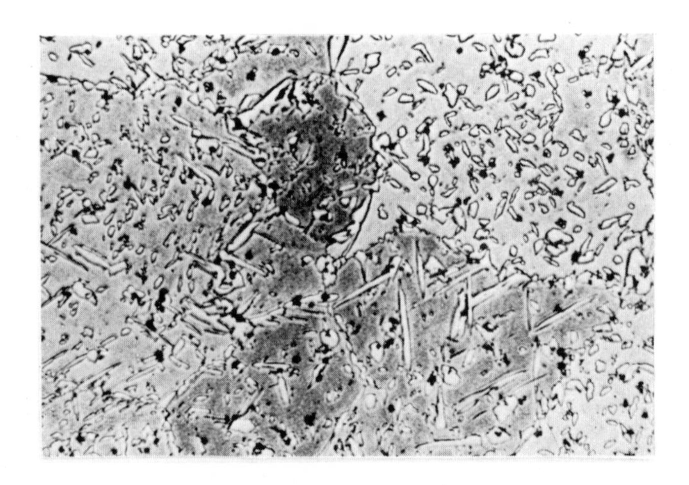

Figure 10.1 *The microstructure of a high tensile brass. The small black areas are an iron-rich phase, and the alpha phase is present as a Widmanstätten structure. (Original slide courtesy of the Copper and Brass Information Centre, Sydney.)*

(a) 70/30 Brass or Cartridge Brass: This is an alpha brass and is soft and ductile in the annealed state. It can be severely cold deformed by drawing, pressing, and extrusion, and work hardens quite severely. Typical applications include the manufacture of cold rolled sheet, wire drawing, tube production, and the manufacture of shell casings.

TABLE 10.2

COMPOSITIONS AND FEATURES OF BRASSES

Name	*Composition*			*Microstructure*	*UTS, depending upon working (psi)*	*Uses*
	copper	*zinc*	*other elements*			
Copper	100	—	—	Pure Metal	22,000 to 70,000	Tubing, Piping, Sheet, Wire Electrical Conductors.
Guilding Metal Tombac or French Gold	90	10	—	Alpha	41,000 to 73,000	Forgings, rivets, jewellery applications.
Low Brass	80	20	—	Alpha	47,000 to 120,000	Drawing and forming operations.
Cartridge or Spinning Brass	70	30	—	Alpha	53,000 to 92,000	Cartridge cases, condenser tubes, sheet fabrication a general purpose brass.
Admiralty Brass	70	29	Sn=I	Alpha	53,000 to 92,000	Condenser tubes exposed to salt water (high corrosion resistance).
High Brass	66	34	—	Alpha	53,000 to 125,000	Stamping and drawing operations.
Muntz Metal	60	40	—	Alpha-Beta	60,000 to 125,000	Suitable for many hot working operations; rolled also cast valves and marine fittings.
Naval Brass	60	39	Sn=1	Alpha-Beta	60,000 to 125,000	As above, but possesses increased corrosion resistance.
Tobin "Bronze"	60	38	Sn=1 AI=1	Alpha-Beta	60,000 to 125,000	Brazing alloy for Naval brasses, etc.
Manganese "Bronze"	62	32	A1=4 Fe = 1.5 Mn=2.25		60,000 to 150,000	High tensile casting metal; Yield Point = 33 tsi UTS = 46 tsi
Brazing Brass	50	50	—	Beta		Brazing rods.

(b) 60/40 Brass or Muntz Metal: This is an alpha-beta brass and is suitable only for hot working applications because of the presence of the hard and brittle β phase. It has higher strength than the 70/30 brass and is commonly used to make hot rolled products, valve stems, condenser tubes, and the like. About 2% lead is often added to this alloy to improve its machinability.

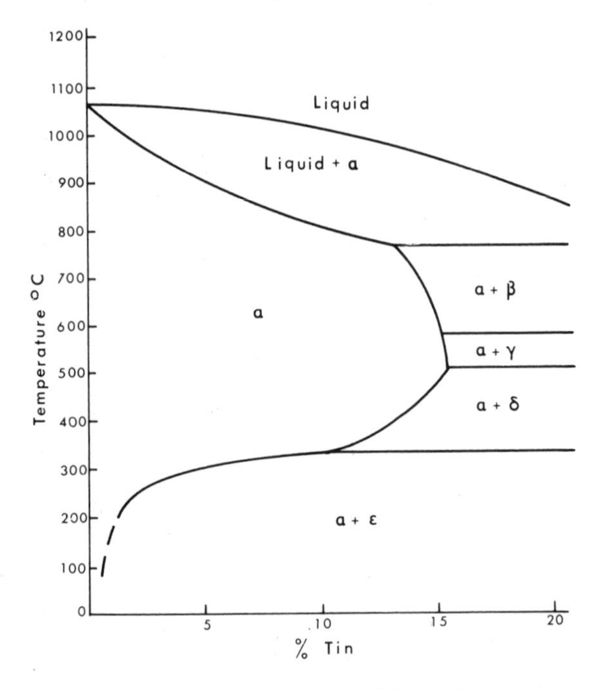

Figure 10.2 *The copper-tin equilibrium diagram.*

(c) High Tensile Brasses: These are all of the basic 60/40 composition with small additions of iron, aluminium, tin, manganese, and nickel, the introduction of which increases strength quite considerably. Typical applications include marine propellers and shafts, pump rods, autoclaves, switchgear, and high-strength fittings of all types. If manganese is the principal strengthening element, the resultant alloy is often erroneously referred to as a "manganese bronze".

(d) Brazing Alloys: These are essentially beta brasses of the 50/50 type, often containing small amounts of other alloying elements such as tin, manganese, or aluminium. They are characteristically hard and brittle.

Table 10.2 summarises the properties and uses of some of the more important brasses.

The Tin Bronzes: Reference to the equilibrium diagram (Figure 10.2) shows that tin, unlike zinc, can only be present in copper to the extent of a few percent at room temperature. The alpha alloys have limited industrial importance, while the "high bronzes", usually containing up to 20% tin, have many applications. Zinc, phosphorus, nickel, and lead are commonly added in small proportions to bronzes in order to improve their mechanical properties or corrosion resistance. The following alloys are of industrial importance.

(a) Coinage Bronze: The composition is usually 95% copper, 4% tin and 1% zinc, the latter acting as a deoxidiser. This alloy is soft and ductile and consists almost entirely of alpha grains in the cast or wrought condition. It is the most commonly used "copper" coinage metal.

(b) Admiralty Gunmetal: Normally containing 88% copper, 10% tin and 2% zinc, the alloy usually contains alpha grains together with the hard alpha-delta eutectoid. It may be hot worked above 600°C, but is usually used in the as-cast condition for steam and water fittings and bearings. The addition of lead improves the pressure-tightness of the alloy.

Figure 10.3 *Microstructure of a leaded gunmetal containing 85% copper, 5% tin, 5% zinc, and 5% lead. The black globules are the lead. The alloy consists entirely of alpha grains which are slightly porous; the pores are readily distinguished from the lead globules since they are surrounded by a dark circle of exuded polishing medium which was trapped in the pores during the polishing process. Magnification x250. (Original slide courtesy of the Copper and Brass Information Centre, Sydney.)*

(c) Phosphor Bronze: These tin-bronzes, containing small percentages of phosphorus, are commonly used in the manufacture of bearings, hard drawn wire, and bronze springs.

The Aluminium Bronzes: Aluminium can dissolve to the extent of about 9% in copper and therefore the aluminium bronzes contain significant amounts of the alpha phase. Above 9% aluminium the alpha-gamma eutectoid appears; this causes embrittlement in the alloy and this effect may be removed by appropriate heat treatment or by the addition of 1–3% of iron.

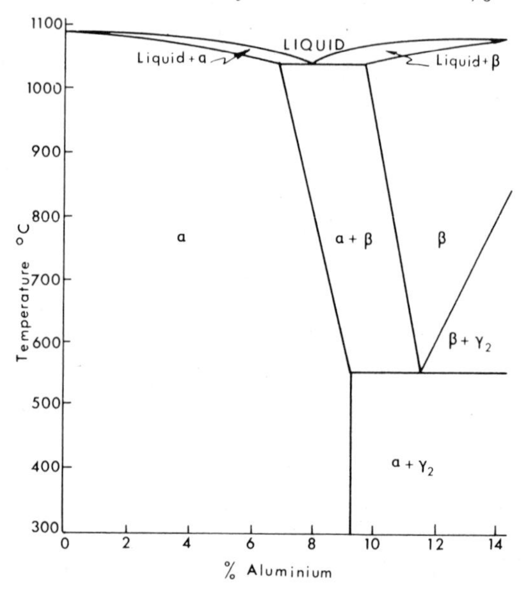

Figure 10.4 *The copper-rich portion of the copper-aluminium equilibrium diagram.*

Copper-aluminium alloys usually possess high strength combined with good resistance to fatigue, corrosion, and mechanical abrasion, and are golden in colour. They can be classified into three broad types:

(a) Wrought Alpha Alloys: Normally containing from 5 to 7% of aluminium, these alloys are suitable for all hot and cold working applications.

(b) Casting Alloys: The 10% aluminium alloy is most common and may be hot worked and heat treated. If the cast alloy is soaked at 570°C the alpha-gamma eutectoid alters to the beta phase which then dissolves the alpha still present. If the alloy is now quenched the beta grains are retained and acicular needles of an alpha[1] phase are formed. These are somewhat analogous to martensite in quenched steels, and strengthen the alloy. If the quenched alloy is now tempered at 350°–550°C, a fine precipitate of the gamma[1] phase appears; this is analogous to sorbite in steels, and toughens the alloy.

(c) Complex Alloys: Small amounts of iron, nickel, and manganese are often added to casting alloys to make them more easily heat treatable.

The Copper-Nickel Alloys: The copper-nickel equilibrium diagram has already been examined in detail (see Chapter 9). It differs significantly from those of the brasses and bronzes in that complete solid solubility occurs between copper and nickel. All alloys exhibit similar microstructures and can be hot and cold worked quite readily. Two alloys are of significance.

(a) Cupro-nickel (nickel silver or German silver): These are extremely malleable and ductile and are used for condenser tubes and other applications where extreme resistance to salt water corrosion is required (70/30 and 80/20 alloys), and also coinage (75/25 alloy).

(b) Monel Metal: Essentially 70% nickel and 30% copper, together with small amounts of iron and other elements, this alloy has high strength and corrosion resistance. It is used extensively in chemical plants, in food manufacturing plants, for valves, turbine blades, and corrosion-resistant bolts, screws, and nails. It has a characteristic silvery lustre.

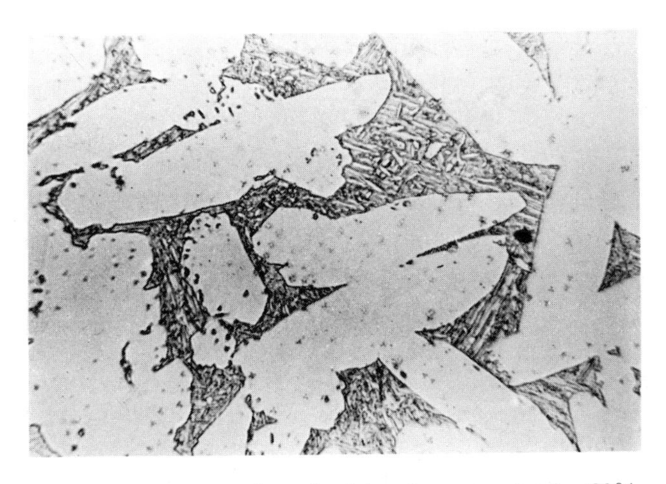

Figure 10.5 *Microstructure of an aluminium bronze containing 90% copper and 10% aluminium. The alloy is cast, and contains large alpha grains surrounded by a mixture of eutectoid plus retained beta phase. Magnification x500. (Original slide courtesy of the Copper and Brass Information Centre, Sydney.)*

ALUMINIUM AND ITS ALLOYS

The main ore of aluminium is bauxite which is the hydrated oxide $Al_2O_3.nH_2O$. The production of aluminium from the bauxite ore involves the following sequence of industrial processes.

(1) Calcining at 600°C to dehydrate the ore.
(2) Treatment with caustic soda at 160°C to form a solution of aluminium hydroxide.

(3) This solution, after standing, precipitates out the aluminium hydroxide.

(4) The precipitate is then calcined at 1150°C forming pure alumina (Al_2O_3).

(5) The alumina is mixed with cryolite (sodium aluminium fluoride AlF_3. $3NaF$) and refined to pure aluminium in the Hall-Herault electrolytic cell.

The aluminium so produced may exceed 99.99% purity and has the following properties: (1) lightness, S.G. = 2.7; (2) high thermal and electrical conductivities; (3) softness and low strength; and (4) good corrosion resistance, due to the development of a resistant oxide layer on the surface after exposure to the atmosphere.

Because of its high electrical conductivity (60% of that of copper), aluminium is used extensively for power cables, but these must be reinforced with a steel core because of the low strength of the aluminium. Actually, on a weight for weight basis, aluminium is a better conductor than copper. Because of its lightness, aluminium is useful for building applications where weight, corrosion resistance, and appearance are more important than strength. Many alloys are used in the aircraft industry since alloying, while it does not greatly alter weight, does improve strength considerably. Foodstuffs can be wrapped in aluminium foil since it is corrosion resistant and the oxides formed are non-toxic.

Alloys can be classified as wrought or cast, both groups containing alloys that are age-hardening.

Wrought Alloys

Wrought alloys are divided into the non heat-treatable and the heat-treatable alloys. The former alloys usually contain manganese up to 2% and/or magnesium from 3½ to 5%, the magnesium alloys having higher strengths and greater corrosion resistance than the manganese-based alloys.

TABLE 10.3

THREE HEAT-TREATABLE WROUGHT ALUMINIUM ALLOYS

Type of Alloy	*Composition (Balance = Aluminium)*
Duralumin	4.5% *Cu*, 0.5% *Mg*, 0.5% *Mn*, 0.5% *Si*
Magnesium-Silicon Alloy	1.0% *Si*, 0.5% *Mg*, 0.5% *Mn*, 0.5% *Fe*
Copper-Magnesium Zinc Alloy	2% *Cu*, 3% *Mg*, 5% *Zn*

Heat-treatable alloys usually contain copper or silicon as the main alloying element. Table 10.3 lists three typical heat-treatable wrought alloys. Of these, the duralumin has the best formability prior to heat treatment, the *Mg-Si* alloy the best corrosion resistance, and the *Cu-Mg-Zn* alloy the greatest strength after heat treatment.

Cast Alloys

Again the basic division into non heat-treatable and heat-treatable alloys can be made, with silicon or copper being the principal alloying element in the former types. Aluminium and silicon form a eutectic at about 13% silicon, the melting point of which is about 660°C. Table 10.4 lists some typical non heat-treatable casting alloys.

TABLE 10.4

NON HEAT-TREATABLE CASTING ALUMINIUM ALLOYS

Type	*Composition*	*Characteristics*
low *Si*	5% *Si*, small % *Cu*, balance *Al*	A low melting-point alloy of high fluidity.
high *Si*	12–13% *Si*, balance *Al*, modified* by 0.5% *Na*	A high shrinkage alloy of low melting point, high surface finish and excellent corrosion resistance.
copper alloys	7–12% *Cu*	Cheaper alloys, but weaker and less resistant to corrosion.

*The sodium prevents silicon segregation and reduces brittleness.

The heat-treatable alloys are similar in composition to the wrought alloys already discussed. Y-alloy, duralumin containing about 2% nickel, deserves special mention. The presence of nickel in this alloy confers strength at high temperatures upon castings that have been suitably heat treated. Y-alloy is therefore useful for pistons and other components in aero engines.

Age-Hardening in Aluminium Alloys

Phases in some alloys occur by precipitation from a single-phase matrix, in which case the precipitated phase may occur as submicroscopic particles distributed both around the grain boundaries and throughout the grains. Certain alloys of aluminium-copper, magnesium-silicon, and beryllium-copper will precipitate such phases after suitable heat treatment, these phases having a strengthening effect upon the alloy. This mechanism is termed age-hardening or precipitation-hardening and is best explained for aluminium alloys if a 4% copper alloy is considered.

Looking at the aluminium-copper equilibrium diagram (Figure 10.6), it is apparent that the solubility of copper in the alpha solid solution decreases steadily and quite considerably with decrease in temperature, so that when the temperature drops to that indicated by point 3, copper in the form of the intermetallic compound, copper aluminide ($CuAl_2$), is deposited as coarse particles in and around the grains of alpha solid solution. This then would be the typical "as-cast" structure of the alloy. The $CuAl_2$ is extremely hard and brittle.

If the alloy is now reheated to about 500°C, a temperature between points 2 and 3 on Figure 10.6, the $CuAl_2$ is reabsorbed into the alpha solid solution, producing a single-phase alloy. If this alloy is now quench-cooled to room temperature there is insufficient time for the copper aluminide to be precipitated out of the solid solution, the copper atoms now being held in a supersaturated solid solution within the aluminium.

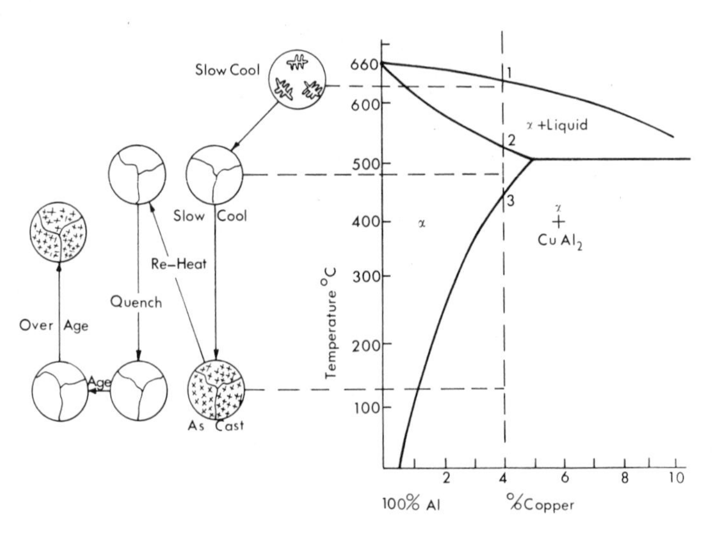

Figure 10.6 *The aluminium-rich portion of the copper-aluminium equilibrium diagram showing the mechanism of precipitation hardening for a 4% copper alloy. Over-ageing causes a coalescence of the $CuAl_2$ particles and a consequent loss of strength in the alloy.*

If the alloy is now allowed to stand at room temperature for from five to seven days, a significant increase in strength properties will result. This is due to the slow precipitation of very fine submicroscopic particles which become more or less evenly distributed throughout the matrix. This process is known as natural ageing, but is often replaced by artificial ageing, when the alloy is reheated to about 120°C, causing the precipitation of this fine copper-rich phase to occur within a few hours.

Close control of both time and temperature is essential in precipitation hardening; this is achieved by using salt baths held at constant temperatures for heating purposes, (solution treatment), plus close time control when artificially ageing the alloy. Duralumin, containing 4% copper, together with 0.5% each of both magnesium and manganese, is a commonly-used age-hardening alloy, favoured for its high strength and its extreme lightness. However, it corrodes readily, and must be clad or otherwise protected in conditions where corrosion could become a problem.

TABLE 10.5

MECHANICAL PROPERTIES AND APPLICATIONS OF COMMON ALUMINIUM ALLOYS

Composition (Balance=Al)	*Condition A=annealed W=cold worked H.T.=heat treated*	*Tensile Strength (ksi)*	*Yield (0.2% proof stress) (ksi)*	*Percent Elongation 2″ gl*	*Brinell Hardness Number*	*Shear Strength (ksi)*	*Endurance* Limit (ksi)*	*Characteristics and Uses*
1.2% *Mn*	A W	16 29	6 27	30 4	28 55	11 16	7 10	A non heat-treatable work hardening wrought alloy used for general sheet metal applications and cooking utensils
4.0% *Cu*, 0.5% *Mg* 0.5% *Mn*, 0.5% *Si*	A H.T.	26 62	10 40	20 20	45 105	18 38	13 18	Duralumin, an age-hardening wrought alloy, hardened by quenching and ageing; used for aircraft construction in the form of Alclad
44% *Cu*, 0.8% *Si* 0.8% *Mn*, 0.4% *Mg*	A H.T.	27	14	12	45	18	11	A strong age-hardening alloy for forging and extrusion
7.0% *Cu*, 2.0% *Si* 1.7% *Zn*	As cast	24	15	1.5	70	—	—	General purpose sand casting alloy that is not age-hardening and has low corrosion resistance
4.0% *Cu*, 1.5% *Mg*, 2.0% *Ni*	As cast	27	18	1.0	70	—	—	A sand casting alloy capable of withstanding high temperatures
12% *Si*	As cast	39	21	2.7	—	—	—	A general purpose non heat-treatable die-casting alloy having excellent corrosion resistance
9.5% *Si* 0.5% *Mg*	As cast	44	27	3.0	—	—	—	A die-casting alloy having high strength and good corrosion resistance

*500 million cycles in reverse bending

Age-hardening alloys containing silicon and magnesium behave in a similar manner to the alloys discussed above, the strengthening sub-microscopic precipitate in these cases being magnesium silicide Mg_2Si. Thus, the age-hardening effect of the copper aluminide in duralumin is reinforced by the presence of Mg_2Si.

Table 10.5 shows some of the more common alloys of aluminium together with their mechanical properties and applications.

ZINC AND ITS ALLOYS

The main ore of zinc is sphalerite (ZnS), which is commonly known as zinc blende; it commonly occurs in association with the ores of silver, lead, and cadmium, and purification processes are usually most complex. However, in essence the extraction of zinc involves the concentration of the ore by flotation, the calcining, and leaching of the concentrate, and then finally the electrolytic extraction of pure zinc from the sulphate solution resulting from the leaching process.

Zinc is a fairly soft metal which exhibits brittleness at room temperature. It has a high resistance to corrosion and this, coupled with its position in the electrochemical series, makes it ideal for the galvanising of iron and steel.

The most important group of zinc-based alloys are those used by the die-casting industry, and it has been the development of such alloys that has resulted in the rapid development of the die-casting industry as a whole. The low melting point of zinc (420°C), its low cost, high-dimensional stability, and its lack of effect on steel dies makes it an ideal die-casting metal. A typical high-strength die-casting alloy would have the following composition and properties:

Composition: $Cu = 1.25\%$; $Al = 4.0\%$; $Mg = 0.08\%$;
$Fe = 0.1\%$; $Pb = 0.007\%$; $Cd = 0.007\%$;
$Sn = 0.007\%$; Zn = remainder

Properties:
Tensile Strength = 47,000 psi
% Elongation 2″gl = 7%
Charpy Impact No. = 48 ft lb
Melting Point = 425°C

LEAD AND ITS ALLOYS

Lead occurs in a large number of complex ores but the principal source is galena, lead sulphide (ZnS). It is extracted in a small reverberatory furnace somewhat like the type used in the first stage of copper refining, the lead bullion which is the product of this furnace containing small amounts of copper, arsenic, antimony, gold and silver. The lead bullion is then purified in a series of fairly complex processes, the 99.95% pure lead thus obtained having the following properties:

(1) softness and low strength
(2) high density
(3) low melting point
(4) low coefficient of friction
(5) high malleability but low ductility
(6) high corrosion resistance

Lead is used in the manufacture of a wide range of chemicals, paints, and chemical containers; lead pipes are used for drainage; cables are sheathed with lead to waterproof them, and lead screens out X-rays and gamma radiation more effectively than any other material. The common alloying elements for lead are tin and antimony; tin lowers the melting point of lead, while antimony increases its strength.

Lead-Antimony Alloys: contain from 1.9% antimony; alloys low in antimony are used for cable sheathing, while 9% alloys are used for plates in storage batteries.

Lead-Tin Alloys: the equilibrium diagram for lead and tin has already been discussed (see Chapter 9). The four main classes of solders are listed in Table 10.6.

TABLE 10.6

SOME FEATURES OF LEAD-TIN SOLDERS

Composition		*Microstructure*	*Applications*
% Pb	*% Sn*		
70	30	α grains + eutectic	lead wiping — the alloy possesses a wide solidification range
60	40	α grains + eutectic	general plumbing applications on galvanised iron
50	50	α grains + eutectic	general soldering applications on non-ferrous metals
40	60	nearly eutectic composition	fine soldering and where a low MP solder is required

Lead-based Babbitt Alloys: these alloys form one of the oldest known groups of bearing alloys, a common composition being 85% lead, 5% tin and 10% antimony. The antimony in these alloys combines with some of the tin to form hard cuboids of *Sb-Sn* solid solution which are distributed throughout the softer lead matrix. In use, the *Sb-Sn* cuboids resist wear while the softer lead matrix acts as a cushion which is fairly friction-free. Such bearings are unsuitable for heavy-duty applications.

TIN AND ITS ALLOYS

Tin is extracted from its principal ore, cassiterite (SnO_2), by smelting in a reverberatory furnace, the impure tin so produced being further refined by further smelting. Tin is a soft white metal of low melting point that has an extremely high resistance to corrosion. It is used in its pure state to make collapsible tubes for foodstuffs and toothpaste and as a coating for "tin cans" that are to contain food. Tin is used in a wide range of alloys, many of which have already been discussed. However, pewter and tin-based bearing metals deserve consideration.

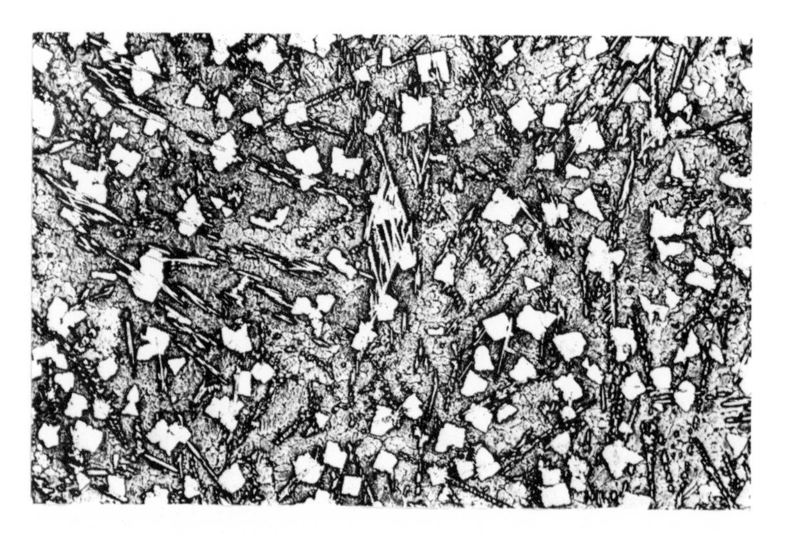

Figure 10.7 *Photomicrograph of a tin-based babbitt alloy containing 75% Sn, 12% Sb, 10% Pb, and 3% Cu. Note the SbSn cuboids and the acicular copper-rich crystals. Magnification x100.*

Pewter is an alloy traditionally used to make beer mugs, plates, and food containers of all types. A typical composition is 91% tin, 7% antimony and 2% copper, the metal being relatively soft and malleable.

Tin-based Babbitt Metals are analogous to the lead-based babbitts but have higher corrosion resistance. However, tin-based babbitts are prone to fatigue failures. A typical composition is 85% tin, 7% copper and 8% antimony.

Both the lead-based and the tin-based babbitt metals are commonly known as "white metals".

NON-FERROUS ALLOYS FOR HIGH TEMPERATURE SERVICE

Many components in jet and rocket engines and in nuclear equipment have to withstand operating temperatures in excess of 1100°C, and a number of highly-specialised alloys have been developed to meet this requirement. Nickel or cobalt usually forms the base metal in this range of alloys, many of which possess strengths in excess of 100,000 psi at room temperature. Also, most of these alloys are precipitation-hardened after manufacture is complete, the usual range being from 250–370 Brinell. Table 10.7 shows some typical high-temperature alloys, most of which are very expensive to produce because of the high cost of the component metals.

TABLE 10.7

COMPOSITION OF SOME HIGH TEMPERATURE ALLOYS

Type of Alloy					*Composition*					
	Ni	*Cr*	*Co*	*W*	*Ti*	*Fe*	*Ta*	*Al*	*C*	*Mo*
Nimonic 80A	Balance	21.0			2.5			1.2	0.04	
Inconel 713C	Balance	12.0			0.5		2.0	6.0	4.5	
Incoloy 910	42.0	13.0			2.4	Balance			0.04	6.0
Hastelloy	45.0	22.0	1.5	0.5		Balance			0.15	9.0
Vitallium	2.5	28.0	62.0			1.7			0.28	5.5

GLOSSARY OF TERMS

Ageing: a change in mechanical properties brought about by either allowing the metal to stand at room temperature for some time, or by reheating it to a certain temperature for a much shorter time; the change in properties is brought about by the precipitation of a new phase from a supersaturated phase.

Calcining: the heating or ores or other concentrates to cause the decomposition of carbonates or hydrates.

Flotation: the concentration of minerals from ores by agitating the ores with water-oil or other flotation mixtures, the minerals being retained on the surface of the flotation tank while the rest of the ore sinks to the bottom.

Gassing: the embrittlement of copper caused by the absorption of hydrogen and carbon monoxide and their subsequent reactions with the copper oxide impurity within the metal.

Leaching: the treatment of ore concentrates with acids or other solutions which dissolve out the valuable constituent of the ore.

Over-ageing: loss of mechanical strength in an age-hardening alloy due to prolonged ageing treatment.

Precipitation hardening: (also age-hardening) the strengthening of an alloy due to the precipitation of a finely dispersed phase from a saturated or super-saturated solid solution.

Solution treatment: a type of heat treatment by which a multi-phase alloy is converted to a saturated single-phase alloy; the usual procedure is to heat the alloy into its single phase region, soak at this temperature, and then quench to retain the single phase structure.

REVIEW QUESTIONS

1. List and discuss some of the important properties of non-ferrous metals and alloys.

2. Compare the properties of tough-pitch and OFHC copper, and give their particular uses.

3. What is the purpose of alloying phosphorus into a tin bronze?

4. How do the copper-nickel alloys differ significantly from the constitutions of almost all other copper alloys?

5. List the properties of aluminium that make it a useful engineering metal.

6. By reference to Figure 10.6, explain the effects of (i) solution treatment and (ii) ageing on the microstructure and mechanical properties of a 4% copper 96% aluminium alloy.

7. List the properties desirable in die-casting alloys. Discuss the relative merits of zinc-based and aluminium-based die-casting alloys.

8. What general features of a 70% lead 30% tin solder make it suitable for "lead wiping"?

9. What is pewter and why is it used for the manufacture of tableware?

10. What general properties are desirable in a metal to be used at high temperatures in an oxidising, corrosive atmosphere?

11. Aluminium alloy rivets are often of such a composition that they can be driven home quite readily and then age-hardened while in place. Describe how a batch of these rivets that had age-hardened prior to use could be reclaimed.

11

The Metallurgy of Iron and Steel

PRODUCTION OF IRON AND STEEL

IN COMMON with many other metals, iron is reduced from its ores by heating the ores in the presence of a powerful reducing agent and a suitable flux, the function of the latter being to cause the rapid removal of impurities as slag. The blast furnace has been used for this purpose for several centuries, but only in the last fifty years have the complex pyrochemical reactions of the blast furnace been brought under sufficient control to bring about high operating efficiencies.

The Blast Furnace

A modern blast furnace consists essentially of a truncated cone-shaped steel shell about eighty to one hundred feet in height, of rigid construction, and lined with heat-resisting firebricks varying from eighteen inches to three feet in thickness, the thicker sections occurring around the hottest zones of the furnace. The hearth diameter of such a furnace is about thirty feet, and furnace operation is continuous, shutdowns only being necessary for internal maintenance such as relining. The charge fed into the blast furnace consists of iron ore of suitable grade, coke, limestone, sinter, and air, the latter being blown in through the tuyeres at about 35 pounds per square inch pressure and often being oxygen-enriched for greater efficiency. The solid ingredients are mixed in their correct proportions by weight before being fed into the top of the furnace via the double-bell system. Sinter is used to a considerable extent in modern blast furnaces, and is made by heating fine coke and iron ore together at red heat, thus causing them to fuse together into small lumps. The sintering process enables these "fines" to be used without danger of clogging the furnace.

The furnace has four main heating zones shown in Figure 11.1, carbon monoxide and solid carbon being the reducing agents acting in these zones. While the exact nature of the blast furnace reactions remains a mystery it is believed that those chemical reactions listed in Figure 11.1 occur in the various heat zones. It should be noted that while carbon dioxide is formed near the tuyeres where the oxygen is abundant, it is soon reduced to carbon monoxide further up the furnace. The exact composition of the slag formed in and around the hottest zone depends upon the impurities present in the raw materials, but, generally speaking, it may be regarded as a highly impure vitreous silicate when cold, stained and coloured by the oxides of manganese, aluminium, phosphorus and magnesium. The main by-product of the blast furnace, known as blast furnace gas, has a heat value of about 115–130 BTU per pound, and is collected, cleaned, and used for heating within the steel-works.

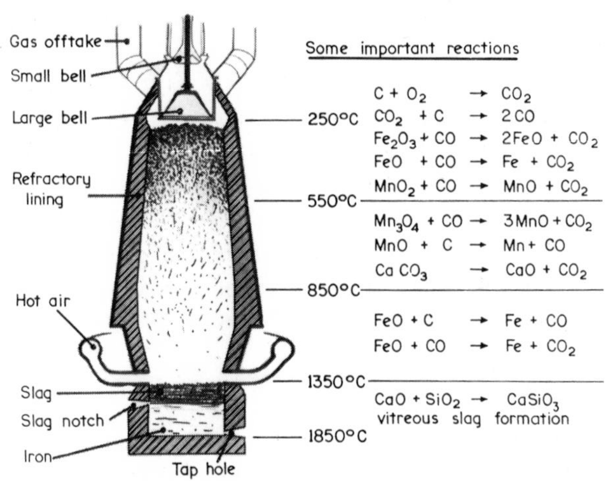

Figure 11.1 *A schematic section through a blast furnace. The important reactions occurring in the various heating zones of the furnace are shown.*

Suitable ores for blast furnace refining include— haematite (Fe_2O_3), magnetite (Fe_3O_4), limonite (Fe_2O_3. x H_2O), and siderite ($Fe\ Co_3$).

Haematite and magnetite are the most common. A modern furnace could consume 1800 tons of iron ore, 400 tons of limestone and 1000 tons of coke in one day with 125,000 cubic feet of air being used every minute. This would yield a daily production of about 1000 tons of high-grade pig iron, the typical composition of which would be—

carbon	3.0 — 4.5%
silicon	1.0 — 4.0%
manganese	0.2 — 1.5%
phosphorus	0.1 — 2.0%
sulphur	0.04 — 0.10%
iron	(balance)

This pig iron is unsuitable for most industrial applications and needs further refining either to steel or cast iron. Steel manufacture is carried out in one of four different types of refining plant; the Open Hearth furnace, the Bessemer converter, the Basic Oxygen converter, and the electric furnace.

The Open Hearth Furnace

A typical open hearth furnace consists of a refractory-lined shallow steel container, the dimensions of a 250 ton furnace being about fifty feet long, twenty feet wide, and three feet deep in its centre portion. Two fundamental types of furnace are in existence; these are known as the "basic" and the "acid" types according to whether their hearths are constructed of basic refractories such as dolomite or acid refractories such as fritted silica bricks. The primary function of any open hearth furnace is the oxidation of impurities. Since the oxide of phosphorus is acidic, excess phosphorus is readily removed in basic furnaces when the basic linings neutralise this acidic component. However, the acid process cannot remove excess phosphorus and sulphur, so the acid process can only be used for pig irons low in these impurities. Both processes remove carbon, silicon, and manganese by oxidation while metals more electropositive than iron, such as copper, tin and nickel, remain unaffected by the refining process. Australian ores being rich in phosphorus demand the basic furnace.

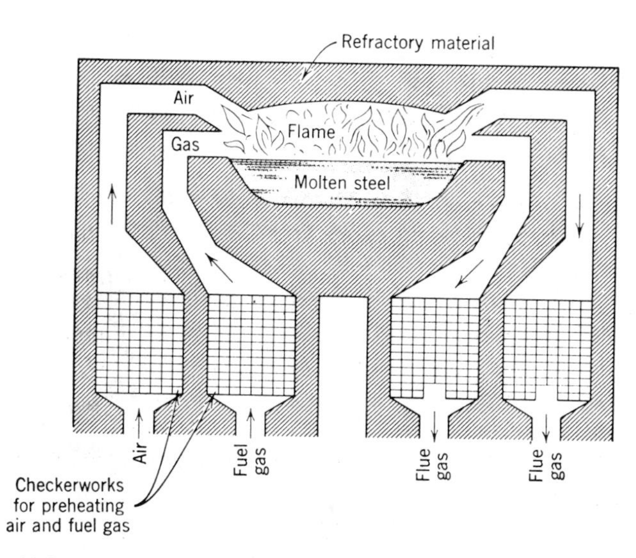

Figure 11.2 *A schematic section through an open hearth furnace. (Reproduced with permission from "Nature and Properties of Engineering Materials" by Z. D. Jastrzebski. John Wiley and Sons Inc., N.Y. 1959.)*

The charge typically consists of steel scrap, molten pig iron, lump iron,* and limestone, with scrap making up to 50% of the total charge. In operation the scrap, the lump iron, and limestone are first added by means of a mechanical ladle, and melted by direct exposure to a naked flame produced

*Lump iron is solid pig iron and iron scrap.

by burning coke oven gas, oil, or liquified tar, after which the molten pig iron is charged in, The gaseous fuel and air are preheated in checkerwork heat exchangers prior to entering the furnace. Two such heat exchangers are always "on heat" while others are used for preheating purposes. A fairly recent innovation is the introduction of pure oxygen into the furnace atmosphere via a long water-cooled lance; this increases production rates considerably.

When impurities have been reduced to acceptable levels the molten steel is tapped off into huge ladles where required alloying elements are added, usually in the form of ferro-alloys. All grades of steel, with the exception of very high-quality special steels, can be produced by this process.

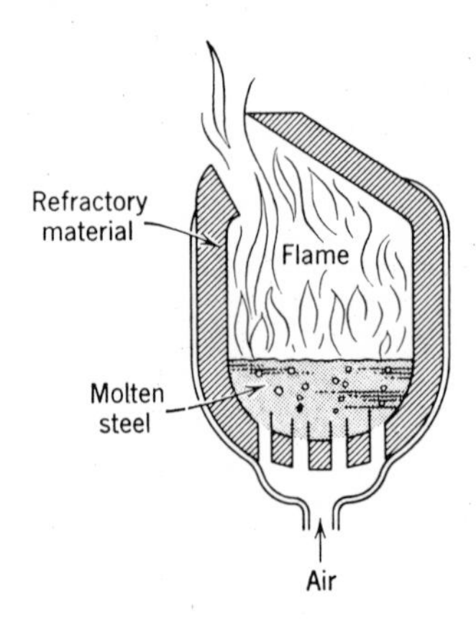

Figure 11.3 *A schematic section through the Bessemer converter. (Reproduced with permission from "Nature and Properties of Engineering Materials" by Z. D. Jastrzebski. John Wiley and Sons Inc., N.Y. 1959.)*

The Bessemer Converter

The Bessemer converter is a much more rapid process than the open hearth furnace, and raw materials of high or low phosphorus content can be refined provided the correct furnace linings are used. The furnace consists of a strong steel shell, refractory-lined, which is supported on trunnions that enable it to tilt through about 90° in either direction from the upright position. At the beginning of the cycle the converter is tilted and the charge of molten pig iron is added. The air-blast is turned on, the furnace rotated upright, and the refining begins. Lime is added at this stage to flux the melt. Carbon, silicon, and manganese are oxidised out first, while phosphorus and sulphur oxidise in the short "after blow" period. The complete cycle for a 25 ton furnace takes only about 25 minutes. Metal loss is high at 12–15%, and the steel produced is not of a high quality due to the lack of control inherent in the process and its high nitrogen content.

Basic Oxygen Processes

Basic oxygen plant has lower installation costs and increased production compared with open hearth furnaces, and the development of such processes represents a major technological advance in the art of steelmaking. Several different types of furnace are being used in different parts of the world; however, the two illustrated in Figure 11.4 are typical. Both the Linz-Donawitz* (L–D) and the Rotor converter have certain similarities in that (1) they consist of steel shells lined with high-grade basic refractories, (2) they take charges consisting of steel scrap, molten pig iron, and limestone, and (3) they both employ oxygen blasts to remove the impurities by oxidation.

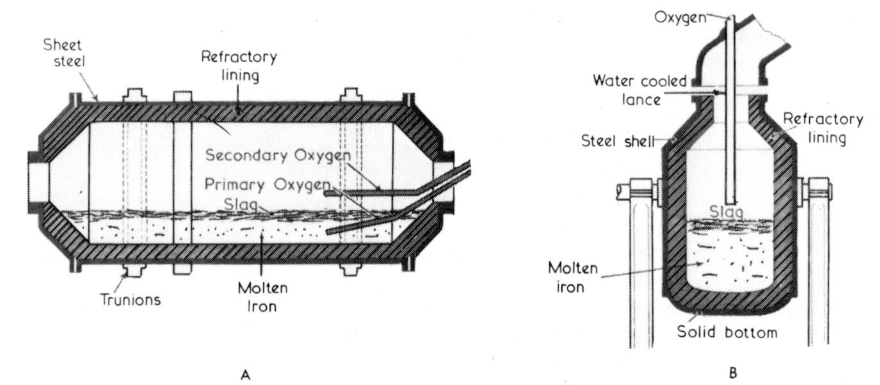

Figure 11.4 *Schematic sections through two types of basic oxygen converters: (A) the rotor converter; (B) the L-D converter.*

However, while the L–D converter remains upright during its working cycle and uses oxygen blown on to the surface of the charge, the Rotor converter is mounted horizontally and slowly rotates while oxygen is blown both into, and on top of, the molten charge.

The furnace reactions are similar to those of the open hearth furnace, but the whole process is much faster and no external heating is required because of the strongly exothermic nature of some of the reactions. Very high-grade steel is produced quite inexpensively by this process.

Electric Furnaces

A relatively small percentage of steel output, consisting mainly of high-grade tool steels and stainless steels, is produced in electric furnaces of one type or another. The main advantage offered by the electric furnace is the neutral nature of the heat source. Perhaps the most common type of electric furnace is the direct-arc type as shown in Figure 11.5, the heat coming from arcing between the long graphite electrodes and the molten charge.

*Named after the two Austrian towns Linz and Donawitz where the converter was developed in the late 1940's.

The final compositions of many grades of steels and irons depend somewhat upon the particular refining processes used for their manufacture. While it is useful to consider plain irons and steels as primarily iron-carbon alloys, the influences of relatively small amounts of other impurities and alloying elements either remaining after refining or deliberately introduced during refining may be quite considerable.

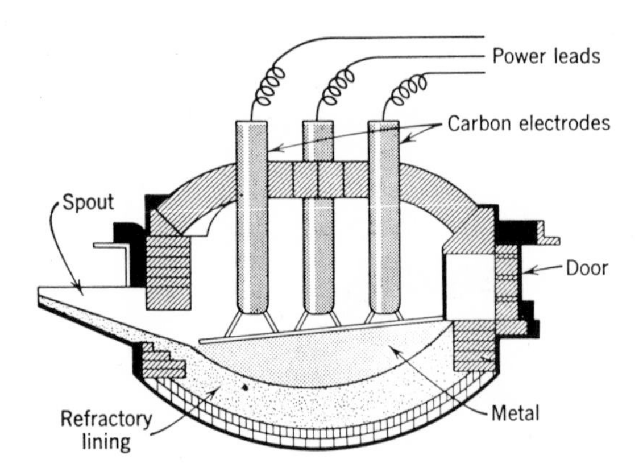

Figure 11.5 *The direct electric-arc furnace. (Reproduced with permission from "Nature and Properties of Engineering Materials" by Z. D. Jastrzebski. John Wiley and Sons Inc., N.Y. 1959.)*

ALLOTROPY OF IRON

Iron exists in three allotropic modifications, each of which is stable over a certain range of temperature. When pure iron freezes at 1540°C, the delta modification forms; this has a body-centred cubic crystal lattice, and is stable down to 1400°C, when, at constant temperature, it alters to the gamma modification, which has a face-centred cubic lattice structure. This crystalline alteration is diffusion-controlled, and sufficient time at 1400°C must be allowed for this reaction to proceed to equilibrium.

The gamma iron thus formed is stable down to 910°C, when it alters to the alpha non-magnetic modification, which has a body-centred cubic lattice similar to that of delta iron. No further allotropic changes occur in the slowly cooling iron, but at the so-called Curie Point of 768°C the alpha iron becomes magnetic. This change from alpha non-magnetic to alpha magnetic is due to electron rearrangements in the outer shells of the iron atoms.

Each of these changes is identified by the presence of a clearly-defined arrestment point on a cooling curve of a sample of slowly-cooled pure iron, such arrestments occurring because latent heat is given out during each allotropic change. It is to be noted that the alpha non-magnetic modification is sometimes called beta iron, but this is not modern nomenclature.

The various allotropic forms of iron have different capacities for dissolving carbon. Thus, while gamma iron can contain up to about 2% carbon, alpha iron can contain a maximum of 0.008% carbon at room temperature. The different solubilities of carbon in iron is a most important aspect of the structure of iron and steel, pertaining most particularly to heat treatment procedures.

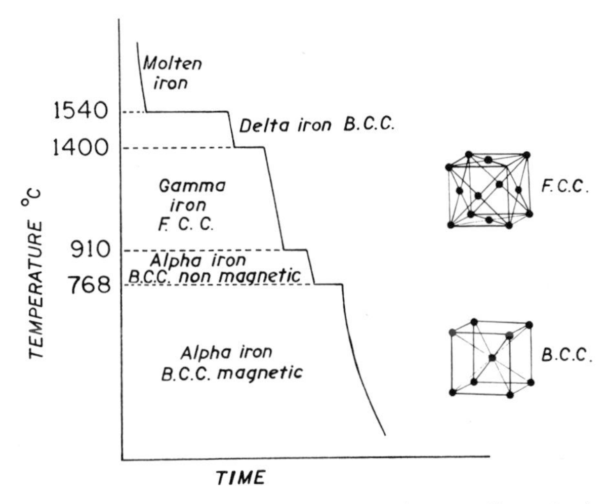

Figure 11.6 *The cooling curve for pure iron showing allotropic changes.*

THE IRON-CARBON EQUILIBRIUM DIAGRAM

Iron and carbon exhibit complete liquid solubility but only partial solid solubility, and thus their equilibrium diagram exhibits a fair degree of complexity. In fact, in that portion of interest to the metallurgist, the phases include several quite different solid solutions, a eutectic, a eutectoid, and an intermetallic compound.

The general effects of carbon in pure iron are:

(1) The progressive lowering of the freezing point of the alloy with increasing amounts of carbon until the eutectic composition of 4.3% carbon is reached, the eutectic temperature being 1130°C.

(2) The disappearance of the high-temperature delta phase after the amount of carbon exceeds 0.5%.

(3) The progressive lowering of the upper arrestment or upper critical temperature, represented on the diagram by the A_3–A_{cm} lines, until a minimum of 723°C is reached for the alloy containing 0.83% carbon, and then a progressive increase in this temperature until it meets the eutectic isotherm at a composition of about 2% carbon.

In this regard it is important to note that both the A_1 and A_3 transformation temperatures vary depending upon whether the alloy is being heated or cooled. Thus, it is possible to have A_{r1}, A_{c1}, A_{r3}, and A_{c3}* temperatures. However, for alloys cooled under true equilibrium conditions these variations are negligible.

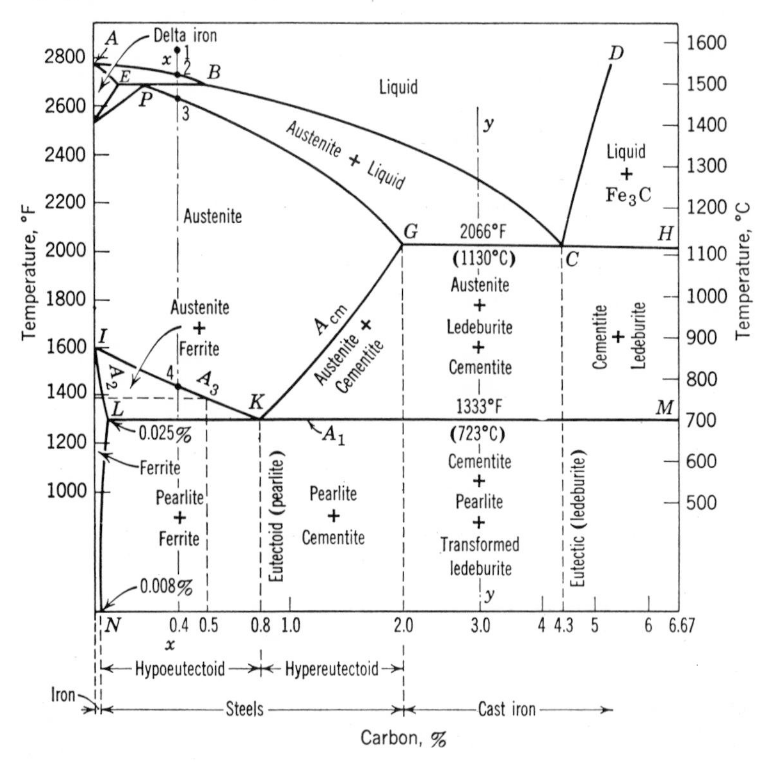

Figure 11.7 *The portion of the iron-carbon equilibrium diagram for alloys containing up to 6.67% carbon. (Reproduced with permission from "Nature and Properties of Engineering Materials" by Z. D. Jastrzebski. John Wiley and Sons Inc., N.Y. 1959.)*

The phases and microconstituents important to the metallurgist and present in the portion of the equilibrium diagram shown in Figure 11.7 are as follows.

Austenite: Essentially gamma iron with or without elements held in solid solution; as shown on the iron-carbon diagram, plain austenite may contain up to about 2% carbon at a temperature of 1130°C. However, other alloying

*From the French nomenclature:
A_c-"chauffage" meaning heating
A_r-"refroidissement" meaning cooling

elements may be held in solution in other than plain carbon steels, and in some few special instances austenite will remain in alloys cooled to room temperature. This is known as retained austenite.

Ferrite: A soft and ductile solid solution of iron containing up to 0.008% carbon at room temperature. Ferrite may also contain other alloying elements in solid solution, these tending generally to strengthen the ferrite and to alter the amount of carbon that it can contain.

Cementite: The hard, brittle, interstitial compound formed between iron and carbon; it contains 6.7% carbon and may be represented by the formula Fe_3C.

Pearlite: This is the eutectoid formed within the alloy system by the simultaneous precipitation of ferrite and cementite from austenite at the lower critical temperature of 723°C. It has a fine to coarse lamellar structure consisting of alternate plates of ferrite and cementite and contains about 87% ferrite.

The eutectic: Known as ledeburite, the eutectic contains 4.3% carbon but is rarely seen in slowly-cooled alloys since it breaks down, due to its unstable nature, to other phases.

TABLE 11.1

GENERAL PROPERTIES OF FERRITE, CEMENTITE AND PEARLITE

Microconstituent	*Tensile Strength Tons/sq in*	*% Elongation 2" gl*	*Brinnel Hardness*
Ferrite	19	50	80
Cementite	very difficult to measure	nil	600—650
Pearlite	50—65	10	200—300

PLAIN CARBON STEELS

A plain carbon steel is an alloy containing carbon as its principal alloying element, other alloying elements such as manganese, sulphur, phosphorus, silicon, and nickel being present in very small amounts only. Several classifications are possible for plain carbon steels, and two common systems are set out below.

1. Plain Carbon Steels Classified by Structure

On the basis of the iron-carbon equilibrium diagram it is possible to describe three classes of plain steels.

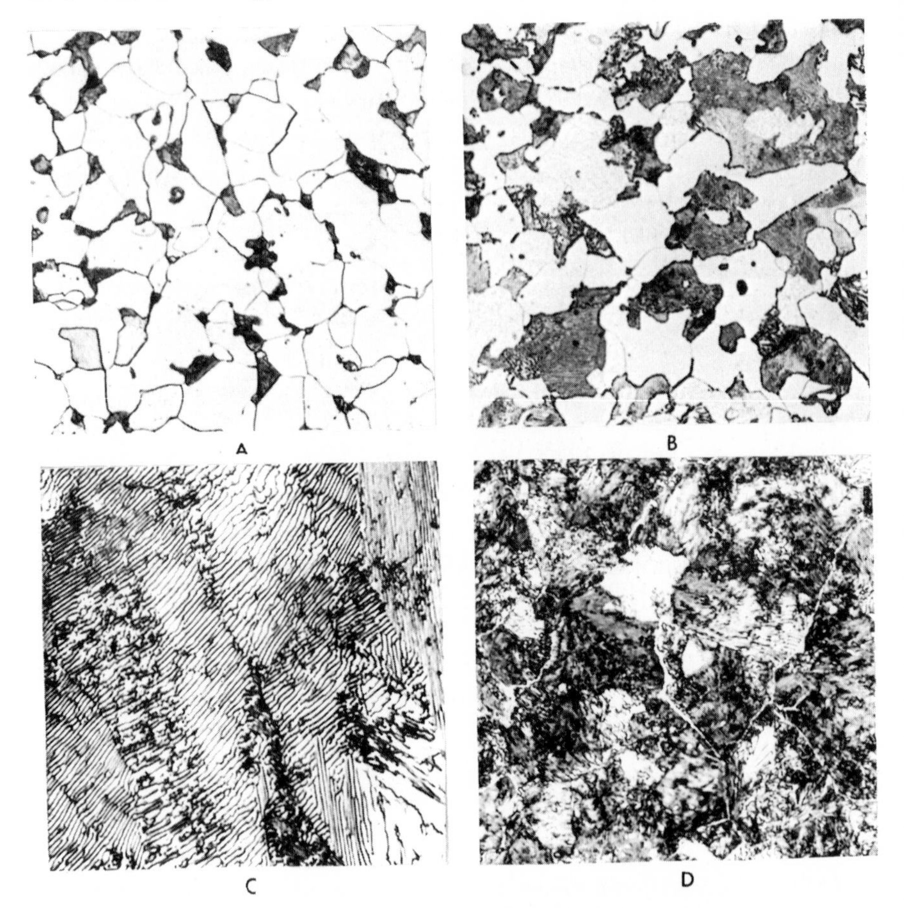

Figure 11.8 *Microstructures of plain carbon steels. (A) 0.16% carbon, showing ferrite (white) with a small amount of pearlite (dark); (B) 0.35% carbon, showing an increasing amount of pearlite; (C) 0.85% carbon, showing coarse and fine lamellar pearlite; (D) 1.3% carbon, showing pearlite grains with an intergranular network of cementite (white). Magnifications: A, B and D x500; C x1000. (Photographs courtesy of the Broken Hill Proprietary Co. Ltd.)*

(a) *Hypo-Eutectoid Steels*: having carbon contents varying from 0.008% to just below 0.83%; these steels have microstructures consisting of grains of ferrite together with grains of pearlite. The strength increases with increasing carbon content due to the increasing proportion of strong pearlite formed, however ductility decreases proportionally.

(b) *Eutectoid Steels*: having carbon contents, ideally, of 0.83%, these steels consist entirely of lamellar pearlite. In practice, fully pearlitic microstructures appear in all steels containing about 0.8 % carbon and the actual eutectoid composition is difficult to determine. Moreover, many alloying elements influence the carbon contents of eutectoid steels; manganese, for instance, if added to the extent of about 1 %, reduces the % carbon in the eutectoid to about 0.7 %.

(c) *Hyper-Eutectoid Steels*: having carbon contents significantly greater than 0.8 %; the steels consist of pearlite and cementite, the latter forming around the grain boundaries of the pearlite in as-cast alloys as an inter granular network. In practice, plain steels containing more than about 1.6 % carbon are rare.

2. Plain Carbon Steels Classified by Applications

This classification of plain carbon steels is of more direct use to the engineer.

(a) *Dead Mild Steels*: 0.07 % to 0.15 % *C*. Such steels are capable of withstanding a large amount of cold working, and may therefore be used for the production of solid drawn tubes and other articles where severe cold deformation is involved.

(b) *Mild Steels*: 0.15 % to 0.25 % *C*. This is the commonest type of steel. It is very weldable and does not harden appreciably when quenched from above its upper critical temperature.

(c) *Medium Carbon Steels*: 0.25 % to 0.55 % *C*. These steels respond to suitable types of heat treatments, the following list being some of the more common applications of medium carbon steels.

General Forging Steel—0.25 % to 0.35 % *C*.
Shafting Steels—0.35 % to 0.45 % *C*. Suitable for shafts, wire, and axles.
Wear-Resisting Steels—0.45 % to 0.55 % *C*. Used in railway rims, rails, cylinders, and die-blocks.
Bright Drawn Stock—0.24 % to 0.4 % *C*.

(d) *High Carbon Steels*: 0.55 % to 0.9 % *C*. Used wherever high strength and/or wear resistance are necessary, e.g. die blocks.

(e) *Carbon Tool Steels*: 0.9 % to 1.6 % *C*. High-duty tooling applications where high hardness and excellent wear resistance is required.

Laboratory (35): *Polish annealed samples of hypo-, hyper- and eutectoid steels and etch with 2 % nital solution. Examine and sketch their microstructures. Now re-etch the hypo- and hyper- eutectoid steels in 1 % picral solution and re-examine. How does picral aid in distinguishing ferrite from cementite?*

Cooling and Solidification of Various Steels in Terms of the Equilibrium Diagram

(a) Consider the case of 0.25 % carbon steel (see Figure 11.7).

Solidification begins at about 1525°C with the formation of a little delta iron. At about 1490°C a peritectic reaction occurs and some austenite is formed as a result of the reaction of the solid delta iron with the residual liquid phase. However, the alloy is still "mushy" and solidification is not complete until the temperature drops to about 1470°C when the alloy becomes fully austenitic. No further change in microstructure occurs until the temperature falls to about 825°C, when the austenite becomes saturated with iron and further small decreases in temperature cause the excess of iron to be precipitated as ferrite. The temperature at which this precipitation of excess iron as ferrite occurs is known as the Upper Critical Temperature (A_3) temperature, and the ferrite so formed is termed pro-eutectoid to distinguish it from that formed in the eutectoid, pearlite.

The pro-eutectoid transformation continues until the Lower Critical Temperature of 723°C (A_1) is reached, when the remainder of the austenite, now being of eutectoid composition, slowly alters to pearlite.

Thus, the microstructure of a 0.25 % carbon steel consists of ferrite and pearlite at temperatures below 723°C, and alterations to the microstructure are minimal during the interval from about 700°C to room temperature. However, the percentage of carbon contained in the ferrite does decrease from 0.025 % at 723°C to 0.008 % at room temperature, so, under equilibrium cooling conditions some carbon from ferrite areas alters to cementite by combining with some iron and the overall percentage of ferrite in the iron decreases slightly.

(b) Consider steel of eutectoid composition, i.e. 0.8 % carbon.

Solidification begins as austenite starts to form from the liquid phase, the metal being wholly solid at about 1330°C. No further change occurs until the lower critical temperature is reached, when the whole mass transforms slowly to the eutectoid, pearlite. Thus, the microstructure consists wholly of one microconstituent, pearlite, and no significant change occurs in the microstructure from the lower critical temperature to room temperature.

(c) Now consider the cooling of 1 % carbon steel.

Again, solidification begins as austenite separates from the molten alloy, solidification being complete at about 1315°C. No further change occurs until the temperature drops to about 810°C, when, because of decreasing carbide solubility, carbon is precipitated with some iron in the form of cementite (Fe_3C). As the temperature falls to the lower critical temperature of 723°C the remainder of the austenite, now at eutectoid composition, transforms to pearlite, the final microstructure being grains of pearlite with an intergranular network of cementite. This cementite is called pro-eutectoid to distinguish it from the cementite within the pearlite.

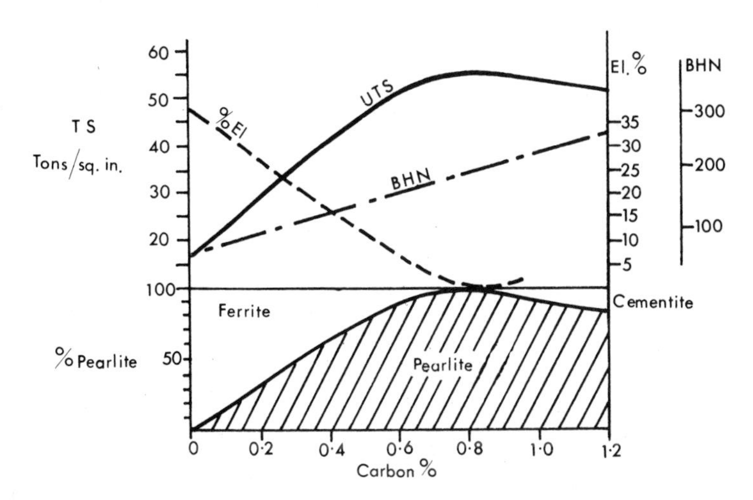

Figure 11.9 *The general effects on hardness, tensile strength, and ductility of increasing the carbon content of steel.*

TABLE 11.2

SOME APPLICATIONS OF PLAIN CARBON STEELS

Type of Steel	*% Carbon*	*Uses*
Dead Mild	0.05 – 0.15	Chains, stampings, rivets, nails, seam-welded pipes, tin plate, automobile body steel, and material subject to drawing and pressing.
Mild Steels	0.10 – 0.20	Structural steels, RSJ's, screws, drop forgings, case hardening steel.
	0.20 – 0.30	Machine and structural work, gears, free cutting steels, shafting and forgings.
Medium Carbon	0.30 – 0.40	Connecting rods, shafting, axles, crane hooks, forgings.
	0.40 – 0.50	Crankshafts, axles, gears, shafts, die-blocks, rotors, tyres, skip wheels.
	0.50 – 0.60	Loco tyres, rails, wire ropes.
High Carbon	0.60 – 0.70	Drop-hammer dies, saws, screwdrivers.
	0.70 – 0.80	Band saws, anvil faces, hammers, wrenches, laminated springs, cable wire, large dies for cold presses.
	0.80 – 0.90	Cold chisels, shear blades, punches, rock drills.
Tool Steels	0.90 – 1.10	Axes, knives, drills, taps, screwing dies, picks.
	1.10 – 1.40	Ball bearings, files, broaches, razors, boring and finishing tools, machine parts where resistance to wear is essential.

ALLOY STEELS

In addition to iron and carbon, all commercial steels contain varying amounts of manganese, silicon, sulphur, and phosphorus, and often also varying amounts of such elements as nickel, chromium, molybdenum and vanadium. If alloying elements other than carbon are present only in small amounts (e.g. manganese up to 0.8 %, silicon up to 0.3 %, etc.) then the steel is usually called a low alloy, or plain carbon, steel.

An alloy steel, or high alloy steel, on the other hand, is a steel containing sufficient amounts of alloying elements such as nickel, chromium, etc., to appreciably alter the nature of the phases present in the steel. Such phase modifications alter the properties of steels quite considerably, causing such changes as increased strength, hardness, shock resistance, and corrosion resistance.

The following four elements commonly occur in very small amounts in plain carbon steels, sulphur and phosphorus being particularly detrimental.

Manganese is the most important alloying element after carbon in most steels; mostly, it exists in solid solution in the ferrite where it has a strengthening effect. However, it may also exist as the carbide Mn_3C which forms part of the pearlite, or as the sulphide MnS. Up to 1 % manganese strengthens the steel but reduces its ductility, while hardenability is increased considerably.

Sulphur normally exists as the sulphide FeS but the presence of manganese alters it to MnS which removes the brittleness caused in steels by the presence of FeS. Excessive sulphur content causes a type of brittleness known as "hot shortness" to occur, the steel tending to crack during hot working operations.

Silicon is a graphitising agent in steels, this effect being most pronounced in high carbon steels. Thus, silicon must be kept to a minimum.

Phosphorus forms iron phosphide Fe_3P which exists in solid solution in the ferrite; above 0.06 % phosphorus becomes detrimental since it causes ferrite segregations to occur during hot working.

Table 11.3 summarises some of the more important effects of common alloying elements in steels.

Some Important Alloy Steels

Structural Steels: Most structural steels are of the low-alloy type and possess high yield stress, good ductility, and high fatigue resistance. Since allowable working stresses are calculated from the yield stress, the high figures for structural steels are of great significance since considerable weight savings are possible if high-yielding steels are used in structures.

A typical low-alloy structural steel would have a composition of 0.12 % carbon, 0.75 % manganese, 0.25 % silicon, and 0.30 % copper, and would

TABLE 11.3

ALLOYING ELEMENTS IN STEELS

Alloying Element	*General Effects in Steels*	*Typical Uses*
Nickel *Ni*	Decreases critical range thus decreasing hardening and tempering temperatures; increases toughness; also increases corrosion resistance when present in massive amounts	Oil hardening tool steels up to 5% *Ni*; Stainless steels from 8 to 25% *Ni*
Chromium *Cr*	Increases hardness by forming hard carbides; increases hardenability; increases corrosion-resistance when present in massive amounts	Structural steels from 0.6 to 1.5% *Cr*; Stainless steels from 12 to 25% *Cr*
Manganese *Mn*	Lowers the upper critical temperature; forms hard carbides; reduces % carbon needed to form eutectoid steel; increases hardenability; stabilises austenite when present in massive amounts	Depth-hardening steel up to 2% *Mn*; Austenite steels from 12% to 14% *Mn*
Silicon *Si*	Strengthens ferrite; causes graphitisation of cementite in high carbon steels (thus forming "self lubricating" steels)	Structural steels contain 1–1.5% *Si*; Electrical steels contain 2.5–4% *Si*; Corrosion-resistant steels up to 14% *Si*; Graphitic die steels 1–1.5% + 1.5% *Si*
Molybdenum *Mo*	Inhibits "temper-brittleness" in *Ni-Cr* steels, helps retain strength of steel at elevated temperatures	Various amounts in high speed steels; Small amounts in 1.5 *Ni* + 0.8% *Cr* steels
Tungsten *W*	Forms stable carbides; slows down diffusion-controlled transformations; inhibits grain growth in heat-treated steels	In tool and die steels from 0.5 to 2% *W*; High speed steels up to 22% *W*

have a yield strength of 52,000 psi and about 15% elongation after hot rolling. The copper increases corrosion resistance, while the manganese and silicon improve weldability by preventing weld embrittlement due to martensite formation.*

Some low-alloy structural steels are used in the quenched and tempered condition and exhibit yield strengths in excess of 90,000 psi. They normally

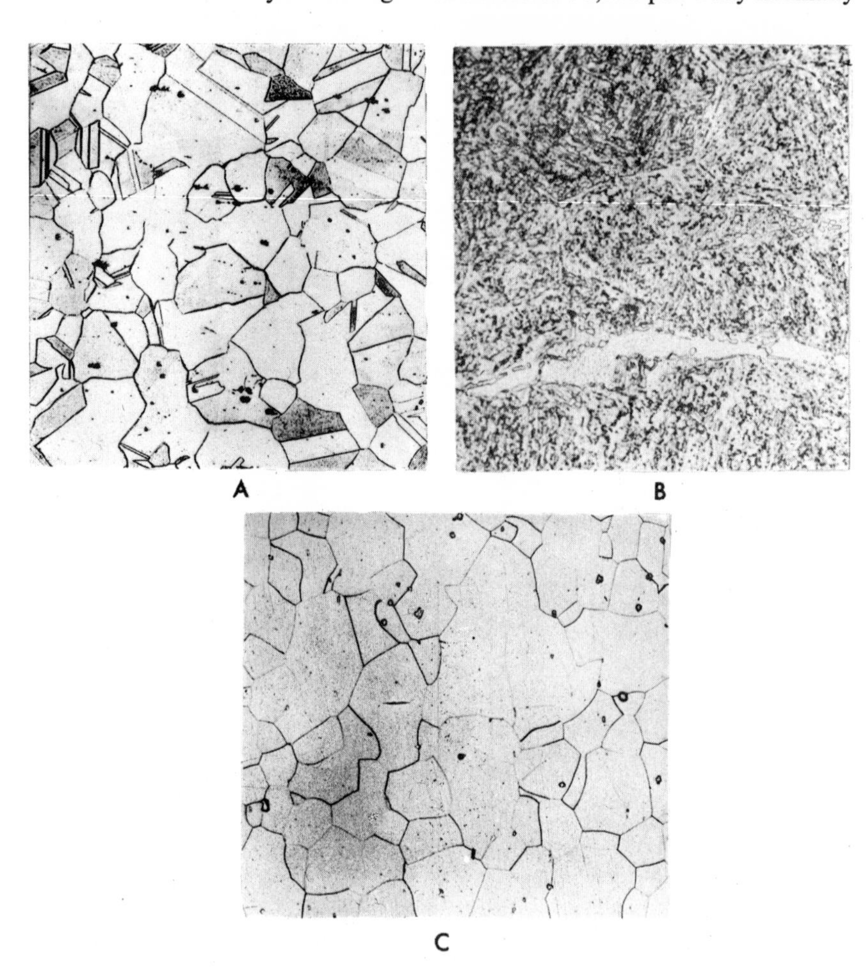

Figure 11.10 *Microstructures of typical (A) austenitic, (B) martensitic, and (C) ferritic stainless steels. The microstructure of the martensitic stainless steel shows a well-defined ferrite band (white). Magnifications x500. (Photographs courtesy of Commonwealth Steel Co. Ltd.)*

*Martensite is the hard and brittle phase that occurs in steels which are quench-cooled from above a certain temperature. See Chapter 12 for details.

contain small amounts of nickel, chromium, and vanadium in addition to carbon, sulphur, silicon, manganese, and copper, and retain good weldability because of their low carbon contents.

Stainless Steels: Stainless steels, first discovered by the metallurgist Brearley in 1913, exhibit excellent corrosion resistance due to the formation of a thin stable protective oxide layer upon exposure to the atmosphere. Although there are many different types of stainless steels, it is possible to classify them as either austenitic, martensitic, or ferritic. Table 11.4 shows the approximate compositions, structures and typical applications of these three groups of stainless steels.

TABLE 11.4

STRUCTURE AND APPLICATIONS OF STAINLESS STEELS

Type	*Structure*	*Compositions*	*Special Applications*
Austenitic	Fully austenitic if quenched from above Upper Critical Temperature	Variations of the 18% *Cr* 8% *Ni* basic alloy; usually about 0.1% carbon	Chemical plant construction; decorative purposes; domestic applications such as saucepans, cutlery, etc.
Martensitic	Martensitic structures if oil quenched from above Upper Critical Temperature	(a) 0.07%–0.10% *C* 13% *Cr* (b) 0.2%–0.4% *C* 13% *Cr* (c) 0.1% *C*, 18% *Cr* 2% *Ni*	Turbine blades, rivets, split pins, cutlery Surgical instruments, springs, ball bearings Pump shafts, regulator valves, aircraft fittings, turbine blades
Ferritic	Ferrite grains plus perhaps free carbides	0.05%–0.15% *C* 16%–30% *Cr*	Where outstanding corrosion resistance and excellent formability are required, e.g. for many deep drawing applications

High Speed Tool Steels: High speed tool steels have been developed to overcome the limitations of plain carbon tool steels at high temperatures when the heat present causes tempering and hence softening of plain steels. The so-called 18.4.1 type, consisting of 18% tungsten, 4% chromium, and 1% vanadium is perhaps the best-known type, these alloying elements forming hard carbides which do not soften at red heat. If 5–12% cobalt is added to such steels they will actually increase in hardness at temperatures of around 600°C, this process being known as secondary hardening.

High Duty Tool and Die Steels: Many different steels of fairly high alloying contents are used for high duty punching tools and dies. However, a most interesting one is the high-carbon—high-chromium die steel of composition 2% carbon, 0.3% manganese, and 12% chromium. Such a steel is shown in its annealed state in Figure 11.11, and consists of small and large hard carbide particles in a matrix of ferrite.

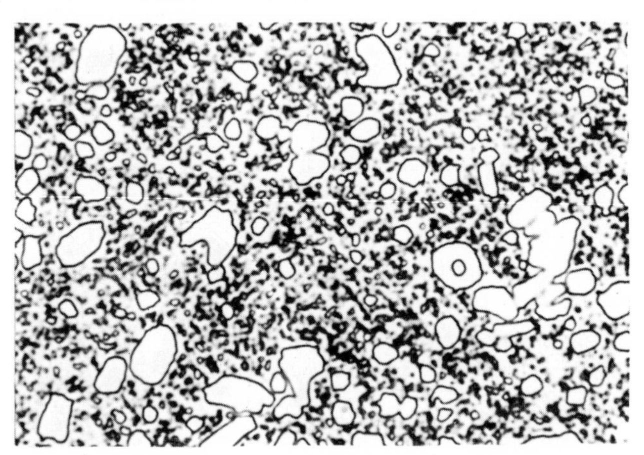

Figure 11.11 *Microstructure of a high-carbon high-chromium die steel containing 2% carbon and 12% chromium. The steel was annealed at 800°C and contains small and large carbide particles dispersed in a ferritic matrix. In this condition the steel can be machined with difficulty, but after heat treatment it becomes extremely hard and tough. Magnification x500. (Photograph courtesy of Bohler Steels Pty. Ltd.)*

Magnetic Alloys: Magnetic alloys are divided roughly into two classes; those that retain their magnetism, and those that do not. The former are usually referred to as "magnetically hard alloys", the latter as "magnetically soft alloys".

The earliest permanent magnet steel was a 1% plain carbon steel in a fully hardened condition, and later developments occurred as tungsten, chromium, and cobalt were added as alloying elements. However, the most useful types of permanent-magnet steels contain very high proportions of nickel, cobalt, and aluminium, with perhaps a small amount of tungsten (e.g. Alnico, 10% *Al*, 18% *Ni*, 12% *Co*, 6% *Cu*, balance *Fe*).

Magnetically "soft" materials possess high magnetic permeability and are readily demagnetised; they should retain as little magnetism as possible after removal from a magnetic field. Early alloys were either ordinary soft iron, or iron-silicon alloys containing about 4.5% *Si*. However, modern high duty alloys are iron-nickel compositions such as Permalloy, containing 78% *Ni*, and Mumetal, containing 75% *Ni*. Such alloys are used as shields for submarine cables and for transformer cores.

CAST IRON

Generally, any material made up primarily of iron with about 2% or more of carbon is considered to be cast iron, with most commercial alloys containing from about 2.5% to 3.8% carbon. Depending upon whether the carbon content is greater or less than 4.3% (the eutectic composition for plain irons), a cast iron is classified as either hyper-eutectic or hypo-eutectic. However, this classification has little industrial significance.

The cupola furnace is the basic melting unit of the cast iron foundry since it produces molten iron very inexpensively and yet allows for good compositional control.

The furnace consists of a cylindrical steel shell standing vertically, lined with high-grade refractories. The bottom of the furnace is closed with drop doors which support the entire charge during firing and allow easy removal of residues left after the firing is complete. The cycle of operation of the cupola is as follows:

(1) The lower portion of the furnace is covered with kindling wood and some coke; this is then lit, and the depth of the coke bed increased until it is from 18 inches to 48 inches deep.

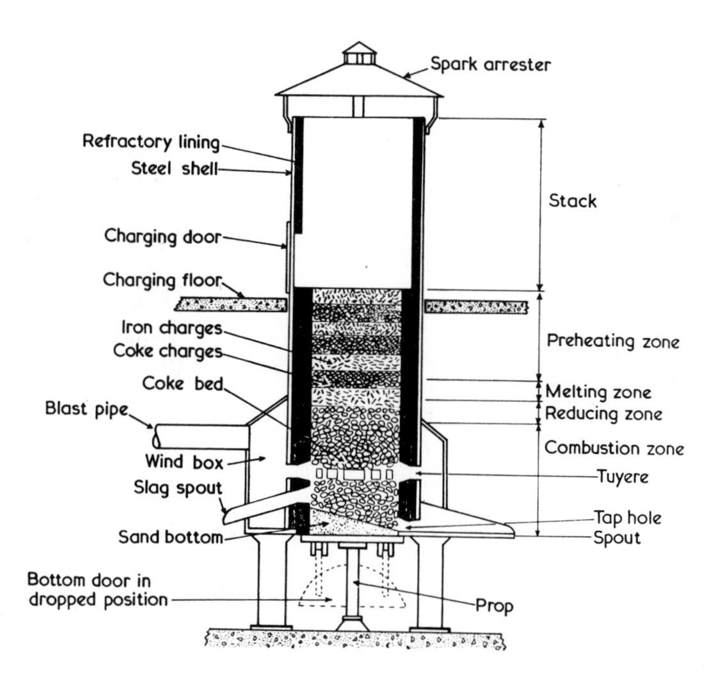

Figure 11.12 *Section through a cupola furnace.*

(2) Alternate charges of pig iron, scrap, and coke are now added via the charging doors half-way up the furnace until the furnace is filled almost to this level. About one part of coke is added for every ten parts of metal, together with a little limestone.

(3) Combustion is assisted using a natural draft through the tuyeres for about half an hour, and then the blower is turned on. The iron melts in the "hot zone" just above the tuyeres and runs down on to the sand bed where it collects ready for tapping.

(4) The limestone forms a basic slag which floats on the molten metal; this is tapped off when necessary.

(5) The molten metal is also tapped off periodically, the air blast being turned off during this operation.

The composition of the molten metal in the ladle is checked prior to pouring into moulds, and special alloying elements are added into the ladle when it is necessary to adjust the composition of the melt.

Types of Cast Iron

Examination of the iron-carbon equilibrium diagram (Figure 11.7) reveals that the slow cooling of cast irons should result in structures that consist essentially of cementite and pearlite. However, in practice, the actual rate of cooling and the presence of other alloying elements such as silicon, manganese, phosphorus, and sulphur play a determining role in the type of microstructure formed in any particular grade of cast iron. The two extremes of microstructure found in cast irons, which are rarely achieved in practice, are (1) all of the carbon is combined as cementite, the resulting microstructure consisting entirely of cementite and pearlite, and (2) all of the carbon exists as free graphite in a matrix of ferrite. The following four basic types of cast irons are usually produced.

Grey Cast Iron: The amount of grey cast iron produced far exceeds that of any other type of cast iron. The microstructure of a typical grey cast iron reveals a large amount of free carbon in the form of both flake and rosette graphite in a matrix of, usually, pearlite and ferrite. When broken, this material exhibits a grey colour on the fracture surface, the brittleness exhibited being due to the weakening effects of the free graphite. Grey cast iron is typically weak in tension, fairly soft, brittle, strong in compression, and possesses excellent casting properties.

White Cast Iron: White cast iron, so named because a fractured surface is white in colour, is a result of fast cooling and has a structure composed entirely of cementite and pearlite. Thus, all of the carbon is in the combined form (cementite) and the iron is extremely hard and virtually unmachinable.

White cast iron is always low in silicon since silicon is the most powerful graphitising agent that can be added to a cast iron melt.

White cast iron is seldom produced except as an intermediate step in the production of malleable irons. However, thin layers of white iron are often produced on the surface of grey iron castings by using metal "chill bars" in the moulding sand close to the metal surface; this produces a casting with a strong and hard surface, and a ductile, shock-resistant core.

Laboratory (36): *Polish small samples of grey iron and chill cast iron. Examine prior to etching, then etch and re-examine. Why is it useful to examine cast irons prior to etching?*

Malleable Iron: The inherent brittleness of grey cast iron limits its applications to those purposes where shock-resistance need not be very high. Malleable irons, however, combine excellent casting qualities with a measure of strength and ductility, and are produced by suitably annealing white iron castings. Two processes are used:

(a) *Whiteheart Iron*: the chilled white iron castings are packed in boxes with a mixture of old and new haematite, heated, and soaked at 875°C for several days, and then slowly cooled in the furnace. Surface carbon is largely oxidised by the Fe_2O_3, giving a surface of ferrite and a core of ferrite and cementite. This process is not commonly used in Australia.

(b) *Blackheart Iron*: the white iron castings are packed in boxes with lime and annealed at 820°C for several days and then slowly cooled. The resultant structure is quite different; the cementite is decomposed to ferrite and free graphite, which exists in nodular or spheroidal form finely dispersed throughout the metal. This type of carbon is often called "temper carbon" since it is obtained by a heat treatment process.

Nodular Iron: Nodular or spheroidal graphite (S.G.) iron is produced by inoculating white cast iron melts with small amounts of such materials as nickel-magnesium (2%), ferro-silicon, or calcium silicide just prior to pouring. The resulting structures may vary from almost completely ferritic to almost completely pearlitic matrixes in which spheroids or nodules of carbon are imbedded. Up to 15% elongation can be present in such irons, which are used where shock-resistance is required. Pearlitic S.G. irons can be surface hardened by a flame hardening technique,* and complete hardening can be achieved by quenching; in the latter case the iron can also be tempered to remove brittleness.

Table 11.5 summarises the major characteristics, mechanical properties, and applications of these main four types of cast iron.

*See Chapter 12.

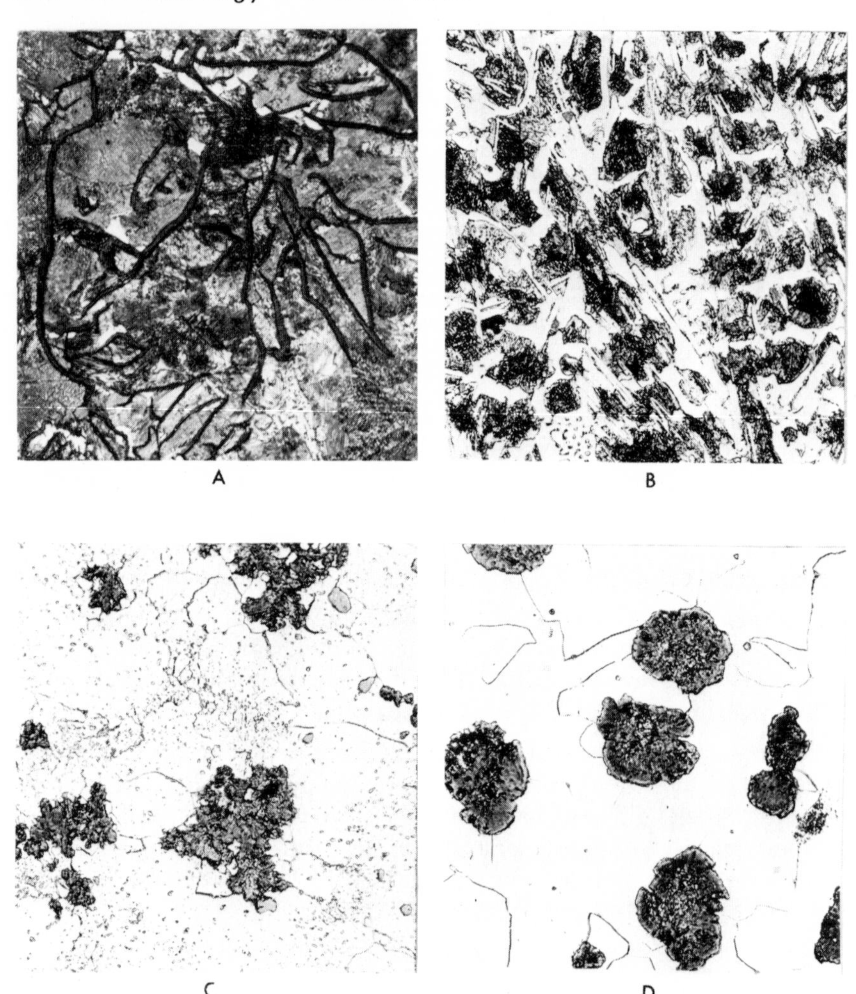

Figure 11.13 *Microstructures of cast irons. The shape and method of dispersion of the carbon determines, in large measure, the mechanical properties of cast iron: (A) grey cast iron, containing flake graphite; (B) white cast iron, in which all of the carbon is combined as cementite; (C) Blackheart malleable iron, containing spheroids of temper carbon; (D) nodular iron, containing carbon in spheroidal form. Magnifications x500. (Photographs courtesy of the Broken Hill Proprietary Co. Ltd.)*

The Damping Capacity of Cast Irons

Cast alloys in general and cast iron in particular possess good damping capacities; that is, they are able readily to absorb vibrations set up by succes-

TABLE 11.5

PROPERTIES, CHARACTERISTICS AND USES OF CAST IRON

Type	*Form of Carbon*	*Composition and Formation*	*Tensile Strength (ksi)*		*Compressive Yield (ksi) Formation*	*Elastic Modulus (psi × 10^6)*	*% Elongation (tensile)*	*B.H.N.*	*Typical Applications*
			Yield	*Ultimate*					
Grey Cast Iron	Flakes	Slow cooling of a high-silicon cast iron melt	—	20	35	15	1	130	Ingot moulds, automobile cylinders, pistons, machine castings, etc.
White Cast Iron	Iron carbide Fe_3C	Quench-cooling of a low-silicon cast iron melt	—	60	100	20	—	400	Ploughshares, chilled rolls, dies, wearing plates, stamping shoes, balls etc.
Nodular or Spheroidal Graphite Iron	Spheroids	Alloying elements added to spheroidise carbon during slow cooling	45	60	80	25	15	180	Machine castings subjected to bending and vibration, and other applications where strength and ductility are important
Malleable Iron	Spheroids	Annealed white cast iron castings	33	50	33	25	14	120	Valve bodies, hinges, manhole covers, machine castings, etc.

sive impact blows or other forces. There are many engineering applications for which it is desirable to have a material with a high damping capacity; for example, large machine beds, automobile crankshafts, engine blocks, etc.

The application of cast iron to crankshaft manufacture is an interesting example. Traditionally, crankshafts have always been forged, but the development of special cast irons which are heat-treatable has removed the need for this expensive forming process. The composition of the cast iron given below is typical of those now used for crankshaft manufacture.

Carbon	= 1.6%	Silicon	= 1.00%
Manganese	= 0.60%	Phosphorus	= 0.06%
Chromium	= 0.40%	Sulphur	= 0.06%
Copper	= 1.50%	(Iron balance)	

GLOSSARY OF TERMS

Austenite: an interstitial solid solution of carbon and often other alloying elements dissolved in FCC gamma iron.

Austenitising: the process of forming austenite in a steel by heating it either into (partial austenitising) or above (full austenitising) its critical range.

Cementite: the interstitial compound Fe_3C, iron carbide, formed in the iron-carbon equilibrium system at 6.67% carbon.

Eutectoid steel: steel composed entirely of pearlite.

Ferrite: a solid solution of carbon and often other alloying elements in BCC alpha iron.

Grey cast iron: cast iron in which an appreciable amount of the carbon exists in the form of graphite flakes.

Hyper-eutectoid steel: steel having a structure of pearlite and pro-eutectoid carbide, this structure being formed due to an excess of carbon over the amount required to produce a fully pearlitic structure.

Hypo-eutectoid steel: steel having a structure of pearlite and pro-eutectoid ferrite, this structure resulting from insufficient carbon to form a fully eutectoid structure.

Ledeburite: the eutectic in the iron-carbide system; it contains 4.3% carbon and forms at the eutectic temperature of 1135°C.

Lower critical temperature: (LCT or A_1) the temperature above which the structure is either fully austenitic, or consists of austenite plus either pro-eutectoid ferrite or carbide.

Nodular iron: cast iron "inoculated" just prior to pouring so that the carbon exists as spheroids.

Pearlite: the iron-carbon eutectoid; it is formed by the simultaneous precipitation of ferrite and cementite from austenite at the lower critical temperature and has a lamellar structure.

Upper critical temperature: (UCT or A_3) the temperature above which the structure of a steel is fully austenitic. The LCT and the UCT for eutectoid steels are equal.

White cast iron: cast iron in which all of the carbon is combined as cementite.

REVIEW QUESTIONS

1. What range of compositions and mechanical properties would you expect in pig iron produced from a high grade ore?

2. What are the advantages and disadvantages of Bessemer steel?

3. List and discuss the advantages of the Basic Oxygen converter compared to the Open Hearth furnace.

4. Pure iron possesses three allotropic modifications. List these, together with their important physical, mechanical, and magnetic properties.

5. Discuss the three main effects of increasing the percentage carbon in binary iron-carbon alloys from zero up to 6.67%.

6. Define the following terms: austenite, ferrite, cementite, pearlite, ledeburite.

7. Outline an engineering classification of plain carbon steels and irons giving typical microstructures of each grade of iron-carbon alloy.

8. Describe the cooling and solidification of iron-carbon alloys containing (i) 0.35% carbon (ii) 0.8% carbon and (iii) 1.5% carbon in terms of the equilibrium diagram shown in Figure 11.6.

9. Calculate, using the Lever Rule, the percentage of (i) pro-eutectoid ferrite, and (ii) ferrite, in a plain 0.35% carbon steel at 765°C, assuming equilibrium cooling conditions.

10. Briefly discuss the general effects of the following alloying elements in steels: sulphur; manganese; nickel; chromium; tungsten.

11. Briefly describe the three important classes of stainless steels in terms of typical microstructures and engineering applications.

12. Distinguish between grey and white cast irons.

13. How do the forms of carbon dispersion in (i) grey cast iron (ii) malleable iron affect the mechanical properties of these materials?

14. What is meant by the damping capacity of a material, and which group of iron-carbon alloys possesses excellent damping properties?

15. Sketch and describe typical microstructure of slowly-cooled samples of iron-carbon alloys containing (i) 0.5% carbon (ii) 0.9% carbon, and (iii) 3.2% carbon.

16. State the approximate compositions of the iron-carbon alloys that would be used for the following applications: auto body pressings; railway lines; bright drawn stock to be hardened by quenching; auto cylinder head castings; lathe cutting tools for light duty work.

17. Explain why grey cast iron is much more machinable than white cast iron.

12

Heat Treatment of Steel

HEAT treating is the controlled heating and cooling of materials in order to deliberately alter their mechanical properties; by far the greatest amount of industrial heat treatment procedures carried out today are done on metals, with ferrous metals, particularly steels, being the subjects for the greater proportion of such treatments. The structure, and therefore the mechanical properties of steel, are profoundly altered by the particular heat treatment used, a fact well known to the metalworkers of centuries long past. The metallurgist must understand and control structure changes brought about by heat treatment if the metals he uses are to be utilised to their fullest advantage. As well, many shaping, forming, and joining processes involve heating and cooling the metal and thus "accidental" heat treatment may result; if these facts are not realised then serious consequences may result.*

All heat treatment processes may be considered to consist of three main parts,

(1) the heating of the metal to the pre-determined heat treating temperature,
(2) the soaking of the metal at that temperature until the structure becomes uniform throughout the section,
(3) the cooling of the metal at some pre-determined rate such as will cause the formation of, or will maintain desirable structures within the metal.

THE HARDENING OF STEEL—THE MARTENSITE REACTION

The hardening of steel depends upon the different solid solubilities of carbon in alpha and gamma iron. If a piece of steel is austenitised by heating it to about 30°–50°C above its upper critical temperature and allowed to cool slowly the austenite present will break down into ferrite and cementite, the final microstructure so obtained being pearlite, or pearlite with either pro-eutectoid ferrite or cementite, depending upon whether the steel was

*A recent example is the Kings Bridge spanning the Yarra River in Melbourne, Victoria, which partially collapsed when subjected to a normal load. This was primarily brought about by weld cracking which occurred during the construction of the bridge. See Chapter 13 for an account of weld cracking.

eutectoid, hypo-eutectoid or hyper-eutectoid in composition. This breakdown of austenite is diffusion-controlled, and will only go to completion if sufficient time is allowed to elapse while the steel cools through its critical range (i.e. through the A_3A_{cm}—A_1 temperature range). If the austenitised steel is cooled very rapidly from above its upper critical temperature the breakdown of austenite is suppressed and the alpha-gamma lattice transformation is unable to go to completion. A completely new phase is formed in the steel; it is known as martensite and has a body-centred tetragonal lattice in which all of the dissolved carbon is held in interstitial solid solution.

Martensite is only a metastable phase and in a sense may be regarded as an intermediate transition product since its structure is broken down by tempering. It is extremely hard and brittle and has a characteristic acicular appearance when examined under high magnifications.

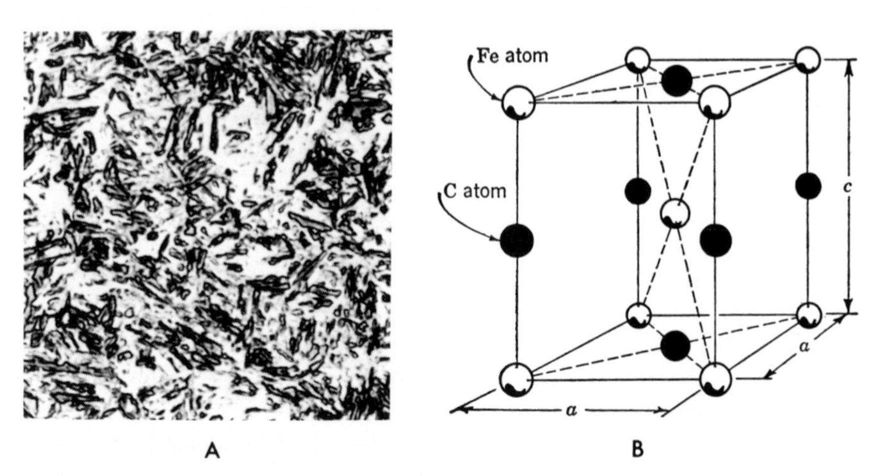

Figure 12.1 *(A) martensite in quenched 0.35% carbon steel. Magnification x500. (Photograph courtesy of the Broken Hill Proprietary Co. Ltd.); (B) The unit cell of the distorted body-centred tetragonal lattice of martensite. (Reproduced with permission from "Metallurgy for Engineers" by Wulff, Taylor, and Shaler. John Wiley and Sons Inc., N.Y. 1952.)*

Generally speaking, in steels up to eutectoid composition the martensite formed by this drastic quenching operation contains the same amount of carbon as did the austenite from which it was formed. However, quenched hyper-eutectoid steels commonly contain some carbide particles as well.

The Effect of Carbon

The amount of carbon present directly determines the hardness of the martensite, a fact well shown in Figure 12.2.

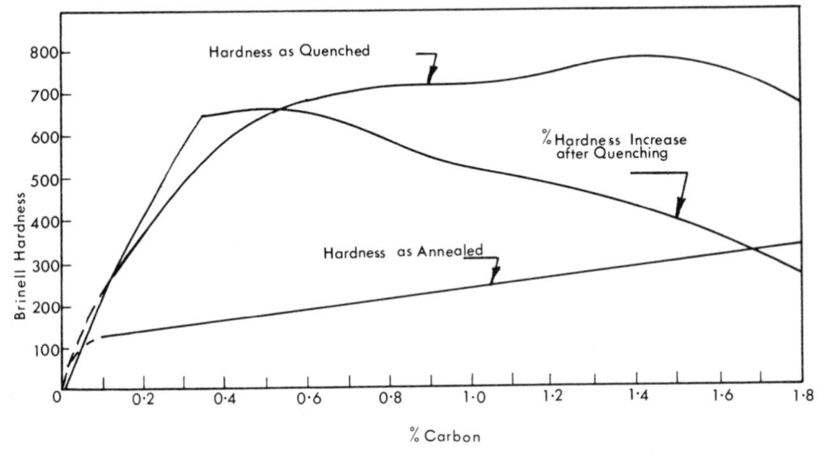

Figure 12.2 *Graphical representation of the relationship between carbon content and the hardness of martensite.*

From this graph it is readily seen that the hardness of quenched plain carbon steels increases rapidly until the eutectoid composition is reached, when the hardness curve levels off and then increases slowly. Martensite formed in eutectoid steels is the hardest form possible, and that in hyper-eutectoid steels has the same hardness. The slight increase in overall hardness in hyper-eutectoid steels is due simply to the presence of free carbide particles (cementite in plain steels) which are themselves extremely hard and brittle.

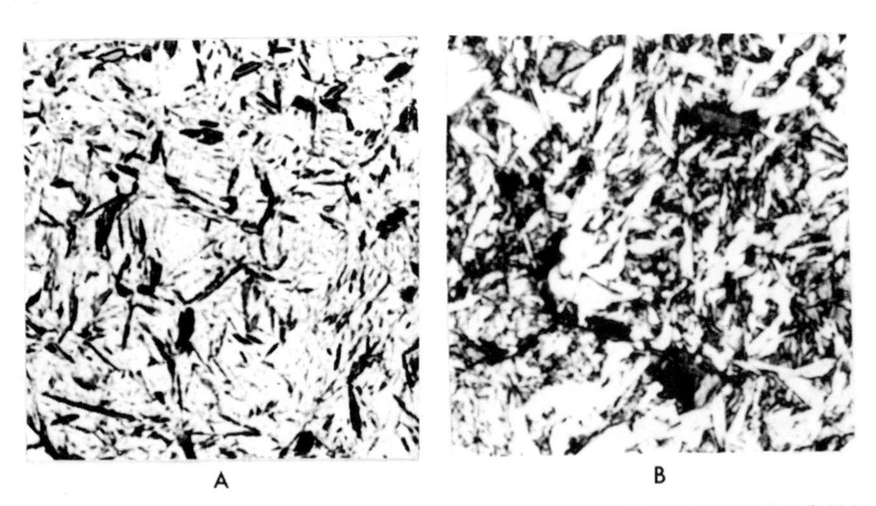

Figure 12.3 *Martensite in austenitised and quenched (A) 0.73% carbon steel and (B) 1.3% carbon steel, the latter showing some free carbide particles (dark). Magnifications x500. (Photographs courtesy of the Broken Hill Proprietary Co. Ltd.)*

The Effects of the Rate of Cooling

In order to understand the results of different cooling rates upon the structure and properties of austenitised steel it is interesting to examine four different samples of eutectoid steel that have been cooled at different rates. The important facts relating to this experiment appear in Figure 12.4.

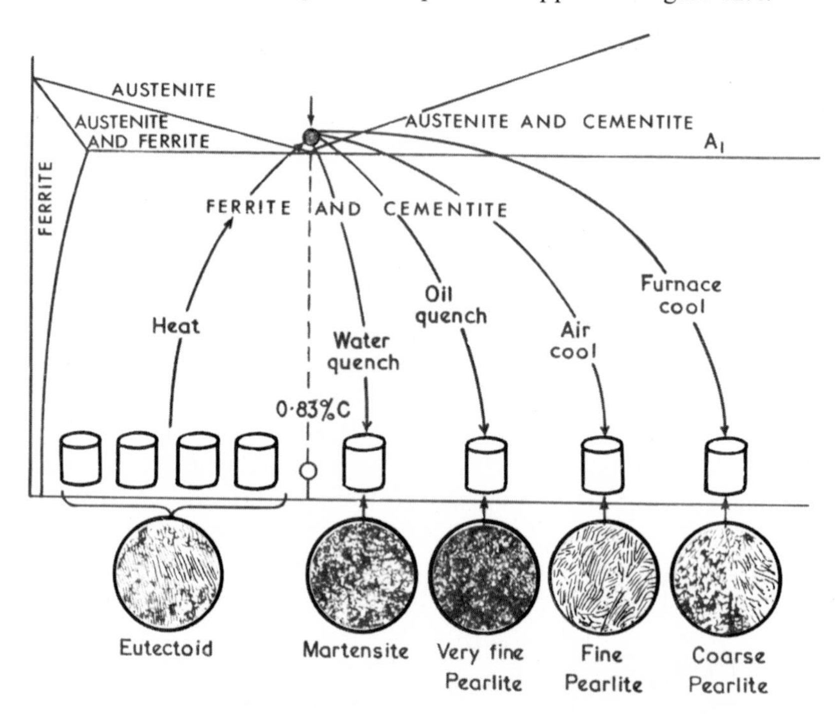

Figure 12.4 *Microstructures resulting from different cooling rates applied to austenitised samples of eutectoid steel.*

Only the drastic water quench produces a fully martensitic structure, the other quenched samples exhibiting different forms of pearlite. The oil quenched sample exhibits a particular form of very fine pearlite known as primary troostite, which is strong and hard, and which etches black under the microscope. It is known as primary troostite in order to distinguish it from secondary troostite which appears in tempered steels.

The mechanical properties and structures resulting from such cooling rates applied to eutectoid steels are set out in the Table 12.1.

Laboratory (37): *Carry out the experiment as described above for eutectoid steel. Polish and etch each specimen after heat treatment, examine microscopically, sketch and describe the microstructures, and carry out Rockwell tests on each.*

TABLE 12.1

Cooling Rate	*Structures*	*UTS (psi)*	*Yield (psi)*	*Hardness (Rockwell "C")*	*% Elongation (2" gl)*	*% Reduction in Area*
Water quench	Martensite	250,000	—	65	low	low
Oil quench	Troostite	160,000	80,000	35	5	30
Air cool	Fine Pearlite	125,000	40,000	25	8	22
Furnace cool	Coarse Pearlite	76,000	20,000	15	12	27

Effects of Different Quenching Media

Different quenching media remove the heat from steel at different rates so that the correct selection of a quenching medium for a particular grade of steel is of the utmost importance. The cheapest quenching medium is plain water, but agitation must be used since there is a tendency for air bubbles to form on the surface of the article being quenched, these having an insulating effect and hence producing "soft spots". Salt water or brine is a more severe quenching medium than water, its main disadvantage being that it causes rapid rusting unless carefully washed away after quenching is complete. For very low carbon steels hydroxide solutions are often used instead of brine since the quench is even more severe. If slower cooling rates are desirable, various grades of oil with high flash points may be used, the grade determining the severity of the quench. Oil would always be used for high carbon steels since water quenching is too severe and results in cracking.

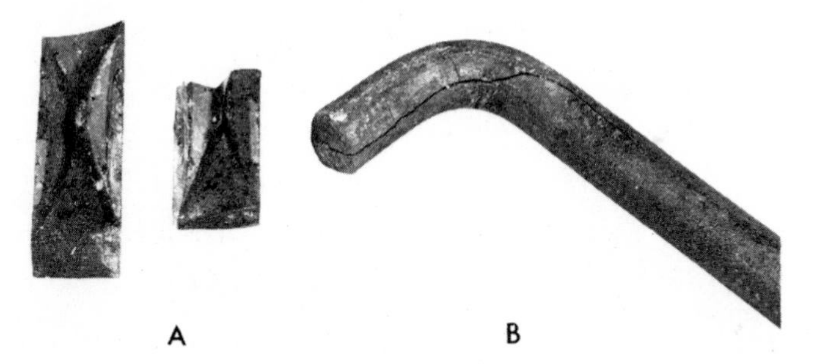

Figure 12.5 *Quench cracking in high carbon steel: (A) in ½" square tool steel; (B) in ⅜" diameter 0.9% carbon steel forged and then quench hardened.*

The Hardenability of Steel

The addition of certain amounts of nickel and chromium to carbon steel produces a steel that can be hardened by relatively slow cooling from above its critical range. For example, steels containing 5% nickel and 1.5% chromium transform to martensite if cooled in still air and are thus known as *air-hardening steels*. However, some steels do not transform to martensite no matter how severely they are quench-cooled. Such steels contain very high amounts of nickel, chromium or manganese, and are known as *quench-*

softening steels since their final microstructures are austenitic. However, these steels work harden very quickly and are used for parts that must resist abrasion, e.g. shovel blades. A typical austenitic manganese steel for this purpose would contain 1.2% carbon and 12% manganese.

In general, alloying elements, unless present in sufficient quantities to stabilise austenite, tend to increase the hardenability of steel, this being clearly shown in Figure 12.6. Alloy steels are always used where depth hardening is required.

Laboratory (38): *Take a piece of about $\frac{5}{16}$th inch diameter tool steel 4″ long (medium carbon will also suffice), heat it evenly to red heat, soak it for several minutes, and then hold about ½″ of it in a bath of cold water, agitating slightly, until the whole piece cools to room temperature. Test for hardness every ¼″ along its length and graph hardness against distance from the quenched end. What does this test indicate about the hardenability of the steel?*

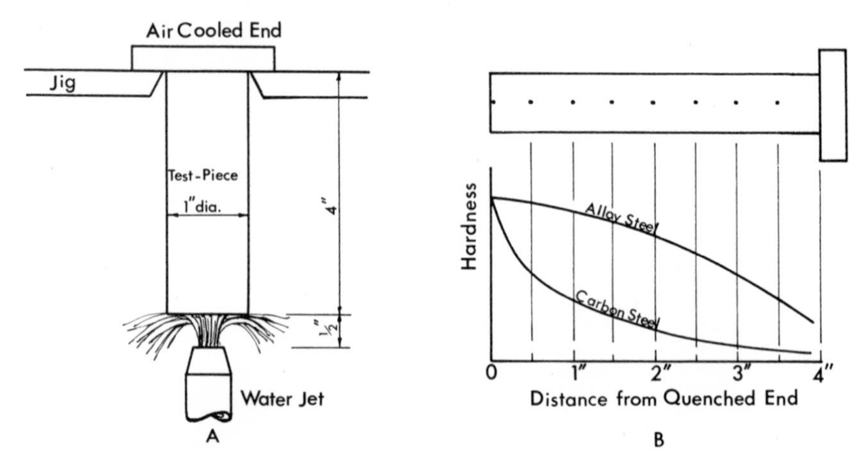

Figure 12.6 *(A) the Jominy end quench test; (B) typical Jominy test results for plain carbon and alloy steels. Note the greater hardenability of the alloy steel.*

THE SOFTENING OF STEEL

Various grades of steel may be softened by different heat treatment processes, and, in general, softening processes may be grouped under the headings of annealing and normalising.

Annealing

Although several different types of annealing processes are in use for steels, their purposes are similar and are:

(1) the relief of all internal stresses within the metal,
(2) the production of a more or less uniform grain structure throughout,
(3) the softening of the metal to some degree or other.

All annealing processes are classified as either full annealing or process annealing, the latter being less expensive than the former and suitable only for low carbon steels.

Full annealing involves heating the steel into the austenitic range, soaking it until the structure becomes fully austenitic, and then cooling very slowly to room temperature. The austenitising temperature used is generally 30°–50°C above the upper critical temperature for the particular grade of steel, and cooling may be done in the furnace, in ashes or sand, or in a special cooling pit, refractory lined, and covered with a refractory lid. If the steel is soaked at the annealing temperature too long, or heated to too high a temperature, the austenite will undergo considerable grain growth, and thus the pearlite grains formed by the transformation of the austenite will be coarse. Such a coarse structure is termed "overheated" and exhibits relatively poor mechanical properties.

Decarburisation and oxidation often occur during annealing; this may be overcome by packing the steel into special boxes or using a "neutral" atmosphere within the furnace. For example, low carbon steel components could be packed into boxes filled with sand, lime, ground mica, or cast iron swarf, while higher carbon components are usually packed into charcoal or other carbonaceous materials.

Full annealing is a slow and costly process and is not used as often as other softening processes since the resulting softness and loss of strength is often both unnecessary and undesirable.

Process annealing, also known as commercial annealing or sub-critical annealing, involves heating the steel to a pre-determined temperature well below the A_1, and then air-cooling or alternatively quenching the metal in a suitable pickling bath. A temperature of between 550° and 650°C is quite common for mild steels, which are usually process annealed to soften them after cold deformation has occurred.

Process annealing results in the complete recrystallisation of the distorted ferrite grains which constitute the major portion of the microstructure, the recrystallisation temperature of pure iron being about 450°C. The pearlite grains, however, do not recrystallise, so that the annealed mild steel would consist of a stress-free ferrite matrix containing a relatively small number of distorted pearlite grains.

Normalising

Normalising involves austenitising the steel at a slightly higher temperature than that used for full annealing and then air-cooling to room temperature. This results in a finer grain structure being formed provided that the steel is soaked at the austenitising temperature for only just enough time to allow the alpha to gamma transformation to go to completion. The finer grain structure increases the yield and ultimate strengths, hardness, and notched-bar impact values, but causes a slight decrease in ductility. Normalising is often applied to castings and forgings as a stress-relieving process. With some grades of steel it is possible to cool in air and produce a troostitic structure, the pearlite in this case being in an emulsified form.

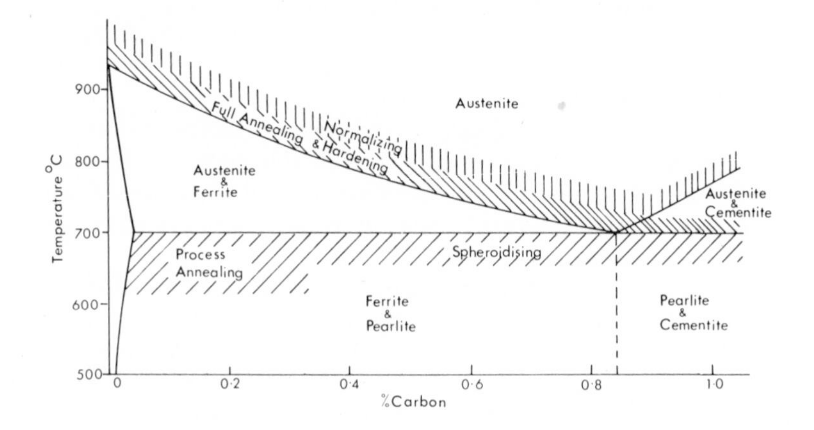

Figure 12.7 *Heat treating temperature ranges shown on the "steels portion" of the iron-carbon equilibrium diagram.*

ISOTHERMAL TRANSFORMATIONS

The iron-carbon diagram does not show time as a variable, and hence it cannot show the effects of different cooling rates upon the structures of various grades of steels. However, another type of diagram, known variously as an isothermal transformation curve, time-temperature-transformation curve (T.T.T. curve) or S-curve, may be used for this latter purpose. An S-curve applies only to one particular type of steel, that for a plain carbon steel of eutectoid composition being shown in Figure 12.8.

The left-hand curve in Figure 12.8A shows the beginning of transformation of this steel for all possible cooling rates, while the right-hand curve indicates the end of the transformation. The two horizontal lines marked Ms and Mf indicate, respectively, the beginning of the martensite transformation and the completion of this transformation. Figure 12.8B shows four cooling curves superimposed on the S-curve, all beginning from above the A_1 temperature (the austenitising temperature for this steel). These curves illustrate full annealing, normalising, a split transformation such as would occur with oil quenching, and full hardening. The following facts are illustrated in Figure 12.8B:

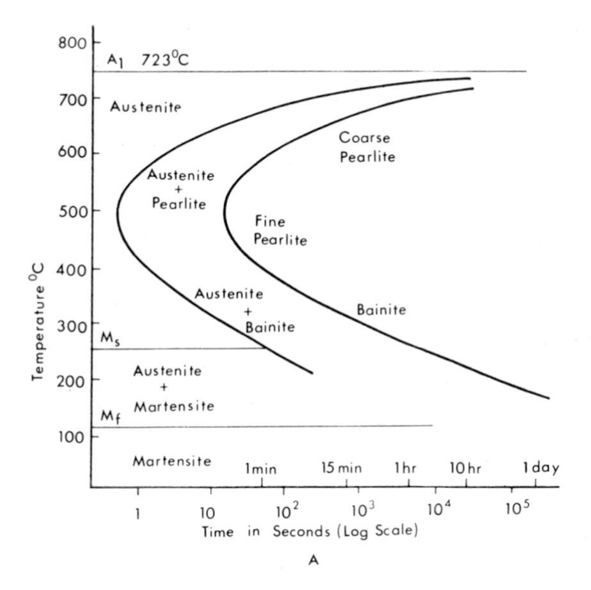

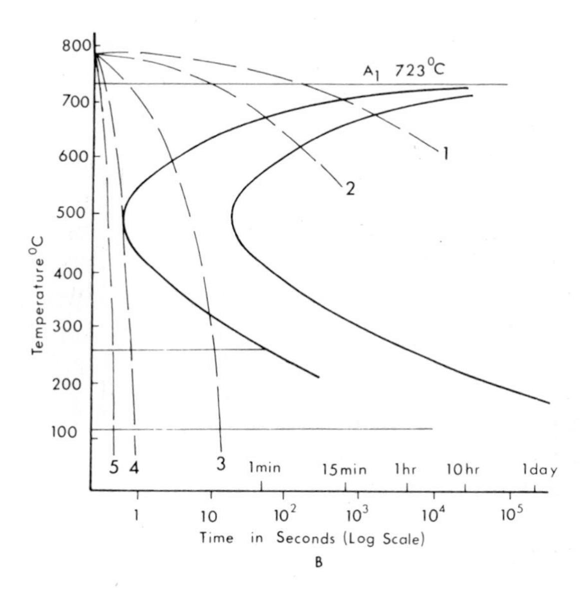

Figure 12.8 *(A) the time-temperature-transformation curve for a steel of eutectoid composition. (B) annealing (1); normalising (2); a split transformation (3); critical cooling velocity (4); hardening (5); shown on the TTT curve for a steel of eutectoid composition.*

(1) If a specimen of this grade of steel follows the cooling curve labelled "full annealing" the transformation from austenite to pearlite will begin at about 700°C and will end at 670°C, the time taken for the transformation having been about 45 minutes. The final structure will be relatively coarse pearlite.

(2) If a similar specimen follows the cooling curve labelled "normalising", it will not start to transform until about 670°C, and it will finish its transformation at about 620°C after a time interval of only about 9 minutes.

(3) If a specimen is to be fully hardened its cooling curve must pass to the left of the "nose" of the left-hand curve, and its transformation to martensite would begin at about 260°C and would end at about 110°C, the whole transformation taking only a fraction of a second. The critical cooling velocity line, which just touches the "nose" of the S-curve indicates the slowest possible cooling rate that will allow a fully martensitic structure to be formed.

(4) If the cooling curve passes through the "nose" area of the S-curve a split transformation will probably occur. As the steel cools through the "nose" area very fine pearlite (troostitic) will be formed, and because of the short time interval involved, some unchanged austenite may remain in the steel. This austenite will transform to martensite when the steel cools through the Ms-Mf zone; thus the final structure will consist of troostite and martensite.

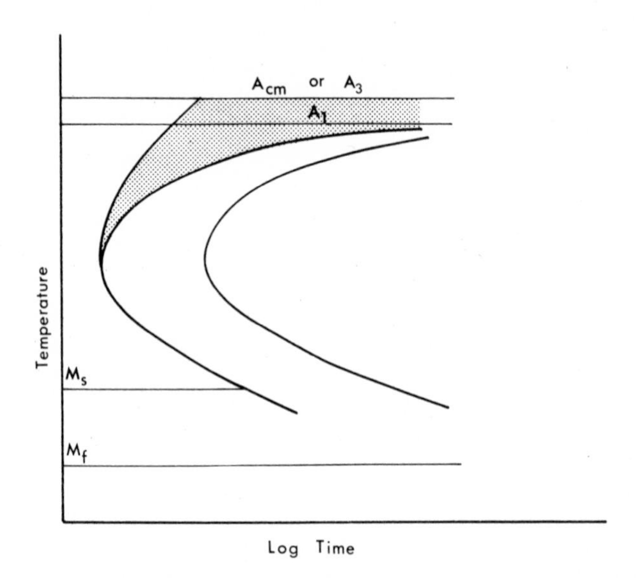

Figure 12.9 *A time-temperature-transformation curve typical of those for plain carbon steels of non-eutectoid composition. The area in which the pro-eutectoid transformation occurs is shaded.*

If S-curves are drawn for non-eutectoid steels another curve makes its appearance; this marks the start of the pro-eutectoid transformation. Such a curve is shown in Figure 12.9, the pro-eutectoid area being shaded. If the steel is hypo-eutectoid, ferrite will begin to precipitate out as the cooling curve crosses this top curve; whereas carbide will precipitate under similar conditions if the steel is hyper-eutectoid.

Under some circumstances, the Mf temperature of a steel can be below the temperature of the quenching medium. Therefore, some of the austenite does not transform to martensite, and is known as *retained austenite*. If no austenite is to be retained, the temperature of the quenching bath must be lowered, and for those steels having Mf temperatures below room temperature, icewater can be used as the quenching medium.

THE MASS EFFECT

The austenite-to-martensite transformation is accompanied by an appreciable expansion due to the need for the carbon atoms to locate themselves interstitially in the distorted alpha-iron lattice. However, this expansion is in direct opposition to the normal contraction resulting from cooling, and since it is almost impossible to avoid some variations in the rate of cooling of different portions of a large article, one section of the article may be expanding while another is contracting. This may result in serious distortion or cracking, rendering the article useless. The greater the mass of the article the more pronounced such differences in cooling rates, and the more likely it is that quench cracking will occur.

Thus, the normal method of quench-hardening is not always advisable so that alternative methods must be found. Two such methods are known as Austempering and Martempering, the former depending upon isothermal transformation for its effectiveness.

Martempering

The part to be hardened is austenitised and then quenched into a lead or salt bath which is being held at a temperature just above that at which martensite would begin to form. It is kept in this bath until its temperature becomes more or less uniform throughout, and is then water-quenched forming a fully martensitic structure. This process successfully separates the cooling contraction from the austenite-martensite expansion, and thus prevents quench-cracking in large articles.

Austempering

Again, the component to be hardened must be austenitised and then quenched into a lead or salt bath held at just above the martensite transformation temperature. However, in this case the component is left in the bath until the bainite transformation is completed, and is then removed and cooled in air to room temperature. The bainite so produced is somewhat softer than martensite of similar carbon content, but has increased shock-resistance.

Bainite is a structure in which the carbide particles are finely dispersed within a ferritic matrix, and is similar to tempered martensite (see the following section "Tempering—The Pearlite Reaction"). The microstructure of bainite varies with the temperature of transformation, higher temperatures producing larger carbide particles and hence softer, more ductile steels. Bainite cannot be formed by continuous cooling.

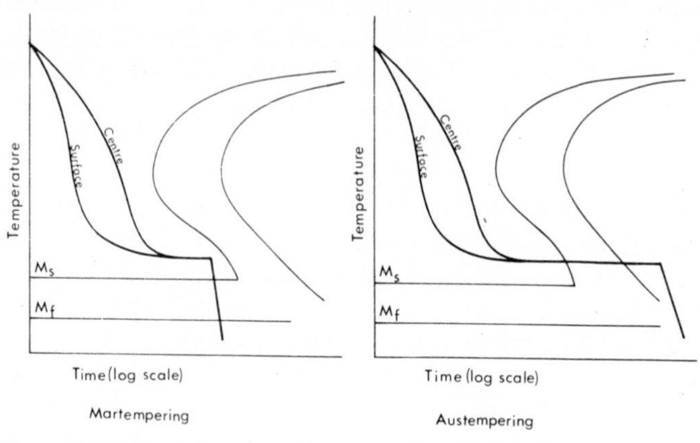

Figure 12.10 *Martempering and austempering shown on the TTT curve.*

TEMPERING—THE PEARLITE REACTION

A fully hardened steel contains internal stresses and is extremely hard and brittle; in this condition even mild shock loads would cause failure and hence the steel must be toughened prior to its use. Tempering is the general name given to that group of heat treatment processes which remove internal stresses from martensitic structures and, while retaining most of their hardness, replace brittleness with toughness. All tempering processes involve reheating the martensitic steel to some temperature well below the lower critical temperature, soaking to remove internal stresses and to allow all structural changes to go to equilibrium, followed by slow cooling to room temperature. Since martensite is itself only a metastable phase, structural alterations induced by tempering proceed fairly rapidly. All structures resulting from tempering processes are referred to as *tempered martensite*.

The following general structural changes occur when medium carbon steels are tempered in the stated temperature ranges for periods of about one hour:

(a) 100°–220°C: very little alteration occurs to the microstructure but considerable stress-relieving occurs; this range is used if maximum hardness is required and brittleness is not a real problem.

(b) 240°–400°C: the martensite decomposes fairly rapidly to an emulsified form of pearlite known as secondary troostite, which is similar to the primary troostite formed by direct quenching. The troostite structure is too fine to be resolved under the microscope and consequently etches black. Most fine-edged tools are tempered in this range.

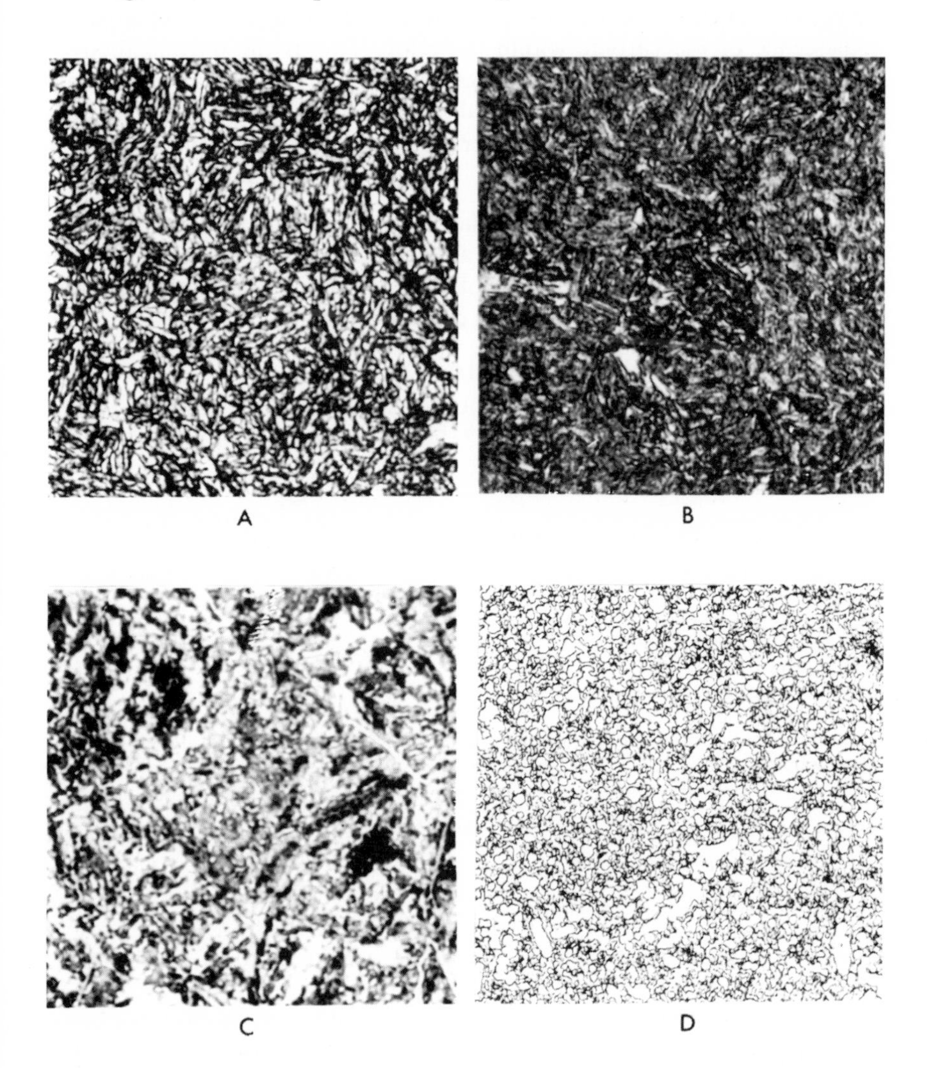

Figure 12.11 *Tempered structures in steels: (A) shows tempered martensite in a 0.35% carbon steel tempered for one hour at 500°C; (B) shows tempered martensite in a 0.73% carbon steel tempered for one hour at 500°C; (C) shows tempered martensite in a 1.3% carbon steel tempered at 500°C for one hour; (D) shows a spheroidised structure in a tool steel. Magnifications x500. (Photographs courtesy of the Broken Hill Proprietary Co. Ltd.)*

(c) 400°–550°C: the precipitated troostite begins to coalesce, forming a coarser form of globular pearlite known as sorbite.* Sorbite appears grey under the microscope and is desirable in such components as beams, springs and axles.
(d) 600°–700°C: spheroidisation occurs if the steel is heated to within this range; the structure is known as spheroidite, and forms because of a further coalescence of the carbide within the alloy. This is the only method of softening high carbon steels and high alloy steels that are air-hardening. Spheroidised steels exhibit fairly good machinability since the hard carbide particles are embedded in the soft ferrite matrix, and consequently do not have to be cut by the cutting tool.

If the spheroidised steel is now heated to just above its lower critical temperature the pearlite present will alter to austenite, and cooling to room temperature will yield a structure of lamellar pearlite plus pro-eutectoid ferrite or cementite, depending upon carbon content.

TABLE 12.2

TEMPERING TEMPERATURES AND TEMPER COLOURS FOR COMMON STEEL PRODUCTS

	Tempering Temperature °C	*Temper Colours*
Springs	300	Dark blue
Screwdrivers	295	Blue
Cold chisels for irons	290	Violet
Planing cutters for soft wood	285	
Axes, gimlets	275	Purple
Dental and surgical instruments	270	
Twist drills for wood	265	Red-brown
Plane irons, stone-cutting tools, moulding cutters	260	
Dies and punches	255	Yellow-brown
Taps, chasers, penknives, picks	250	
Rockdrills, screw-cutting dies, reamers,	245	
Shear blades, bone-cutting tools	240	Golden-brown
Wood engraving tools, paper cutters	235	
Hammer faces, planers for steel	230	Straw
Hacksaws, steel-engraving tools, circular saws for steel cutting	225	
Lathe tools for non-ferrous metals	220	Light straw

It is apparent that tempering causes the alteration of the unstable body-centred tetragonal lattice of martensite to the stable body-centred cubic lattice of ferrite, the visible changes in microstructure indicating this by the change to a structure in which the carbide exists as spheroids, the sizes of

*The terms "sorbite" and "troostite" are now not generally used by metallurgists who prefer to consider all tempered structures as various forms of tempered martensite. Refer to Figure 12.11.

TABLE 12.3

VARIATIONS IN MECHANICAL PROPERTIES OF FERROUS METALS CAUSED BY MECHANICAL WORKING AND HEAT TREATMENT

Material and Treatment	*Tensile Strength (ksi)*		*Compressive Yield Strength (ksi)*	*Tensile Modulus of Elasticity Psi × 10^6*	*% Elongation 2″ g.l.*	*Brinell Hardness*
	Yield	*Ultimate*				
Ingot iron annealed 0.02% C	24	42	21	30	45	70
Steel 0.2% C						
Hot rolled	40	60	40	30	35	120
Cold rolled	60	80	60	30	15	160
Annealed castings	35	60	35	30	25	130
Steel 0.4% C						
Hot rolled	42	70	42	30	25	135
Heat treated for fine grain	60	90	60	30	25	190
Annealed castings	35	65	35	30	15	130
Steel 0.6% C						
Hot rolled	63	100	63	30	15	200
Heat treated for fine grain	78	120	78	30	15	235
Steel 0.8% C						
Hot rolled	73	120	73	30	10	240
Oil quenched	125	180	125	30	2	360
Steel 1% C						
Hot rolled	83	135	83	30	10	260
Oil quenched	140	220	140	30	1	430
Steel 3.5% Ni 0.4% C						
Heat treated for good machinability	150	170	150	30	12	350

which increase with increasing tempering temperatures and soaking times.

Figure 12.11 summarises the changes occurring in eutectoid steels subjected to various types of heat treatment.

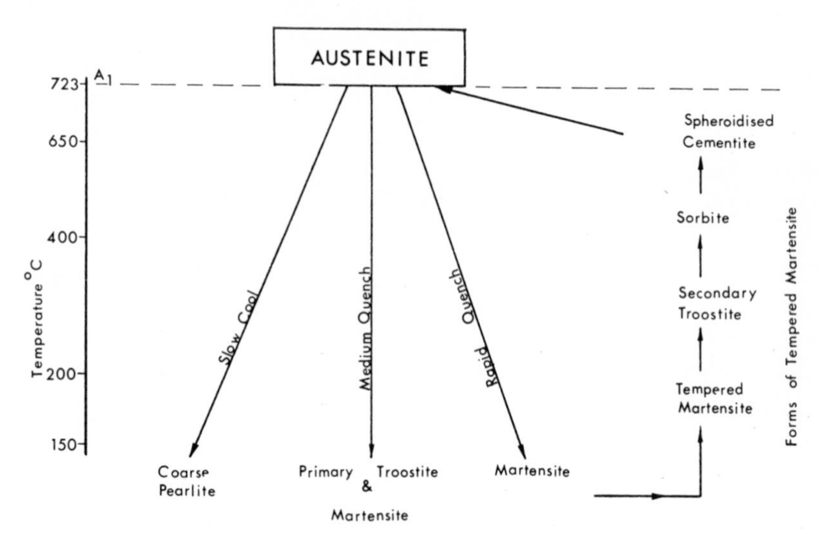

Figure 12.12 *Changes in the structure of eutectoid steel caused by slow cooling, quenching, and tempering.*

Laboratory (39): *(a) Take three pieces of mild steel. Polish and etch one as a control. Heat the other two to red heat and quench in cold water. Now temper one of these hardened pieces at 500°C for one hour. Polish and etch these heat treated pieces and examine all three together.*
(b) Repeat this experiment with three pieces of carbon tool steel (about 1% carbon).
Sketch and describe the microstructures of the six specimens.

THE SURFACE HARDENING OF STEEL

Very often it is desirable for the surface of a steel component to be hard while its core or centre portion is relatively soft; this gives a combination of good wear resistance on the surface combined with good shock resistance. In general, surface hardening is divided into the three basic procedures of carburising, nitriding, and selective hardening.

Carburising

The basic aim is to impregnate the surface of low carbon steel with sufficient carbon to raise its composition to the eutectoid level. The steel commonly used has a carbon content of 0.15% or less, this low figure preventing any possibility of embrittlement in the core after subsequent heat treatment.

Examination of the iron-carbon equilibrium diagram (Figure 11.7) shows that, while ferrite has a maximum solid solubility for carbon of 0.025% at 723°C, austenite can dissolve up to 2% of carbon at 1135°C. Thus, the absorption of carbon will be best achieved if the steel is in the austenitic condition, and, in practice it has been found that a carburising temperature of 900°–950°C produces quite an appreciable layer of carburised steel in 3–6 hours.

The basis of the case carburisation process lies in the dissociation of carbon monoxide into carbon dioxide and carbon under favourable conditions:

$$2CO \rightleftharpoons CO_2 + C$$

The carbon monoxide necessary for this reaction, itself formed by the combustion or breakdown of the carburising agent being used, diffuses into the surface of the steel where it breaks down and the carbon is taken into solid solution in the austenite.

Various materials are used as carburising agents and are classified as solid, liquid, or gaseous.

(a) Using Solid Reagents: the articles to be hardened are packed into mild steel boxes containing a carbon-rich substance such as charcoal, leather, or petroleum coke together with a promoter which assists in the absorption of carbon into the steel. The boxes are then placed in a furnace and soaked at about 900°C for several hours, the length of time varying with the desired depth of case. Portions which are not to be hardened are coated with a mixture of asbestos in fireclay or may be copper plated.

(b) Using Liquid Baths: the articles are suspended in baths of molten sodium cyanide plus sodium carbonate for sufficient time to allow a deep enough case to be developed. The cyanide decomposes to form carbon. Advantages include ease of operation and rapid carburisation. Direct quenching from the bath is possible, thus eliminating the need for further heat treatment.

(c) Using Gaseous Reagents: town gas may be used to produce case depths of up to 0.15″; however, the treatment is not often used even though it does prevent the formation of undesirable salts on the surface of the steel.

After carburising some form of heat treatment is necessary in order to develop martensite on the high-carbon case. *The Double-Quench Method* is often used to produce both extreme hardness in the case and good ductility in the core. The process involves:

(1) heating the metal to just above its upper critical temperature, soaking to fully austenitise the steel, then quenching in oil to retain the fine-grained core.

(2) reheating the metal to just above its lower critical temperature, soaking and requenching. This causes the pearlite in the case to transform to martensite but does not materially alter the fine-grained core.

Laboratory (40): *Take two pieces of 22 gauge mild steel wire about 2″ long. Bend one in your fingers to assess its ductility. Heat the other piece in a bath of sodium cyanide solution for half an hour, water quench, and repeat this rudimentary ductility test. Mount small pieces of each length of wire in a cold setting resin, polish, etch and examine each piece microscopically. Relate the change in ductility to the differences in the microstructures.*

Note: Sodium cyanide is a dangerous poison, and must be handled very carefully.

Figure 12.13 *Macro-photograph of a case hardened knife-edge, original size ½″ across section. The outer surface zone is slightly hyper-eutectoid. (Original specimen from Avery Australia Pty. Ltd.)*

Nitriding

Iron, aluminium, and chromium form nitrides of high hardness. However, as the nitrides of iron are exceptionally brittle, plain carbon steels are seldom nitrided. Special "nitralloy" steels, consisting of 1% aluminium, 1.5% chromium and 0.2% molybdenum in a low carbon steel have been developed; if such a steel is heated at 500°C for 40 to 90 hours in a gas-tight box in which ammonia is circulated, its surface will become very hard due to nitride formation while its core will remain unaffected. Quenching is not necessary after nitriding, and, thus, nitrided components are usually free from distortion or cracking and retain a high surface finish.

Selective Hardening

By using either induction heating or a high intensity oxy-acetylene flame, it is possible to bring the surface of a piece of steel into the austenitic range while its core remains below the critical range. If the surface is then quenched rapidly, usually by a water jet following the flame or induction hardening coil, the surface will become martensitic while the core remains soft. In both cases a medium or high carbon steel must be used. However, surface finish is excellent and the component remains distortion-free.

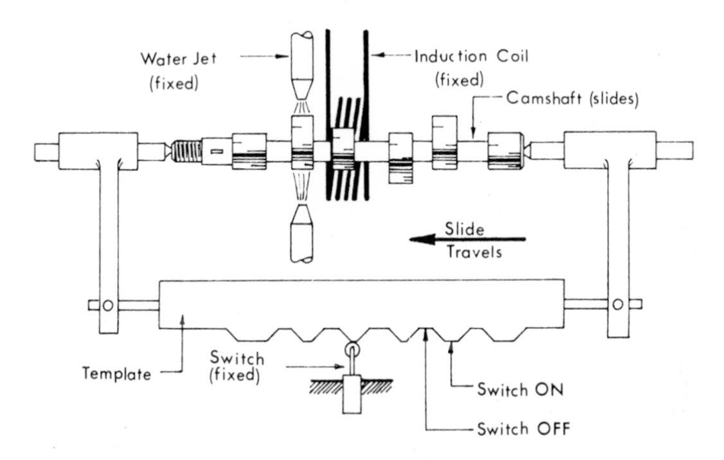

Figure 12.14 *The automatic induction hardening of a camshaft.*

While induction hardening is usually applied to small components, such as axles or camshafts, flame hardening is usually applied to sections of larger objects, such as the teeth of large gear wheels or the "ways" of large machine beds.

GLOSSARY OF TERMS

Air-hardening steel: alloy steel that transforms to martensite when air cooled from above its upper critical temperature.

Annealing: heating and cooling a metal to produce the softest possible state.

Austempering: the isothermal transformation of steel from austenite to bainite.

Austenitic steel: high alloy steels that retain their austenitic structures at room temperature due to the stabilising effects of the alloying elements.

Bainite: a structure in steels formed by isothermal treatment in which carbide particles are finely dispersed in a ferrite matrix.

Carburising: a process in which the carbon content of the surface of a piece of low carbon steel is raised to eutectoid or slightly hyper-eutectoid composition so that it can be hardened by quenching.

Decarburisation: the removal of carbon from the surface of a piece of steel or iron caused by heating it in an oxidising atmsophere.

Heat treatment: the controlled heating and cooling of materials in order to alter their properties.
Induction hardening: a hardening process in which the steel is heated by induced eddy currents caused by placing it in an induction coil.
Martempering: an interrupted quenching technique used to prevent quench cracking in steels that are quenched to martensite.
Martensite: a metastable saturated solid solution of carbon in body centred tetragonal iron formed by quenching from within or above the critical range at a rate that exceeds the critical cooling velocity; martensite is the hardest form of steel known.
Nitriding: a surface hardening technique applied to steels containing *Ni*, *Cr* and *Mo* which causes the formation of hard nitrides on the surface of the steel.
Normalising: a softening process in which steel is austenitised and then air-cooled.
Spheroidisation: (spheroidisation annealing): heating a martensite high carbon steel to a temperature below its LCT so that the structure alters to a dispersion of coarse carbide particles in a ferrite matrix.
TTT curve: an isothermal curve; a graph showing the relationships between phases in a particular steel, time, and temperature.
Tempering: heating a martensitic steel to some temperature well below its LCT so that stress relieving can occur; this removes brittleness, increases ductility, and reduces hardness.
Tempered martensite: the end product of a tempering process as applied to a martensitic steel; the older terms "sorbite" and "troostite" were used to refer to particular forms of tempered martensite as seen under the microscope.

REVIEW QUESTIONS

1. Define the term "heat treatment", and list the three main stages of any heat treatment procedure.
2. How is a piece of steel austenitised?
3. Briefly outline the martensite reaction in steels, and discuss the effects of carbon content on the properties of martensite.
4. Samples of low carbon, medium carbon, and high carbon (plain) steels are to be quench hardened. State which quenching media would be used for each type of steel, giving reasons for your choice.
5. Distinguish between the hardness and the hardenability of a steel, and briefly outline the general effects of alloying elements on hardenability.
6. Explain the meanings of the following terms: annealing; process annealing; normalising; tempering.
7. Distinguish between quench-softening steels and air-hardening steels.

8. Why is the mass effect a problem when large components are to be hardened? How can its adverse effects be prevented?
9. Explain the meaning of the term "isothermal heat treatment" and briefly discuss one example of this type of treatment.
10. Briefly discuss the difference in structures and properties existing between martensite and bainite.
11. What are the general effects of tempering martensitic steels?
12. Two pieces of 0.6% carbon steel are heated to 30°C above their upper critical temperature and are then quenched to room temperature at a rate exceeding the critical cooling velocity for this grade of steel. One piece is reheated to 450°C, soaked for one hour, and then slowly cooled to room temperature; the other is reheated to 680°C, soaked for two hours, and cooled to room temperature. Sketch and describe the final structures of both pieces of steel and list their probable mechanical properties.
13. A piece of high carbon steel is examined microscopically and found to have a structure in which carbide particles are dispersed within ferrite. Outline the most economical heat treatment by which a pearlite matrix could be developed in the steel.
14. Give two reasons why a piston pin should be made from case hardened low carbon steel and not from fully hardened medium carbon steel.
15. A 1″ diameter piece of 0.2% carbon steel is packed into an airtight box with charcoal and the box is heated to 950°C and soaked for 5 hours. The box is cooled and the steel is removed, reheated to 920°C, and is quenched in water. Sketch and describe the structures of the outer portion and of the core of the steel (i) prior to any heat treatment; (ii) after the initial heating and cooling; (iii) after reheating to 920°C and quenching.
16. List and discuss the advantages of nitriding compared to a normal carburising process.
17. What is retained austenite? How could the quenching process be altered to prevent the presence of retained austenite?

13

The Joining of Metals

METALS may be joined in a variety of ways, some of which disturb the structure of metal to a considerable extent, while others create only minimal disturbance. Bolts, studs, machine screws, and rivets fall into the latter class since the only disturbance to the grain structure occurs during drilling and tapping operations; however soldering, brazing, and welding are of more interest to the metallurgist. In the following discussion it must be borne in mind that the best possible joint between two pieces of metal would be one having complete continuity of grain structure throughout. This in practice is almost impossible to achieve, particularly when two dissimilar metals are being joined.

The main techniques for permanently joining metals together are: (1) soft soldering, (2) brazing and silver soldering, and (3) welding.

SOFT SOLDERING

If two pieces of metal are to be soft soldered together they must first of all be thoroughly cleaned of all scale and dirt and then a suitable flux brushed along the joint. The pieces are then heated and some molten solder is run into the joint by one means or another. In the case of lead-tin solders a copper soldering bit is usually used to heat the work, to melt the solder, and then to apply it to the joint. However, solder may also be applied using a heating torch such as those run by propane or other L.P. gases. All soft solders have the following general properties:

(1) they have melting points considerably lower than those of the metals being joined,
(2) they readily "wet" the surfaces of the joint and then flow freely into the joint by capillary action,
(3) they form a sound and well-attached film within the joint, and,
(4) they have *adequate* mechanical strengths.

The structure of the soldered joint will vary depending upon the type of metal being soldered and the composition of the solder itself. However most soldered joints reveal some surface alloying between the film of solder and the metal being joined, the actual amount of such surface alloying depending upon the solubilities of the metals involved. If surface alloying results in a gradual change in structure and composition from the parent metal into the solder film and then back into the parent metal again the joint will be a strong one.

While the lead-tin solders are most commonly used for soft soldering, lead-antimony solders are also in use; such "antimonial" solders are less expensive than the lead-tin types but cannot be used on galvanised iron.

FUNCTIONS OF FLUXES

Continuity of grain structure across a soldered or welded joint can only be obtained if the metals are brought into atomic contact, and this is not possible if the metals are coated with oxide layers, grease, corrosion products, or other surface films. Mechanical cleaning can only remove the bulk of such surface films, some form of chemical cleaning being necessary to complete the cleaning operation. Fluxes perform some or all of the following functions:

(1) they chemically clean the surfaces to be joined,

(2) they prevent the formation of new oxide layers during the heating cycle of the joining process,

(3) they assist the filler metal to run freely into the joint, and,

(4) they assist the "wetting" process by which surface alloying occurs.

TABLE 13.1

FLUXES FOR SOFT SOLDERING

Flux	*Suitable Metals*	*Action*
"Killed Spirits" Zinc Chloride $ZnCl_2$	Iron, tinplate, copper, brass, bronze	Chemically cleanses surface and prevents further oxide formation. Causes corrosion if not thoroughly washed away
"Spirits of Salts" dilute hydrochloric acid HCl	Zinc, galvanised iron	Cleanses the zinc and forms $ZnCl_2$ by chemical reaction. Must be washed away
Sal Ammoniac Ammonium Chloride NH_4Cl_2	Copper, iron	Has a cleansing action but not as good a flux as $ZnCl_2$
Resin	All metals	Non-corrosive, thus used for all electrical work
Tallow	Lead and lead-rich alloys	Has a protective action only, and does not cleanse the metal

BRAZING

Brazing is also known as "hard soldering" since it is done with solders of considerably higher melting points than those of the soft solders, and also because the joints formed are harder and stronger than soft soldered joints. The solder, or spelter as it is commonly known, is melted directly into the joint using a propane or oxy-acetylene flame.

Brazing alloys are classified as:

(1) *silver solders*: commonly silver-copper-zinc alloys with or without cadmium that melt in the range 600°–800°C.

(2) *brazing brasses (or spelters)*: these are copper-zinc alloys varying from about 50–60% copper and may contain up to 2% tin. They melt in the range 850°–900°C.

(3) *phosphorus-bearing brazing alloys*: these are usually self-fluxing alloys that usually contain at least 75% copper and between 4–8% phosphorus.

TABLE 13.2

CHARACTERISTICS AND USES OF SOME BRAZING ALLOYS

Type	*Composition*						*Characteristics and Uses*
	Ag	*Cu*	*Zn*	*Cd*	*P*	*Sn*	
Brazing Brass		50	50				Melts at about 865°C and forms a very hard joint
Brazing Brass		60	39			1	Melts at 890°C and runs more freely due to the tin content
Silver Solder ("Eazy-Flo")	50	15	16	19			Melts at about 640°C; it is very free-flowing and freezes rapidly since it is almost eutectic in composition
Silver Solder	42	38	20				Melts at 770°C and is stronger than "Eazy-Flo"
Phosphorus-bearing alloy ("Silfos")	15	80			5		Self-fluxing due to the formation of a fluid compound upon oxidation; melts at about 630°C; must not be used upon ferrous or nickel-base alloys since the solder alloys with the base metal forming brittle compounds that weaken the joint

For those solders melting between 600–750°C a fluoride-type flux is required while those melting above 750°C require a borax-base flux. Borax, an excellent flux for brazing all types of metal, cannot be used below 750°C because it remains too viscous and does not clean and protect the metal during the heating cycle.

The metallurgical aspects of all types of brazing are similar in some respects to those of soft soldering. For instance, surface alloying effects between the base metal and the brazing alloy are similar to those in a sound soft soldered joint. However, the higher temperatures used mean that structural changes can quite readily occur in the base metal itself. Copper and copper alloys, for instance, are annealed during brazing operations, and considerable grain growth may occur in the vicinity of the weld itself.

Laboratory (41): *Take four pieces of brass strip about ½″ wide by 1″ long and join 2 pieces with a soft solder and the other two with silver solder. Take two similar-sized pieces of mild steel and braze them together. Cut sections across each joint, mount carefully in a suitable cold-setting plastic, polish, etch, and microscopically examine each in turn. List the significant differences between each type of joint.*

WELDING

Welding is carried out at much higher temperatures than brazing and thus the metallurgical aspects are much more complex and diversified. Looked at in one way all welding operations must be seen as heat treatment processes in which metals are permanently joined together either with or without the application of additional filler or weld metal. All welding operations are classified as either:

(1) *Pressure welding processes*: in which the metals, although usually heated, do not melt, welding being effected mainly by pressure. No additional filler metal is used in pressure welding processes;
(2) *Fusion welding processes*: in which the metals to be joined are locally melted and the extra metal required to fill the joint is added in the form of filler wire or consumable flux-coated electrode. No pressure is used in these welding processes.

Figure 13.1 illustrates the common welding processes used in industry.

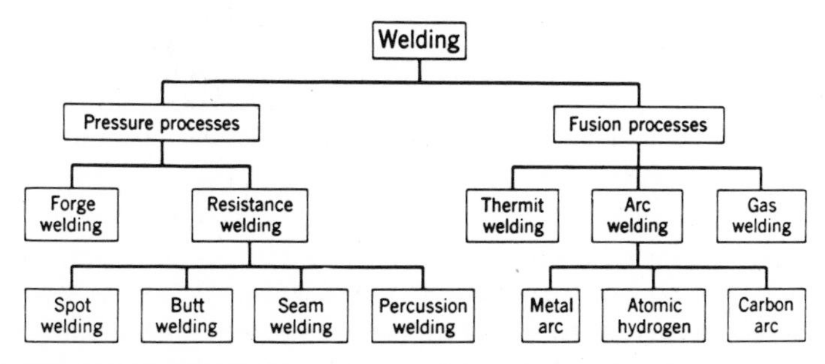

Figure 13.1 *Classification of common welding processes. (Reproduced with permission from "Textbook of Engineering Materials" by M. Nord. John Wiley and Sons Inc., N.Y. 1952.)*

Forge Welding

Forge welding is part of the traditional art of the blacksmith and has little place in modern industry, although some forge welding is still done on large components using power hammers instead of hand-driven hammers. The steel pieces to be welded must be heated to just below their melting point, descaled by hammering, fluxed with fine sharp sand, then brought together and hammered so that the joint is completely closed. Any slag formed during the welding process is hammered out of the joint, and continued hammering promotes recrystallisation which strengthens the joint. Unless done very carefully, the heating will damage the parent metal by "burning" it, and so forge-welded articles are often not very strong in the parent metal adjoining the weld itself.

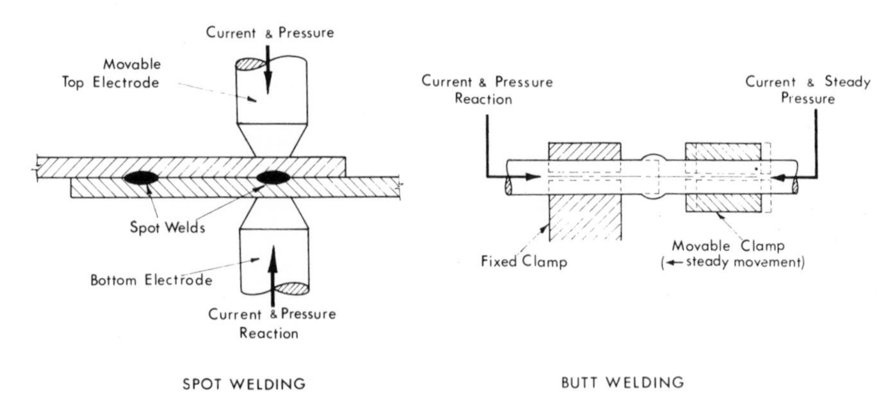

Figure 13.2 *The principles of spot welding and flash butt welding.*

Resistance Welding

In resistance welding processes the heat necessary to cause welding is generated by a low-voltage high-amperage current which is passed through the components to be welded together. No additional filler metal is added, no flux is required, but considerable pressure is exerted in the area of the weld by means of the electrodes used. Most metals can be resistance-welded; however, unlike metals to be arc-welded, they must be clean and free from scale or rust, the presence of which would seriously weaken any resistance weld. The following types of resistance welding techniques are common:

(a) Spot Welding: This is used to weld sheet and thin plate material together when a gas-tight joint is not required. The pieces to be welded together are pressed between two water-cooled copper electrodes which pass the welding current through the sheets. The current melts the metal in the area of the electrodes and thus a good strong weld is effected. Obviously many spot welds would be needed if a fairly long seam had to be effected.

(b) Seam Welding: If the single copper electrodes used in spot welding are replaced by rotating wheel electrodes then a seam weld can be effected. The materials to be welded are clamped between the electrodes, the welding current is turned on, but this time the current is left on while the materials are pushed along in between the electrodes, a continuous seam weld resulting from this operation.

(c) Butt Welding: The most common process is flash-butt welding which may be used to join rods of similar cross-section together. The pieces to be welded are clamped in the welding machine and their ends are brought close together. The welding current is turned on and an arc is generated between the pieces which are, in effect, the electrodes. This arc causes localised melting on the ends of the pieces to be welded, which are forced together once they reach welding heat. The weld is strong and the slightly increased section of the weld can be machined away if necessary. This process is used in the manufacture of high performance engine valves which have a wear-resisting stem flash-butt welded to a heat-resisting head section.

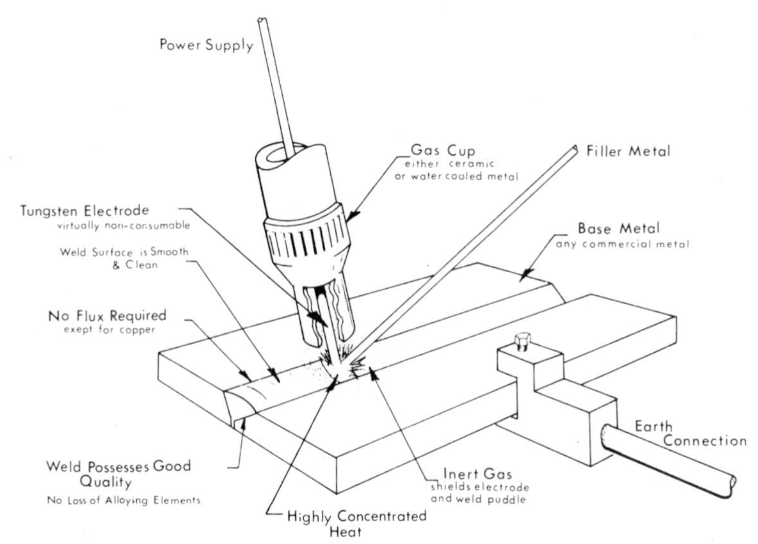

Figure 13.3 *Shielded inert gas metal arc (SIGMA) welding.*

Gas Welding

In gas welding the edges of the material to be welded are heated by some sort of blowpipe and the filler metal is run into the heated area where it melts and mingles with the metal being welded. The most common gas used is acetylene, which is stored in high pressure cylinders where it is dissolved in acetone to minimise the danger of an explosion. Oxy-acetylene welding may be used on sheet steel, cast iron, and on most non-ferrous metals. However, when welds between cast iron or dissimilar metal pieces are effected, a bronze filler rod is used and the process is really one of brazing.

Electric-Arc Welding

Electric-arc welding uses the heat generated by an electric arc to effect the weld. The arc is usually struck between a flux-coated consumable metal electrode and the piece to be welded (metal-arc welding), between two carbon electrodes and the object to be welded (carbon-arc welding), or between a tungsten electrode and the parent metal (shielded-arc processes). If carbon-arc or shielded-arc processes are being used the filler metal is supplied in the form of a continuous consumable electrode (wire of suitable composition). In the case of the carbon-arc process this wire is flux-coated; however, in the shielded-arc process the weld is protected from oxidation by the presence of the inert gas argon which is blown around the joint during welding.

METALLURGICAL ASPECTS OF FUSION WELDING

Every fusion-welding process involves the following four steps:

(1) the heating (including preheating if necessary) of the parent metal,
(2) the correct manipulation of the torch flame or electrode to deposit the weld metal,
(3) the cooling of the weld,
(4) in some cases, the reheating of the weldment in order to effect stress-relieving within the weld itself or the parent metal or both.

Microscopic examination of a polished and etched section through a fusion weld reveals the existence of the following zones or areas:

(1) the unaffected parent metal, either side of the weld,
(2) the weld deposit itself,
(3) the "heat-affected" zone, existing along the boundaries of the weld deposit, being the areas in which the heat and alloying effects of the welding operation have caused changes in the structure of the parent metal.

The Weld Metal

The weld deposit is, in actual fact, a miniature casting which has cooled quite rapidly from a high temperature; it may be likened to a small casting made in a warm metal mould. The actual crystalline structure present depends upon a number of factors, one of the most important being the number of "runs" made to deposit the weld metal.

If a single-run weld is examined, long columnar crystals will be seen growing outwards from the sides of the weld. If the welding temperature is too high these columnar crystals will meet at the centre of the weld deposit forming a plane of weakness, this tending to cause inter-crystalline cracking within the weld. However, if the welding temperature is correct, equi-axed grains will form at the centre of the weld before the columnar crystals can meet; this results, of course, in an appreciably stronger joint.

A multi-run weld joining two pieces of steel exhibits quite a different structure. The first run exhibits the typical structure as described for a single-run weld. The second run normalises the first, causing a considerable degree of grain refinement and the destruction of the original columnar structure. Each successive run thus normalises the preceding run, so that only the final deposit (*i.e.* the top run) exhibits the coarse cast structure typical of a single-run weld. However, the possibility of defects such as slag and gas inclusions occurring is increased in a multi-run weld.

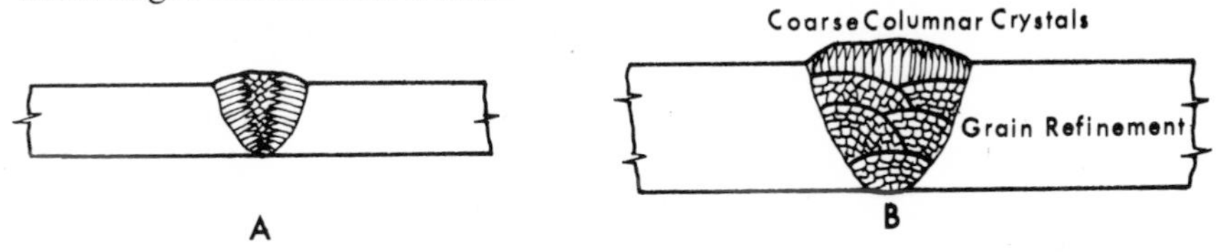

Figure 13.4 *Typical structures of (A) a single-run arc weld and (B) a multi-run arc weld.*

Non-metallic inclusions within the weld metal are usually avoided by the use of a suitable flux, or by shielding the weld with an inert gas such as argon, as in the argon-arc process. Both of these agents prevent the formation of oxide and nitride inclusions within the weld. Modern flux coated electric-arc electrodes usually provide good quality weld deposits substantially free from harmful inclusions. The flux itself forms a slag, and care must be taken to remove this each time a run is laid down during a multi-run weld, otherwise slag inclusions will occur in the weld metal. Slag removal is readily accomplished by chipping.

Gas porosity is caused by gases escaping, or attempting to escape, from the weld metal as it solidifies and cools. The solubility of hydrogen, the gas principally responsible for gas porosity in welds in ferrous metals, decreases sharply as temperature decreases. Hydrogen may be dissolved in the molten weld metal in large quantities, being generated in the welding flame used in gas welding, or by the flux coatings of many electrodes. Hydrogen dissolved in this way may react with oxides present in the metal to form steam, which may increase gas porosity in the weld. This is the case when tough-pitch copper is welded. If the metal cools before all of the dissolved gases reach the surface, small areas of trapped gases, known as *gas pockets*, will occur in the weld metal, further weakening the welded joint. The small holes and craters left on the surface of the weld by escaping gases are called *blowholes*; these do not appreciably weaken the weld.

Weld metal cracking often occurs in a welded joint that is held under restraint due to the contractional strains set up in the metal as it cools to room temperature. This "hot cracking" is usually intercrystalline in character. *Crater cracks* often form down the centre of an electric-arc weld deposit due to hot shrinkage within the weld metal; this form of cracking can be prevented by correct manipulation of the electrode during the welding operation.

The Heat-Affected Zone

The extent of any changes caused in the area of the parent metal which adjoins the actual weld deposit will depend on a large number of factors. Some of the more important are the dimensions of the article being welded, the thermal conductivity of the parent metal and its composition, and the length of time during which the metal is heated to above critical temperatures while welding is being carried out.

In general, non-ferrous metals and alloys are softened in the heat-affected zone, due to the annealing of the metal in this area. This localised softening is particularly apparent when cold rolled (work hardened) or age-hardened alloys are being welded. Grain growth will also occur in this zone, being most pronounced in the area adjoining the weld itself. This also weakens the metal to some extent, but is not often critical.

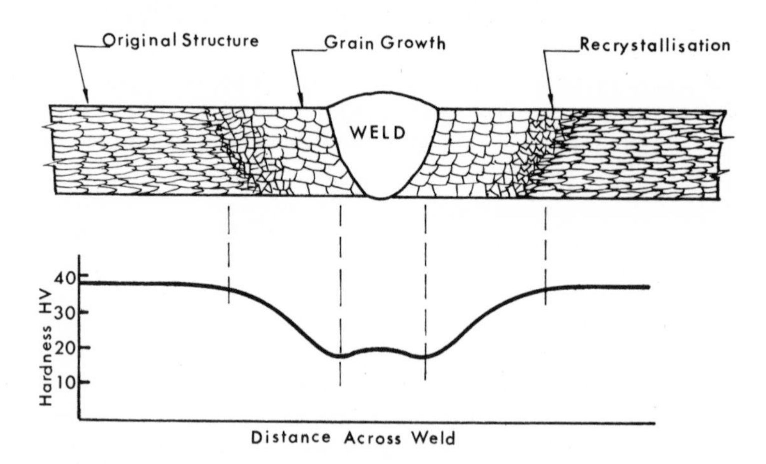

Figure 13.5 *The effects of welding on the structure and hardness of cold rolled aluminium plate; note the coarse grains in the heat-affected zone nearest the weld and the fine grains (recrystallised zone) further away from the weld.*

Mild steel is the easiest ferrous metal to weld, the typical heat-affected zone ranging from an overheated area near the weld metal to an under-annealed structure further away from the heat source (i.e. the weld itself). Medium and high carbon steels are more difficult to weld, and complex structures are formed within the heat-affected zone due to the rapid rate of cooling; this generally results in an increase in hardness across the weld. For this reason such steels are often cooled at a controlled rate, so that martensite formation is prevented.

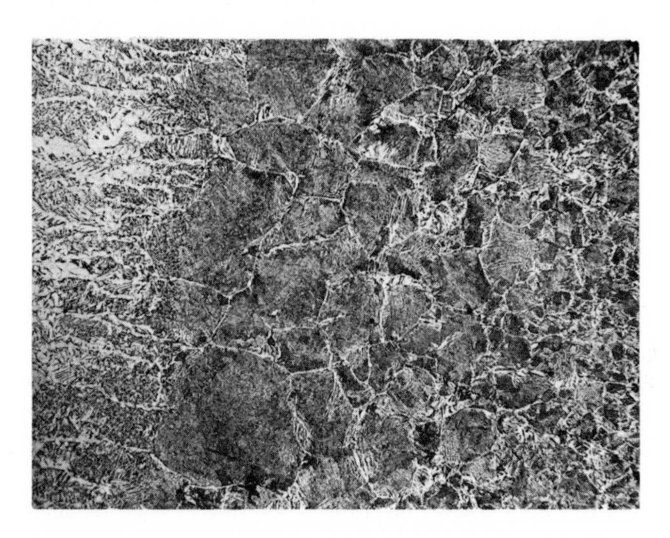

Figure 13.6 *A photomicrograph showing the heat-affected zone of an arc weld in mild steel. The long columnar grains of the weld metal are clearly seen adjoining the very large polygonal grains of the heat affected zone. (Photograph courtesy of the New South Wales Department of Railways.)*

Laboratory (42): *Arc weld two pieces of 3/8″ mild steel plate together using several "runs". Section the weld, polish, etch and examine microscopically.*

Hard-zone cracking may occur in many high carbon steels due to the formation of a martensitic structure in the heat-affected zone; this is very pronounced if the joint is prepared under restraint. It appears to be related to the presence of hydrogen in the weld; the hydrogen, being insoluble in the martensite, collects in microfissures within the metal and exerts sufficient pressure to cause the metal to crack through this brittle zone.

Preheating is often employed when welding high carbon or alloy steels that tend to form martensite when cooled fairly rapidly in air since preheating slows down the cooling rate of that area of the parent metal close to the weld itself. Many weld failures have occurred due to the failure to preheat the parent metal.

Figure 13.7 *(A) Poor weld penetration in arc welding (B) welded specimens broken in tension show the strength of the weld. (Photographs courtesy of the New South Wales Department of Railways.)*

GLOSSARY OF TERMS

Blowholes: small surface imperfections seen in welds caused by escaping gases that were dissolved in the weld metal.
Brazing: the joining of metals by running in molten non-ferrous "brazing" alloys that melt in the range, 800°–900°C.
Crater cracks: cracks formed down the centre of a weld because of hot shrinkage within the weld metal.
Flux: a substance used when soldering or welding to clean the surface of the metal, to assist the weld metal to flow into the joint, and to prevent oxide and scale formation during heating.
Fusion welding: the metals to be joined are locally melted and extra filler metal is added; no pressure is required.
Gas porosity: unsoundness in welds caused by dissolved gases such as hydrogen, nitrogen, and oxygen.
Hard-zone cracking: cracking in welds in high carbon steel due to martensite formation in the heat-affected zone.
Heat-affected zone: the area of the parent metal adjoining the weld in which structural changes have occurred because of the heat used in the welding operation.
Pressure welding: welding in which heat and pressure are used to cause fusion; generally, no filler metal is required.
Silver soldering: a form of brazing in which silver-copper-zinc alloys are used; silver solders have lower melting points and greater fluidity than normal brazing alloys.

REVIEW QUESTIONS

1. What are the essential requirements of an alloy for use as a soft solder? State the composition, uses, and phases present in (i) plumbers' solder (70%

lead, 30% tin), and (ii) tinman's solder (50% lead, 50% tin). The lead-tin equilibrium diagram is given in Figure 9.10.

2. What is silver soldering and how does it differ from a normal brazing operation?

3. List the principal functions of a flux in a soldering operation.

4. Discuss the essential differences between pressure welding and fusion welding processes.

5. Sketch and discuss the structures that would occur in the weld and in the heat-affected zone of (i) a single-run arc weld in ¼″ mild steel plate, and (ii) a multi-run arc weld in ½″ mild steel plate.

6. Two pieces of hard rolled aluminium sheet are welded together using a single run. Describe the resulting metallurgical structures that occur in the weld and in the heat-affected zone, and discuss the changes in mechanical properties that occur across the welded area.

7. What is hard zone cracking and how does it occur in steels?

14

Wood as an Engineering Material

WOOD is probably the oldest structural material and is organic in origins and structure; unlike most other engineering materials, it is harvested as a "crop" rather than produced as a raw material. The early part of this century saw the beginning of the gradual replacement of wood as a structural material by other manufactured materials. However, the realisation that wood offers many unique advantages over other materials has grown over the last few decades, and today wood and wood products occupy a most important place in engineering. The following are some of the important unique properties of wood that are of fundamental importance to the engineer.

(1) It is subject to shrinkage, swelling and warping due to alterations in moisture content.

(2) It is combustible, a property shared by virtually all other organic materials. However, wooden beams burn slowly and do not distort and soften as do structural steel girders.

(3) Its strength properties are quite different when measured in different directions; for instance, it is stronger in tension along the grain than across the grain.

(4) Different species (types) of wood vary considerably with respect to properties; *e.g.* balsa has a density of 6 lb/cu ft, while grey box has a density of 70.5 lb/cu ft.

(5) There is considerable variation of properties among individual members belonging to the same species.

All of these properties result from the basic structure of wood which in turn is determined by the life cycle and growth conditions of the tree from which the wood is cut.

THE STRUCTURE OF WOOD

If the cross-section of a tree is examined closely the following general features are clearly visible.

(a) The pith, which is the soft and often decayed centre portion, usually darker in colour than the rest of the wood, about which the rest of the tree has grown. The pith may be thought of as the original shoot of the tree, and it extends out into all the branches. It has no economic significance.

(b) The heartwood is the darker zone of wood surrounding the pith, it is of great economic importance since it is more resistant to decay and insect attack than the next layer which is called the sapwood. Heartwood is also known as truewood.

(c) The sapwood is the zone of lighter coloured wood cells surrounding the heartwood; in general, these cells are more open in structure since the processes of secondary thickening are not advanced in the sapwood. Sapwood is less dense than truewood for this reason, and is less resistant to decay and insect attack, but is generally as strong as the heartwood.

(d) The bark of a tree really consists of two fairly distinct layers, the inner bark and the outer bark. While the latter is purely a protective layer, the former is essential to the life cycle of the tree since food passes downwards from the leaves through the inner bark to the roots where osmosis occurs and the tree gains the minerals essential to its further growth.

(e) The cambium layer is a unicellular layer existing between the inner bark and the sapwood, and it is here that all growth in girth takes place. Although not normally visible on an ordinary cross-section, the cambium layer can be distinguished if the bark is pulled away since it exists as a "sticky" film on the outer portion of the sapwood. Continual division of the thin-walled cambium cells gives rise to growth in girth, division on the inner side forming new sapwood cells, and division on the outer side giving rise to fresh layers of bark.

(f) Growth rings occur in trees due to the irregular nature of growth in girth. In the colder months there is an almost complete cessation of growth; however, the onset of spring causes cellular growth to increase rapidly, the "springwood" so formed being lighter in weight and colour than the "summerwood", the growth of which proceeds at a much slower rate during the hotter and drier summer months. This means that following the zone of light, quickly-grown springwood there is a smaller zone of denser, darker summerwood. Repetition of this growth pattern year after year results in the formation of definite growth rings. In sub-tropical and tropical climates the growth rings of a tree are usually not prominent since growth occurs more or less regularly throughout the year.

(g) Medullary rays are zones of cells radiating from the inner heartwood out towards the sapwood at right angles to the vertical cells of the tree. They provide conducting channels from the heartwood to the sapwood.

Heartwood and Sapwood

In a very young tree all the wood is sapwood, the cells of which give support to the tree and at the same time conduct mineral solutions ("sap") up from the roots to the leaves where photosynthesis occurs due to the action of sunlight and the presence of chlorophyll. When the sapwood cells are first laid down by the cambium layer they consist of very thin-walled tubes of cellulose. As growth proceeds these cells grow rapidly and their walls become thicker and a change known as lignification occurs. The growth processes associated with the thickening of the cell walls are collectively

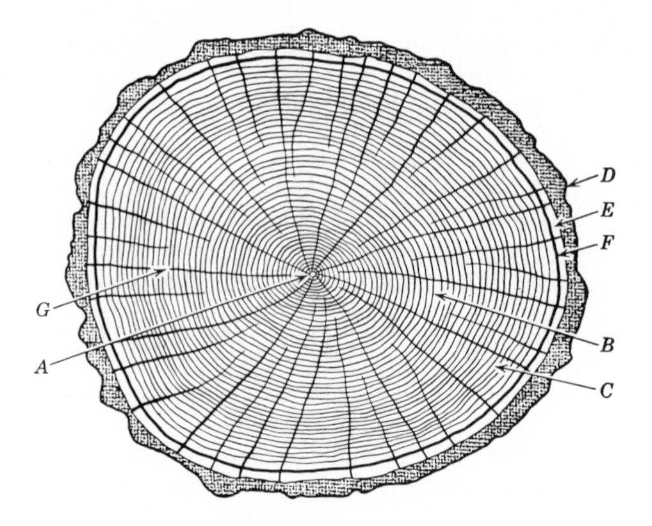

Figure 14.1 *Cross-section of a tree trunk. (A) pith, a soft tissue about which wood grows; (B) heartwood; (C) sapwood; (D) outer bark; (E) inner bark; (F) cambium layer; (G) ray cells. (Reproduced with permission from "Handbook of Engineering Materials", edited by Miner and Seastone. John Wiley and Sons Inc., N.Y. 1955.)*

known as *secondary thickening*, and mark the beginning of the change of sapwood to heartwood. Associated with secondary thickening is the deposition of mineral matter, gums, and resins inside the cells, which eventually die in the sense that they are no longer capable of acting as conducting channels from the roots to the leaves.

Thus, heartwood is dead wood, and plays no further role in the life cycle of the tree except to act as storage cells in some instances.

Laboratory (43): *Examine the cross-section of a small tree or branch. Sketch it and label the important zones that you can see.*

Types of Cells—Hardwoods and Softwoods

Reference has already been made to the fact that structural differences, and not mechanical properties, determine the classification of a timber as either a hardwood or a softwood (see Chapter 2), and that, in terms of hardness and strength, this classification is sometimes misleading. Briefly, the main feature of a hardwood is that is contains very large cells known as vessels or "pores", while a softwood does not contain such cells.

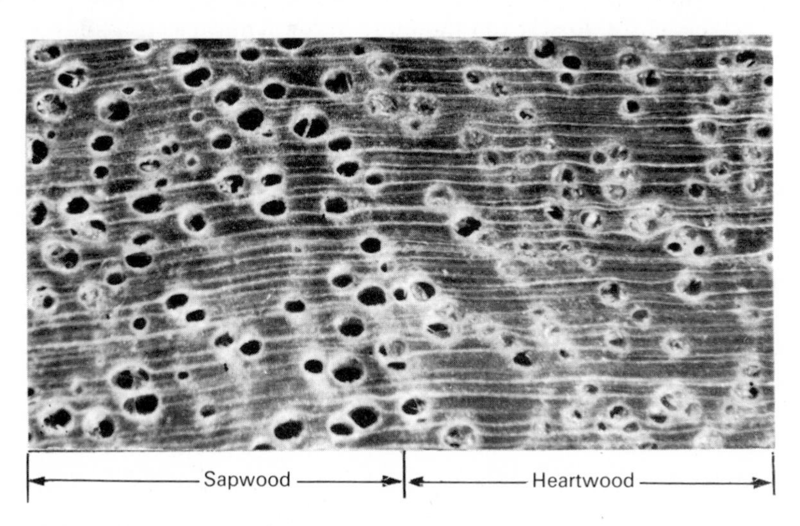

Figure 14.2 *A reflected light photograph of the hardwood Flooded Gum clearly showing the secondary thickening of the vessels in the heartwood. Magnification x20. (Photograph courtesy of the New South Wales Division of Wood Technology.)*

The cells commonly found within a hardwood are:

(a) Vessels: these are the large tubular wood cells running vertically up the tree. In the first place the vessels are made up of many cells joined end-to-end. However, the cross-walls are quickly broken down either completely or partially so that long tubes up to several yards in length are formed. Vessels often occur in groups and are the main conducting cells within a hardwood. They are often large enough to be seen on a cross-section by the naked eye.

(b) Wood fibres: these are also vertically-orientated cells, but unlike vessels they are small in diameter, short, and thick-walled. Their sole function is to support the tree during its lifetime. Wood fibres vary in structure from springwood to summerwood and undergo secondary thickening.

(c) Wood parenchyma: thin soft-walled cells often occur in areas surrounding the vessels, and may contribute to the "figure" of the wood. They are

collectively known as "soft tissue" since they do not undergo secondary thickening and they make no contribution to the strength of the tree.

(d) Ray Cells: these are similar to wood parenchyma cells but radiate outwards horizontally from the heartwood. They act as storage and conducting cells and in sawn wood often contribute towards "figure".

In contrast to the somewhat complex structure of a hardwood, a softwood exhibits a simpler cellular structure in which one type of cell combines both the functions of the vessels and the wood fibres. Known as *trachieds*, these softwood cells both support the tree and act as conducting cells; they are similar in appearance to the wood fibres of a hardwood, but have end connections typical of vessels. Rays may be present but are never pronounced, while wood parenchyma is largely non-existent. Thus softwoods exhibit structures dominated by trachieds, which are intermediate in size between hardwood vessels and wood fibres.

Laboratory (44): *Examine longitudinal, radial, and transverse sections of pinus radiata, oregon, silver ash, and red meranti using a 10x hand lens. Sketch their cellular structures and label the different types of cells. Classify each as a hardwood or a softwood.*

CHEMISTRY OF WOOD

Wood is a fairly complex material chemically. Wood substance itself is made up of two main chemical constituents, cellulose and lignin. The greater portion of the lignin present occurs in a layer between the wood cells and may be considered to be the cement holding the wood cells together. However, some lignin occurs in intimate association with the cellulose within the cell walls themselves.

When wood is broken down chemically it is found to consists of the following:

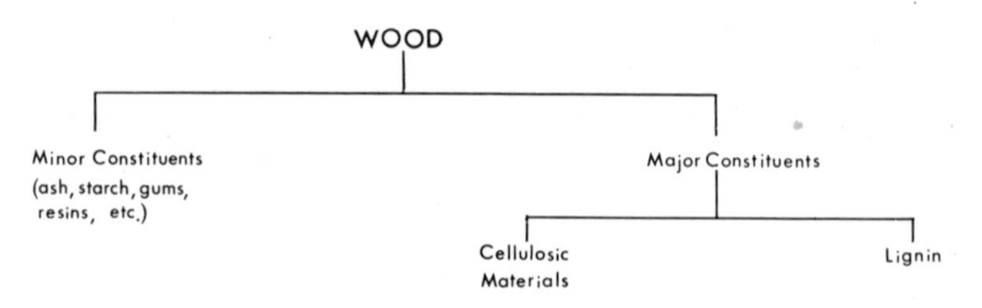

The Cellulosic Constituents are of course carbohydrates and their formation is begun during the process of photosynthesis, in which the essential reaction may be written as:

$$\begin{matrix}\text{carbon}\\ \text{dioxide}\end{matrix} + \text{water} + \text{energy} \xrightarrow[\text{catalyst}]{\text{chlorophyll}} \text{sugars} + \text{oxygen}$$

The sugars resulting from photosynthesis within the leaves are dissolved in the sap and are altered by little-understood processes into the very complex carbohydrates known collectively as cellulosics.

Cellulose is a most important derivative of wood, being used as raw material for the manufacture of artificial silk, celluloid, cellulose-based lacquers and plastics, and explosives.

Lignin, unlike cellulose, has little commercial value, and is mostly run to waste after removal by concentrated acid solutions. Wood to be used for white paper or artificial silk manufacture must be completely delignified. Some lignin is now being used in the manufacture of lignin-based plastics, while lignin from certain types of trees is used to make artificial "essence of vanilla".

Table 14.1 lists the properties and uses, if any, of the so-called minor constituents of wood; that is, those constituents existing in small quantities and which do not form an essential part of the life cycle of the living tree.

THE PHYSICAL AND MECHANICAL PROPERTIES OF WOOD

Density

The density of wood varies considerably with moisture content, and is an excellent indication of the strength of a particular piece of wood. The specific gravity of wood substance itself is virtually constant irrespective of species, being about 1.53. Density may be measured as:

(a) basic density: the mass of the bone-dry piece of wood divided by its volume. The volume is obtained by displacement after the specimen has been soaked for at least eight days, or boiled for six hours, in water.

(b) air-dry density: the density at 12% mosture content, the condition that most wood is in after thorough air drying.

(c) green density: the density of freshly-felled wood, and varies depending upon the season in which the felling occurs.

Since nearly all woods have apparent specific gravities of less than one because of the empty spaces within the wood cells, they will float upon water. However, unsealed wood will become "water-logged" due to the absorption of moisture, thus losing its buoyancy. Grey Box is the densest wood in Australia at 70.5 lb/cu ft; balsa, which is not grown in Australia, has a density of only 6 lb/cu ft.

TABLE 14.1

PROPERTIES AND USES OF THE MINOR CONSTITUENTS OF WOOD

Constituent	*How Extracted or Detected*	*Characteristics*	*Industrial Applications*
Starch	a blue colour indicates starch if the wood is smeared with a solution of iodine in potassium iodide in water.	promotes attack by the powder post borer (*Lyctus*) which lives on starch.	
Resins	may be dissolved out of wood shavings by alcohol or ether.	protects wood from insect and fungus attack.	may be extracted by "tapping" the trunk of the tree and broken down into resin and turpentine.
Fats and Waxes	shavings boiled in water, and fats and waxes float to surface upon cooling; they are also soluble in ether.	unpleasant odours and may cause "wood taint" in foods such as butter if they are boxed in woods containing fats and waxes.	
Tannins	extracted by boiling; indicated by the addition of iron chloride or other iron salts to the residual water.	cause staining of nails driven into unseasoned wood.	tannins extracted from wood, bark and nuts used by the leather industry in "tanning processes".
Terpenes and Other Oils	separated by steaming the wood and collecting and cooling the vapours.	some oils are toxic to fungi and thus make the wood very durable.	turpentine, sandlewood oil and Huon pine oil commercially extracted from appropriate types of wood.
Mineral Matter	remain in the ash when a piece of wood is burned.	used in the conversion of sugars to cellulose and starch in the living tree; frequently cause blunting of tools during the working of timber.	

Moisture Content

Under normal conditions of growth wood may contain considerable amounts of water either in the form of "free water" existing within the cell cavities or as "combined water" which is adsorbed into the cell walls. While the amount of combined water is always about 25% of the oven-dry weight of the wood substance itself, the amount of free water varies considerably, 100% free water being quite common.

If fallen wood is allowed to stand in air for some considerable time the moisture content of the wood will reach equilibrium with that of the surrounding air, this moisture content being known as the "equilibrium moisture content". The water is removed in two distinct stages.

(a) The free water diffuses out first, and when all the free water has been removed the wood is said to be at its "fibre saturation point" since none of the water from the cell walls has as yet been dried out. No shrinkage of the cellular structure of the wood occurs during this stage of drying; however, cell collapse can occur, in which case reconditioning must be carried out in steam chambers.
(b) Continued drying results in the removal of some of the adsorbed (combined) water and this results in shrinkage. Air-dried timber commonly reaches its equilibrium moisture content at 12–15% moisture. However, kiln-dried timber is normally dried to 8–10% moisture, then allowed to stand in the air until it takes in enough moisture to reach equilibrium moisture content.

Moisture content is usually determined by either oven-drying or electrical resistance methods. Oven-drying is a slow but reliable method applicable to small samples, and is carried out as follows:

(a) the sample is weighed in its moist condition and is then placed in an oven held at 103±2°C for one hour,
(b) it is then removed and re-weighed, dried for a further hour, re-weighed again, and so on, *until a constant weight is obtained,*
(c) the moisture content is found from the following relationship:

$$\%\text{M.C.} = \frac{\text{weight in air} - \text{oven dry weight}}{\text{oven dry weight}} \times \frac{100}{1}$$

Electrical resistance meters are more applicable to the testing of pieces which cannot be cut and the method is reliable only on pieces under two inches in thickness.

Calorific Value

This is only of importance when wood is to be used as a fuel, and varies with moisture content and the presence of minor constituents such as resins, gums, and oils. The calorific value of pure wood substance is about 8,300 BTU/lb.

TABLE 14.2

MECHANICAL TESTS APPLIED TO WOOD

Test	*Properties Assessed*	*Test Details*	*Applications of Test Data*
Hardness	Indentation Hardness	The force in pounds weight required to press a 0.444″ diameter steel ball to half its depth into the surface of the specimen.	Hardness figures are directly related to strength values within species, and hardness also varies consistently with density and moisture content.
Izod or Charpy	Notch-toughness	Conducted on $\frac{7}{8}$″ square clear specimens $5\frac{1}{2}$″ long containing a standard keyhole notch.	Test values useful for design calculations where impact loads are involved; for instance in the selection of wood for axe handles or aeroplane propellers.
Denison	Brittleness	Conducted on unnotched clear specimens 10″ long × $\frac{5}{8}$″ square, the specimens being broken as beams by an impact load.	Similar applications as for Izod or Charpy, except that the Denison figure relates more directly to an unnotched member.
Static Bending	(1) Transverse strength (2) Fibre strength at proportional limit (3) Elasticity	Clear specimen 2″ × 2″ × 30″ is tested over a 28″ span.	Data essential in all calculations where timber is to be used as a beam.
Compression Tests	(1) Compressive strength along the grain (2) Compressive strength perpendicular to the grain in the (a) radial (b) tangential directions	Tests along the grain use 2″ × 2″ × 6″ specimens, or 2″ × 2″ × 8″ specimens with a 6″ gauge length if the proportional limit is to be found; tests across the grain use 2″ cubes. All specimens must be clear.	Compressive strength along the grain is useful when the behaviour of short columns is being calculated (long columns have a beam action); compressive strength across the grain is important in applications where the wood itself must resist crushing in this direction, e.g. in frameworks where faces or edges are load-bearing.
Tension	(1) Tensile strength along and across the grain (2) Elasticity	Tensile tests along the grain are difficult to conduct but may be done of 24″ specimens $\frac{3}{4}$″ square which are shaped down to an $\frac{1}{8}$″ square cross-section in the central portion; across the grain tests are donducted by the tensile splitting technique.	Tensile strength along the grain is very high but cannot be utilised due to the difficulties involved in securing the ends of the member, however elasticity is important in all structural members.
Shear	Shear strength along and across the grain	Clear 2″ cubes tested in single shear.	Shear strength is an important consideration in structures and is related to bending and compressive strength values.
Torsion	Torsional strength	Tests conducted on clear 2″ square × 14″ long specimens machined down to $1\frac{1}{2}$″ diameter in their centre sections.	Torsion tests are rarely conducted today but torsional strengths are important in design calculations for aircraft propellers

Thermal and Electrical Properties

Both thermal and electrical conductivity increase with moisture content since water is a better conductor than air. Air-dry wood is an excellent thermal insulator, the lighter woods being superior to the dense woods due to their larger cellular cavities. It is unusual, however, to use wood for electrical insulation since the absorption of only small amounts of moisture can increase its electrical conductivity considerably. The very dense woods like *lignum vitae* are sometimes used for insulators since they absorb very little moisture after drying.

Mechanical Properties

Table 14.2 lists details of tests which may be carried out on prepared wood samples.

Laboratory (45): *Carry out transverse bending tests and compression tests perpendicular and parallel to the grain on small samples of red meranti, oregon, and hoop pine. Determine the moisture contents of each specimen after testing has been completed. Explain how the test results show that oregon is "tough" while hoop pine is "brittle".*

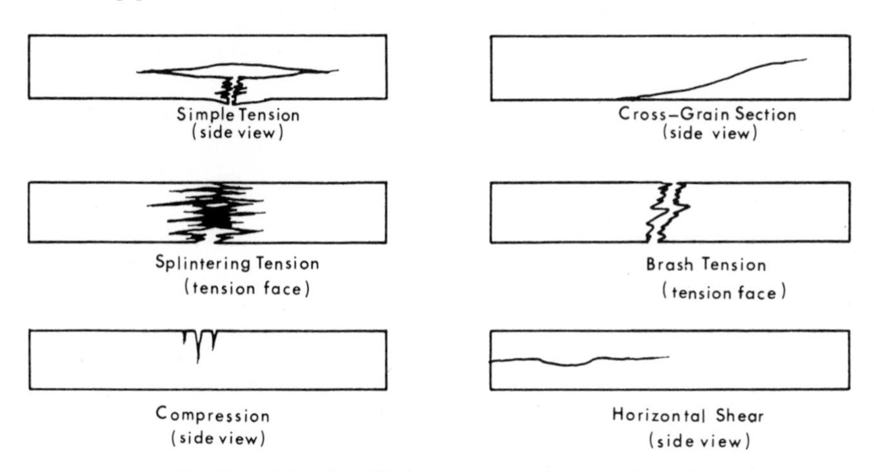

Figure 14.3 *Modes of failure in wooden beams subjected to centre-point loading in standard transverse tests.*

FACTORS AFFECTING THE STRENGTH OF WOOD

Moisture Content

The strength of "green" wood is not appreciably affected until the fibre saturation point is reached. However, once this point is passed the removal of combined water strengthens and stiffens the wood since the cell walls themselves become stronger.

Temperature

In general, temperature increases are accompanied by corresponding increases in plasticity. This is useful if wood has to be bent or moulded, particularly since the wood regains its original strength and stiffness once the temperature is lowered.

Duration of Stress

Wood is a material in which deformation occurs fairly slowly and therefore the duration of load application is important. Structural members may not fail if their ultimate strengths are exceeded for very brief periods only, but may fail in time if subjected to constant loads well below those causing failure in short-term tests. Creep, then, is an important property of wood, and allowable working stresses have to be low enough to prevent creep strains from becoming significant.

Physical Defects

The following physical defects occur in seasoned timber and, in general, exert undesirable influences upon strength properties.

(a) Knots: formed where the stem of the tree branches.
(b) Checks: cracks caused by uneven shrinkage during seasoning.
(c) Shakes: splits occurring around or across the growth rings formed when internal growth stresses are released by felling or seasoning.
(d) Sloping Grain: is probably the most important defect in sawn timber; a slope of 1:14 along the grain reduces bending strength by 25% and longitudinal compression by 13%.
(e) Reaction Wood: compression wood or tension wood may occur if parts of a leaning tree are in tension or compression. Compression wood exhibits reduced toughness while tension wood lacks shear or compressive strength.

THE GRADING OF STRUCTURAL TIMBER

Tests on many different trees all belonging to the same species have shown a wide variation in mechanical properties, and it has been found useful to group all timber cut from trees of the same species into one of three different grades. The following grades are commonly used for structural timbers in Australia:

Select Grade: free from all major defects and having at least 75% of the strength of absolutely clear test specimens.
Standard Grade: can contain some physical defects but strength must be at least 60% of that of clear specimens.
Common or Merchant Grade: only used where strength is not all-important, but must exhibit at least 30% of the strength of clear test specimens.

GLOSSARY OF TERMS

Cambium: the unicellular layer in a growing tree between the inner bark and the sapwood in which all growth in girth occurs.
Cellulosics: complex carbohydrates; in wood they are formed by photosynthesis and make up the major portion of wood tissue.
Checks: cracks caused by uneven drying shrinkage.
Growth rings: the rings seen on a cross-section of a log that mark each year's growth.
Heartwood: wood in which all cells have undergone secondary thickening.
Knots: irregularities of structure, formed in growing trees where the stem branches.
Lignin: the amorphous material found both within and between the walls of wood cells; it is thought to be the "cement" holding the cells together.
Pith: the remains of the original stem of a growing tree; usually seen as a small, dark, dense core in a log.
Reaction wood: wood from those parts of a tree that were subjected to either tensile or compressive stresses during growth.
Sapwood: that part of a tree containing wood cells that have not undergone any appreciable amount of secondary thickening.
Shakes: splits occurring when stresses within a tree are released by felling or seasoning.
Trachieds: the supporting and conducting wood cells found in softwoods.
Vessels: the conducting, thin-walled wood cells characteristic of hardwoods.

REVIEW QUESTIONS

1. List and discuss four properties of wood that make it a rather unique engineering material.
2. Distinguish between softwoods and hardwoods in terms of their typical cellular structures.
3. Sketch the cross-section of the trunk of a tree and label the important zones.
4. What are the essential differences between heartwood and sapwood?
5. How are the strength properties of wood affected by (i) moisture content, (ii) density?
6. How significant is the creep behaviour of wooden beams when designing (i) a small bridge to carry light traffic only, and (ii) a heavy two-storey building. In the latter instance, suggest some reasons why steel would be a desirable substitute for the main structural members.
7. How is structural timber normally graded?
8. List some of the uses of the minor constituents of wood.

15

Polymers

POLYMERS form a large group of naturally-occurring and synthetic materials which are playing increasingly important roles in modern technology. Most polymers are organic materials and have the following characteristics:

(1) they consist of combinations of carbon with oxygen, hydrogen, nitrogen and other inorganic or organic substances,

(2) they are generally non-crystalline "solids"* at ordinary temperatures, but pass through a viscous stage during their formation when shaping is readily carried out,

(3) they exhibit distinctive macromolecular or long-chain structures in which individual molecules are large, being composed of perhaps tens of thousands of similar small molecules all covalently bonded together.

The common naturally-occurring polymers include protein, cellulose, resins, starch, shellac and lignin, the latter being the amorphous "cement" holding the cellulose fibres together in wood tissue. All polymers are characterised by structures quite unlike those in metals. However, like metals their structures in every particular instance determine their physical, thermal, electrical and mechanical properties.

THE MECHANISMS OF POLYMERIZATION

The characteristic macromolecular or long-chain structure of all polymers is formed by the chemical combination of many smaller and simpler molecules which may or may not be of the same kind. This is the process of polymerization, and two mechanisms are known to operate. These are *addition polymerization* and *condensation polymerization*, and each results in the formation of a certain type of polymeric structure.

Addition Polymerization

If a large number of simple molecules are chemically "added together" without waste then the process of addition polymerization has occurred. An

*The term "solid" is restricted to those materials having a definite crystalline structure and hence non-crystalline polymers are not true solids.

example is polyethylene which is formed from the monomer, ethylene, by subjecting the latter to heat and pressure in the presence of a suitable catalyst (see Figure 15.1).

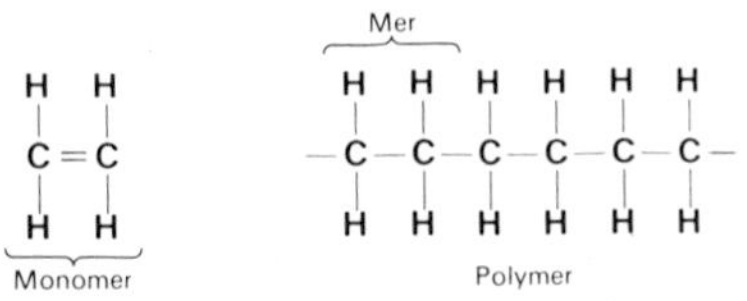

Figure 15.1 *The addition polymerization of polyethylene (polythene).*

This polymerization reaction involves the breaking down of the double covalent bonds between the two carbon atoms within the ethylene molecules and the subsequent redistribution of these bonds as single bonds between a whole series of carbon atoms. The carbon-hydrogen bonds remain unaffected by the reaction. This is the simplest form of polymerization that can occur.

Condensation Polymerization

In contrast to addition reactions in which a simple molecular summation occurs, condensation reactions result in the "splitting out" of simple non-polymerizable molecules which are considered to be by-products of the process. The formation of the polymer, phenol-formaldehyde, more commonly known as Bakelite, is typical (see Figure 15.2).

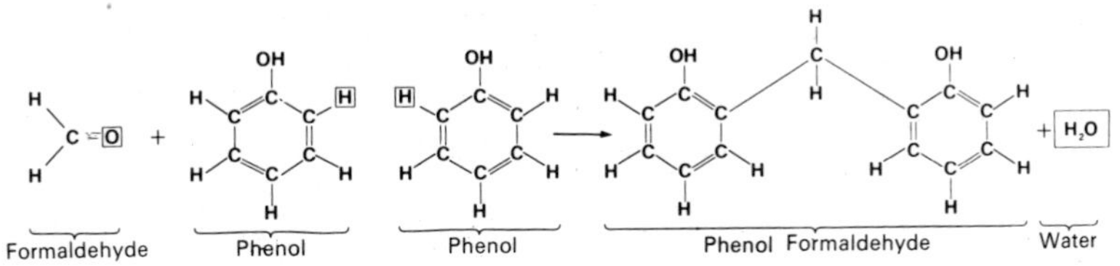

Figure 15.2 *Condensation polymerization of phenol-formaldehyde.*

In this case molecules of water are split out during the reaction; however the formation of dacron (or mylar) results in the condensation of methyl alcohol (see Figure 15.3).

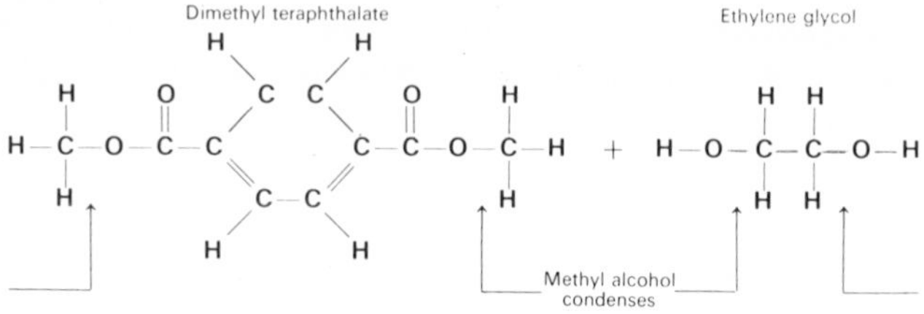

Figure 15.3 *Condensation polymerization of Dacron (mylar).*

Functionality

The ethylene is considered to be a bifunctional monomer since it possesses two reaction sites where polymerization can be effected. In other words, when the double bonds are broken two single bonds become available; other monomers may be trifunctional or tetrafunctional. Still higher functionality, while theoretically possible, does not generally occur in practice.

Importance of Unsaturated Hydrocarbons

In general, if polymerization is to occur at least one double bond must be present in the monomer. Thus, the large group of organic materials known collectively as unsaturated hydrocarbons provide raw materials for most polymers. In contrast to saturated hydrocarbons in which all covalent bonds are single, unsaturated hydrocarbons contain at least one multiple covalent bond.

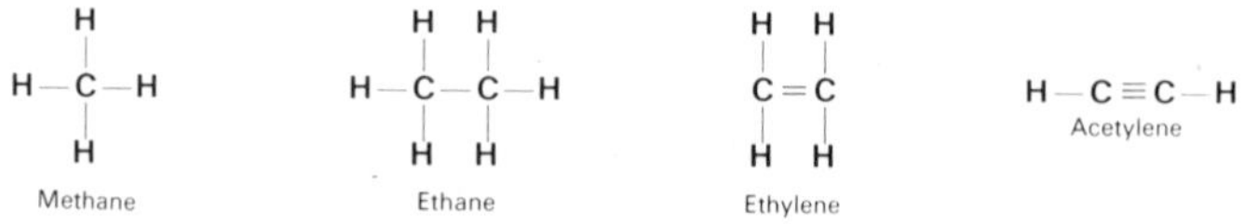

Figure 15.4 *Typical saturated and unsaturated hydrocarbons.*

It is important to note that not all double or triple bonds present must be broken down to single bonds during polymerization. The polymerization of butadiene, for example, into a form of synthetic rubber involves the breaking down of only one of the two pairs of double bonds present in the monomer. However, the other double bond must shift its position in order to maintain four covalent bonds around each carbon atom (see Figure 15.5).

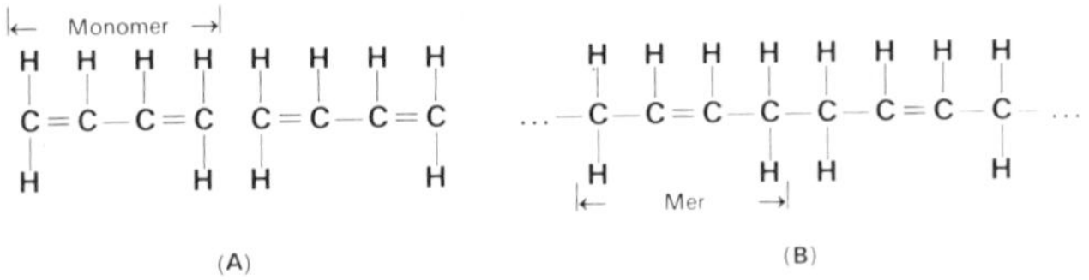

Monomer

Mer

(A) (B)

Figure 15.5 *The polymerization of butadiene into an elastomer.*

Polymerization Conditions

The breaking down of double bonds within monomers requires considerable energy. For example, the energy required to break the C=C bond is equal to 146,000 cal/6.02 $\times$ 10^{23} bonds. However 83,000/6.02 $\times$ 10^{3} calories of energy are released every time a single C–C bond is formed so that external energy is required to start the polymerization reaction rather than to sustain it. However, polymerization reactions of this simple type are not self-sustaining because the diffusion processes by which the molecules of the monomer move to the ends of the growing long-chain molecules operate very slowly, and thus chain formation must eventually cease due to the unavailability of further simple molecules.

Laboratory (46): *Some monomers such as methyl methacrylate can be polymerized by the addition of a chemical catalyst. Take a small sample of such a monomer, place into a suitable metal mould, and mix in the required amount of catalyst. Hold the mould in your hand while polymerization takes place. What is the significance of the heat that is evolved?*

Copolymerization

If an addition reaction utilises two different types of monomers, then the process is known as copolymerization and the resulting material is termed a copolymer. A good example is butadience-styrene rubber, the molecular structure of which is shown in Figure 15.6.

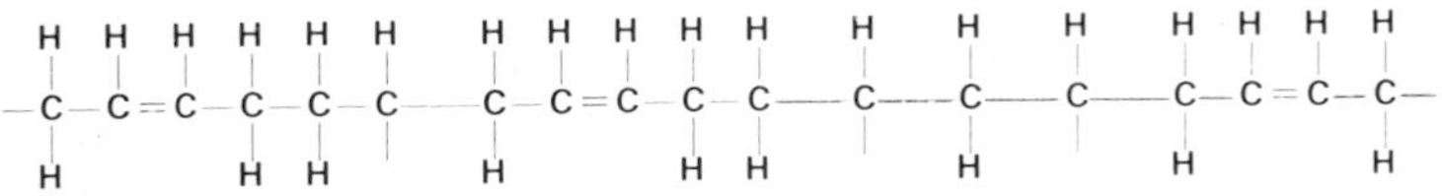

Figure 15.6 *Copolymerization of butadiene-styrene rubber.*

The physical and mechanical properties of copolymers are often more desirable than those of either of the constituent mers, the effect being comparable to that seen in solid solutions formed between two pure metals. Table 15.1 shows a range of copolymers formed between vinyl chloride and vinyl acetate.

TABLE 15.1

VINYL CHLORIDE-ACETATE COPOLYMERS

Material	*% Vinyl Chloride*	*Number of Chloride Mers/Acetate Mer*	*Molecular Weight*	*Applications*
Polyvinyl Acetate	0	0	5,000–15,000	Adhesives
Chloride-Acetate Copolymers	(a) 85–87	8–9	9,500–10,500	Injection moulding plastics of excellent strength and solvent resistance.
	(b) 95	26	20,000–22,000	Extrusion-moulding plastic for electrical insulation.
Polyvinyl Chloride	100	—	—	Limited commercial applications.

ADDITIONS TO POLYMERS

In most instances several to many different additives are compounded into the monomer material either before or during the polymerization process. Such additives are usually one of the following different types of compounds:

(1) *plasticisers*: usually complex organic compounds, essentially "oily" in nature, which have high boiling points and low evaporation rates. These substances act as internal lubricants, particularly during moulding operations, improving material flow and imparting toughness and even flexibility to the cured material.

(2) *fillers*: these are inert materials that do not change chemically during the curing process; they may be likened to the sand or aggregate added to cement to make concrete. Typical fillers include wood flour, asbestos, dry wood pulp, mica and slate powders, and cotton fibre. Fillers may be added in high proportion to many plastics, particularly to those thermosetting materials to be used as moulding powders.

Both plasticisers and fillers disrupt the tendency of many polymers to crystallise during cooling, crystalline polymers being undesirable due to their greater brittleness.

TABLE 15.2

COMMON FILLERS AND THE PROPERTIES THEY IMPART TO PLASTICS

Filler	*Properties Imparted*
Wood flour	Improves strength and imparts good mouldability for low cost
Cloth fibre	Improves impact strength but results in poorer mouldability
Macerated cloth	Imparts very high impact strength
Glass fibre	Imparts high strength and improves dimensional stability
Asbestos fibre	Improves heat resistance and dimensional stability
Mica	Reduces moisture absorption factor and also reduces electrical conductivity

(3) *accelerators and hardeners*: usually added to promote faster or more complete polymerization, being catalysts of one type or another.

(4) *dyes and pigments*: added to colour the material to the desired shade.

(5) *initiators*: usually added to allow polymerization to begin. They stabilise the ends of the molecular chains, such end stability being essential if long molecular chains are to form freely. Hydrogen peroxide is a common initiator, as are organic peroxides and many other simple compounds which will supply terminal radicles or atoms for the molecular chains (see Figure 15.7).

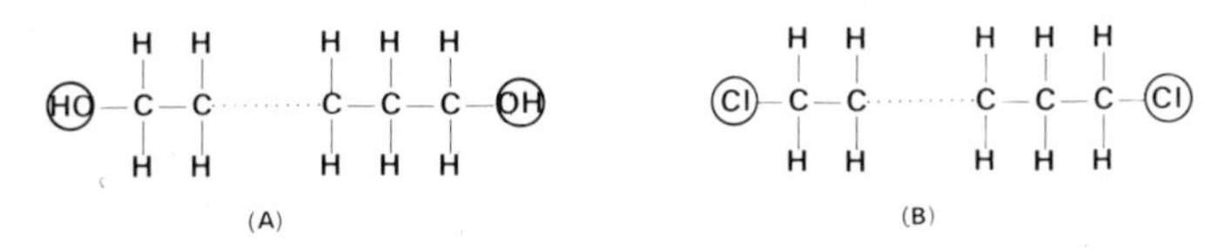

Figure 15.7 *The action of H_2O_2 and Cl_2 as initiators.*

STRUCTURE OF POLYMERS

The structure of a polymer affects its properties in definite and well-defined ways, the following five aspects of structures being most important in this respect.

Linear and Framework Structures

A bifunctional monomer such as ethylene produces a linear polymer, that is, a polymer in which the individual long-chain molecules are quite separate from one another and are held together only by relatively weak secondary bonds. This is not to say that the chains have to be straight; on the contrary, it is quite common for linear polymers to have a tangled structure in which some of their mechanical strength comes from the inter-twining of the molecules.

Tri- and tetrafunctional monomers usually form polymers with framework structures (or network structures) in which individual molecular chains are difficult to distinguish. Such is the case in phenol-formaldehyde which, in contrast to polyethylene, is a hard, strong plastic of relatively high melting point.

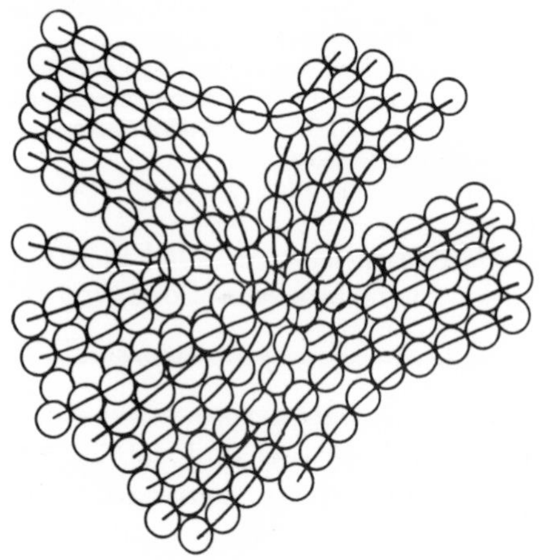

Figure 15.8 *Schematic of a linear polymer (polyethylene). The circles represent the individual mers in the chains.*

Cross-Linked Chains

If a linear polymer becomes cross-linked the movement of individual molecular chains becomes quite restricted and thus mechanical properties are considerably altered. Cross-linking may be done deliberately to increase strength and reduce plasticity, or it may occur naturally. Both situations can occur with natural rubber. If rubber is vulcanised by the addition of about 5% sulphur, the sulphur atoms form covalent bonds with carbon atoms in adjoining molecular chains; in this case the cross-linking is done deliberately and results in improved tensile strength, abrasion-resistance, and elasticity for the rubber. However, if rubber is placed in an oxidising atmosphere or alternatively exposed to the atmosphere for a long period of time, oxygen atoms will diffuse into the rubber and form extensive cross-links between the molecular chains; in this case, the rubber loses its elasticity and "perishes".

The Shapes and Distribution of Molecules within the Polymer

Some linear polymers are composed of simple, uniform, atomic groups, while others are composed of large and small atomic groups unevenly distributed along the molecular chains. Polyethylene is an example of the

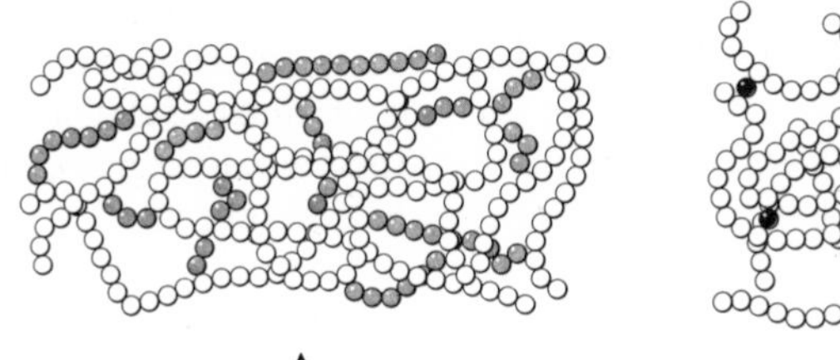

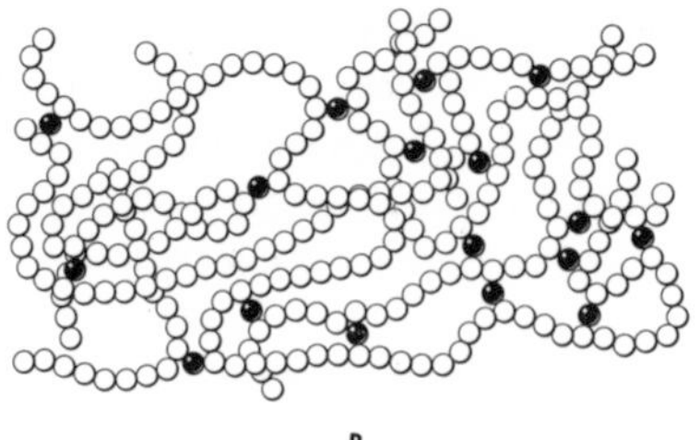

Figure 15.9 *Cross-linked polymer chains: (A) cross-linked chains in which small chain segments act as cross-links; (B) cross-linked chains in which foreign atoms or molecules are the cross-links. An example of this latter type of cross-linking is found in vulcanised rubber, in which sulphur atoms act as the cross-links. (Reproduced with permission from Volume I of "The Structure and Properties of Materials" by Moffatt, Pearsall, and Wulff. John Wiley and Sons Inc., N.Y. 1964.)*

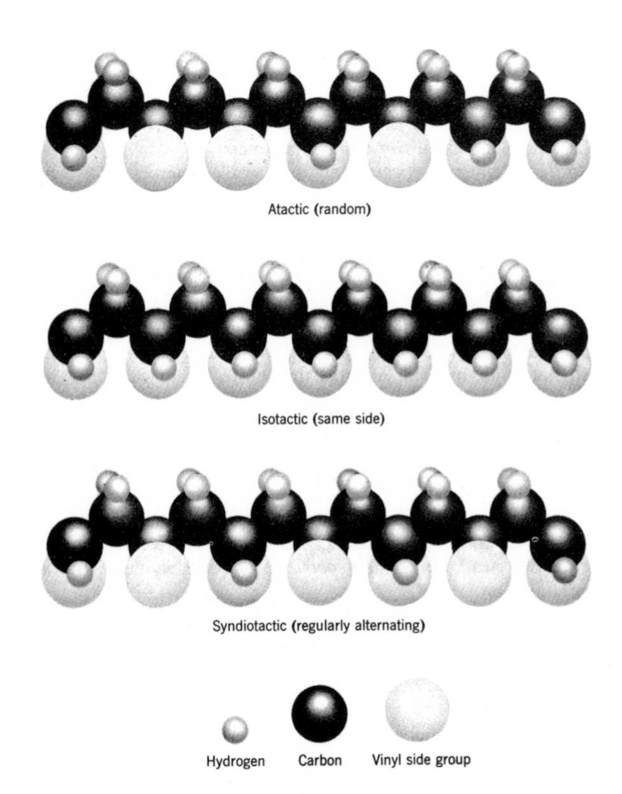

Figure 15.10 *The possible arrangements of side groups in a simple vinyl polymer: the side group may be a single atom, such as chlorine as in polyvinyl chloride; or it may be a group of atoms, such as the benzene ring in polystyrene. (Reproduced with permission from Volume I of "The Structure and Properties of Materials" by Moffat, Pearsall, and Wulff. John Wiley and Sons Inc., N.Y. 1964.)*

former type of structure, while polyvinyl chloride is an example of the latter; polyvinyl chloride is a tougher, stronger plastic than polyethylene since intermolecular movements are restricted by the irregularities of the chains themselves. The three different arrangements of "side groups" of atoms along the long-chain molecules in a vinyl plastic are shown in Figure 15.10.

Crystallinity of Polymers

Many polymers exhibit a tendency to exist in the crystalline state. However, few if any polymers can be perfectly crystalline. If linear polymers have their molecular chains arranged such that they lie side by side over a definite area then they are considered to be crystalline; however, they are rarely crystalline throughout their entire structure. It is quite normal for single linear polymers to have crystalline and non-crystalline areas, and this kind of structure is commonly termed microcrystalline. Network and cross-linked polymers rarely crystallise even to this limited extent and thus may be considered amorphous.

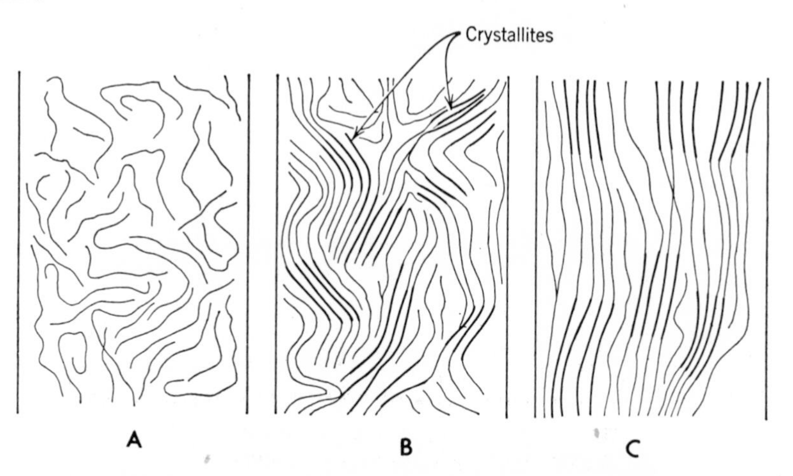

Figure 15.11 *Schematic representations of amorphous and crystalline polymers. (A) an amorphous polymer; (B) a crystalline polymer with some amorphous areas; (C) an oriented crystalline polymer. (Reproduced with permission from "Mechanical Behavior of High Polymers" by A. Turner. Interscience Publishers Inc., 1948.)*

Branched Chains

It is possible, under well-defined conditions, to cause simple linear polymer molecules to branch into bifurcated chains. If this branching is extensive, the polymer will become stronger and less plastic due to the simple interlocking actions of the branched chains with each other. In many polymers it is easier to cause branching than cross-linking, since the former requires less energy than the latter, and for this reason controlled branching within polymers is of great industrial importance.

Figure 15.12 *Branching in a linear polymer. The side branches are formed from the second mer in the copolymer. (Reproduced with permission from Volume I of "The Structure and Properties of Materials" by Moffatt, Pearsall, and Wulff. John Wiley and Sons Inc., N.Y. 1964.)*

CLASSIFICATION OF POLYMERS

Plastics and Rubbers

It is very difficult to draw a clear-cut distinction between those polymers known as plastics and those known as rubbers. Natural rubber, of course, is very readily distinguished from other polymers because of its distinctive origins and structure. However, many synthetic polymers are known variously as plastics or rubbers. The distinction is clarified to some extent if the concept of an elastomer is introduced. An *elastomer* is a material that can be repeatedly stretched to at least twice its original length and, upon release, return to its original length. Rubbers, unless highly vulcanised, are elastomeric materials, while that group of polymeric materials termed "plastics" are commonly rigid or semi-rigid in character.

Thermosetting and Thermoplastic Polymers

Plastic materials all soften when first heated, being shaped by one means or another while in this soft state. However, a fundamental distinction can be made on the basis of their behaviour upon cooling and subsequent reheating.

Thermosetting plastics harden quickly while in the soft state, the curing process usually being assisted by pressure. These plastic materials cannot be softened and reworked once this curing process has gone to completion.

Thermoplastics, on the other hand, only set when the temperature is lowered below a certain limit, and may be softened and reworked by reheating to a suitable temperature. Thus, thermoplastics are reclaimable while thermosetting plastics are not.

This difference in behaviour is directly related to structural considerations. Thermoplastics are linear polymers with or without some degree of cross-linking or branching within their structures. Heating such a structure breaks the secondary bonds between the individual chains, thus allowing plastic flow to occur quite readily. Cooling the polymer allows it to harden again since the secondary bonds re-establish themselves. Thermosetting plastics, on the other hand, have three-dimensional network structures in which bonding is predominantly covalent. Upon heating, these bonds retain their strength and prolonged heating simply causes charring of the polymer. Extensively cross-linked polymers behave in the same way and are also thermosetting.

There are some resins that set at ordinary temperatures because of the chemical action of a hardener. These polymers may be thermosetting or thermoplastic and require no pressure during polymerization.

Thermosetting plastics are usually harder, stronger, and more brittle than thermoplastics. Because they do not progressively soften when heated thermosetting plastics have applications where moderate service temperatures are encountered: for example, saucepan handles.

PROPERTIES OF POLYMERS

The following properties are typical of "solid" plastics:

(1) Low specific gravities
(2) Good thermal and electrical resistance
(3) Good surface finish direct from the forming dies
(4) Availability in a wide colour range or transparent if required
(5) Low strengths compared to metals
(6) Unsuitable for service conditions where temperatures in excess of several hundred degrees centigrade exist
(7) Poor to fair dimensional stability, particularly in moist conditions.

Tables 15.3 and 15.4 give the properties, characteristics, and uses of eight common plastics.

Laboratory (47): *Take pieces of polythene, nylon, and bakelite and weigh each piece (about 20g is sufficient). Place these pieces in a water bath for 24 hours. Remove them, soak off excess surface water on blotting paper and reweigh each piece in turn. Repeat this procedure after resoaking for another 24 hours. Calculate the percentage moisture absorption factor for each polymer and on the basis of this and other known properties state which one would be most suitable for (a) an electrical panel and (b) sheathing for electrical cables.*

TABLE 15.3

COMMON THERMOSOFTENING PLASTICS—PROPERTIES AND USES

Name	*Specific Gravity*	*Tensile Strength (ksi)*	*Izod Number ft lb/in*	*% Water Absorption (in 24 hrs)*	*Top Working Temperature °F*	*Colours*	*Characteristics and Typical Uses*
Cellulose Acetate (celluloid)	1.23–1.5	3–8	0.75–4	2.0–6.0	145–260	Transparent, translucent; and all colours	Slow burning, excellent moulding qualities, tough, machines well; lampshades, tool handles, packaging, camera parts, etc.
Poly-methyl Methacrylate (Acrylic, 'Perspex')	1.15–1.18	4–8	0.3–0.6	0.4–0.6	195	Clear and transparent; also in all colours	Strong, tough, very weather resistant, high optical clarity; used for windows, observation domes on aircraft, light fixtures, display case fronts, etc.
Polyamides (Nylon)	1.06–1.19	5–8.5	—	7.6	450	Translucent; all colours possible	Strong; good abrasion resistance, resistant to oils and greases, good electrical insulator; used to make bristles, rope, parachute cords, tyre fabric, bearings, small gears, etc.
Polyethylene or Polythene	0.96	4	10	0.003	200	Transparent; wide range of colours	Odourless, tasteless, non-toxic, good chemical resistance; used to make "unbreakable" containers such as squeeze bottles, flexible and rigid pipes, plastic bags, etc. It can be heat sealed.

TABLE 15.4

COMMON THERMOSETTING PLASTICS—PROPERTIES AND USES

Name	*Specific Gravity*	*Tensile Strength (ksi)*	*Izod Number ft lb/in*	*% Water Absorption (in 24 hrs)*	*Top Working Temperature °F*	*Colours*	*Characteristics and Typical Uses*
Polystyrene	1.05–1.15	5–9	0.3–0.6	less than 0.2	150–190	All transparent and opaque colours	Good moulding characteristics, unaffected by temperatures commonly encountered in freezing chambers, inert; commonly used in refrigerators, wall tiles, tumblers, light shades, radio cabinets, etc.
Phenol-formaldehyde (Bakelite)	1.25	4–7	0.2–0.4	0.25	250–300	Mainly black and brown	Strong, scratch resistant, very good thermal and electrical resistance; used for handles, vacuum cleaner parts, switches, knobs, coatings, laminates, etc.
Melamine-formaldehyde	1.76–1.98	5–8	—	0.1	350	All colours; translucent to opaque	Tough and scratch-resistant, tasteless, odourless, chemically inert; used to make paper and fabric water-resistant, light fittings, auto ignition parts, dinnerware, etc.
Urea-formaldehyde	1.4–1.8	4–8.5	0.2–0.5	1–3	160–185	All opaque and translucent colours	Durable, water-resistant, good strength, good insulator; used to make switches, plugs, buttons, buckles, glues for plywood, etc.
Silicones	2.0	3–5	0.2–3.0	0.5	550	—	Resistant to oils and greases, high stability, water repellent, high resistance to acids; used as resins, as moulding compounds, coatings, greases, and fluids, and also used in silicone rubber.

One of the outstanding properties of plastics is their light weight as shown by the fact that their specific gravities are usually much less than 1.75, which is the value for magnesium, the lightest of the engineering metals. This property combined with the ease of formability and high surface finish that can be obtained direct from the die, mould, or extrusion press, makes plastics very suitable for the mass production of articles and components requiring only low strengths and low impact resistance. Another advantage is that, unlike metal articles, the colour in a plastic permeates the whole material and thus cannot be worn off. Contrary to popular belief most plastics are not inexpensive materials, the cost per pound often being higher than that for the cheaper engineering metals. However, plastics are competitive with metals for certain applications because of their particular properties, even though the total tonnage of plastics produced today does not exceed that of magnesium, the least used of the engineering metals.

Table 15.5 lists some of the more important industrial applications of plastics.

TABLE 15.5

APPLICATIONS OF PLASTICS

General Applications	*Specific Examples*
Packaging and container materials	Polythene for plastic bags; polypropylene for flexible sealed containers; Phenolics for radio cabinets.
As thermal and electrical insulators	Phenolics for iron and other "hot" handles, electrical switchboards, and switchgear; polystyrene used as expanded foam insulator in refrigeration plant; polythene and polypropylene used for the sheathing of electrical cables.
Flexible foams	Rubber "latex" foam for mattresses, pillows, cushions, etc.; other synthetic foams may be used for similar purposes, e.g. polyurethanes.
Engineering applications	Gears are often made from laminated plastics, e.g. nylon; both light and heavy duty bearings are made from suitable plastics such as nylon or polytetrafluorethylene and will not "sieze" in uses; laminated plastics can be used for lightweight building panels; and clutch and brake linings are made from asbestos-reinforced phenolics.

MECHANISMS OF DEFORMATION IN POLYMERS

Elastic Deformation

Elastic deformation varies considerably between linear and network structures since it depends upon the amount of bond-straightening and bond-lengthening that can occur. Naturally both of these mechanisms are very restricted in network structures, which consequently have higher elastic moduli than linear polymers. Also, if the polymer has a highly tangled and kinked structure which lacks extensive cross-linking, the elastic deformation due to chain straightening will be very large; such is the case in elastomers where the major part of their large amount of elastic deformation is due to chain movements rather than bond adjustments.

Plastic Deformation

Permanent or plastic deformation is due to the "slip" that occurs between the molecular chains when sufficient load is applied to overcome the initial elastic deformation. Linear polymers without cross-linking or branched molecules exhibit the greatest degree of plastic deformation, while network structures lack plastic deformation and are consequently brittle. As well, the plasticity of thermoplastics increases with temperature, molecular movements becoming easier due to the breakdown of the secondary bonds by the heat.

Stress Crystallisation

Stress or deformation crystallisation occurs when the molecules in a basically linear polymer are pulled into close alignment as a result of applied stress. Elastomers exhibit a very high degree of stress crystallisation, resulting in an increase in the elastic modulus as the breaking point is reached.

Laboratory (48): *Take similarly sized pieces (a good size is ½" diameter by 1" long) of ebonite, polythene rod, and lightly vulcanised rubber and subject each to a standard compression test. Plot the curves to failure, and sketch the mode of failure of each sample. How does the elastic and plastic behaviour of each of these polymers differ from that of a ductile metal?*

METHODS OF FORMING AND WORKING PLASTICS

Compression Moulding

Compression moulding is the most common method of forming thermosetting materials, but is not generally used for thermoplastics. Moulding powder or pellets mixed with such materials as fillers and pigments is placed directly into the open mould cavity. The mould is then closed, the plunger pressing down on the plastic and causing it to flow throughout the mould. The plastic material polymerizes under the influences of heat and pressure, generally requiring little curing time in the mould. The three factors of temperature, pressure, and the time during which the mould is closed vary with the design of the article and the type of moulding powder used. Since thermosetting plastics are used the mould can be maintained at its operating temperature all the time.

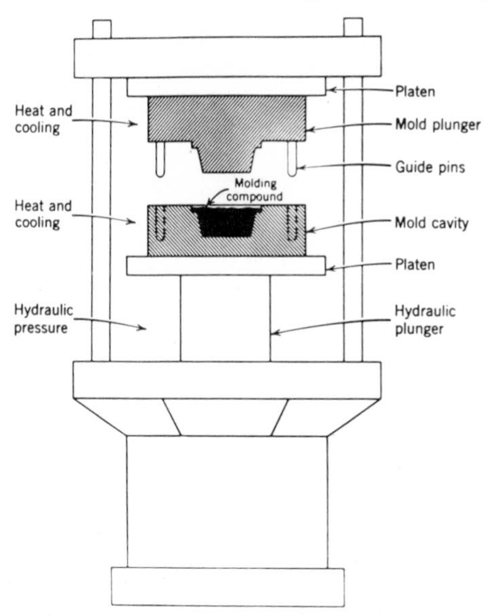

Figure 15.13 *The compression moulding of a powdered or granulated thermosetting polymer. (Reproduced with permission from "Textbook of Polymer Science" by F. W. Billmeyer. John Wiley and Sons Inc., N.Y. 1962.)*

Transfer Moulding

Transfer moulding is like compression moulding in that the plastic is cured in the mould under heat and pressure. It differs in that the plastic is heated to the point of plasticity before it is forced into the mould chamber by a hydraulically operated plunger. Transfer moulding is most commonly used for thermoplastics, particularly in the manufacture of small parts having either small deep holes or cavities or metal inserts, since compression moulding is unsuitable for such applications. Dies must be cooled after each article is formed to allow for the hardening of the thermoplastic material.

Injection Moulding

In injection moulding, plastic material in the form of granules or powder is put into a hopper which feeds the heating chamber. A plunger then forces the plastic through this long heating chamber and around the torpedo where it is thoroughly mixed, the material leaving this chamber in a fluid state. This fluid plastic is then forced under high pressure into the cold mould via a special nozzle that abuts the end of the heating chamber. As soon as the plastic cools to a solid state the mould opens and the part is ejected. This method is used principally for thermoplastics; however, modern injection moulding equipment is available for thermosetting resins. This is a most economical production method for long production runs.

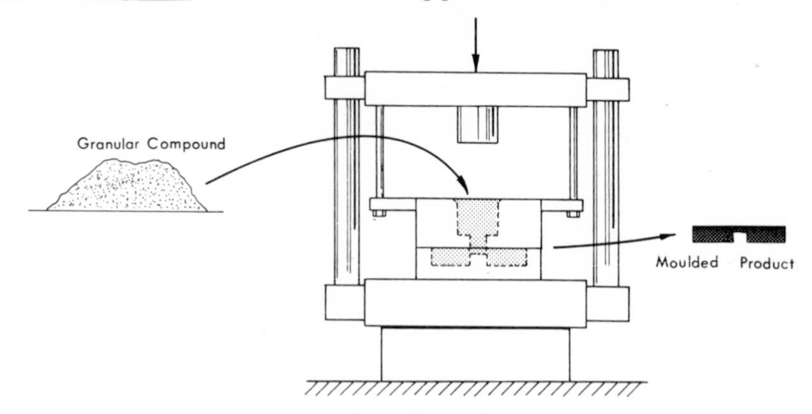

Figure 15.14 *Transfer moulding using a press and a granulated polymer.*

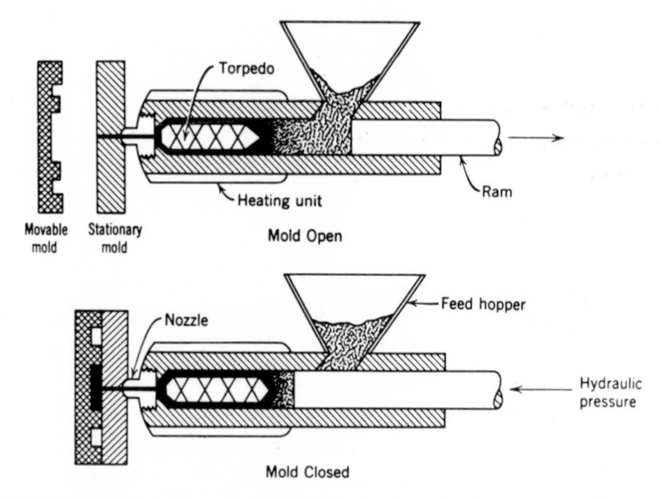

Figure 15.15 *The injection moulding of a thermoplastic polymer. (Reproduced with permission from "Textbook of Polymer Science" by F. W. Billmeyer. John Wiley and Sons Inc., N.Y. 1962.)*

Blow Moulding

Blow moulding consists of stretching and then hardening a sheet of plastic against a mould. In the method as illustrated in Figure 15.16, a gob of molten thermoplastic material has been roughly shaped by extrusion so that it will fit correctly into the mould. Air is then blown into the "bag" of plastic forcing it against the sides of the mould, where it is allowed to cool.

This is the most common method of making plastic bottles. It is analogous to the forming of glass-blown bottles.

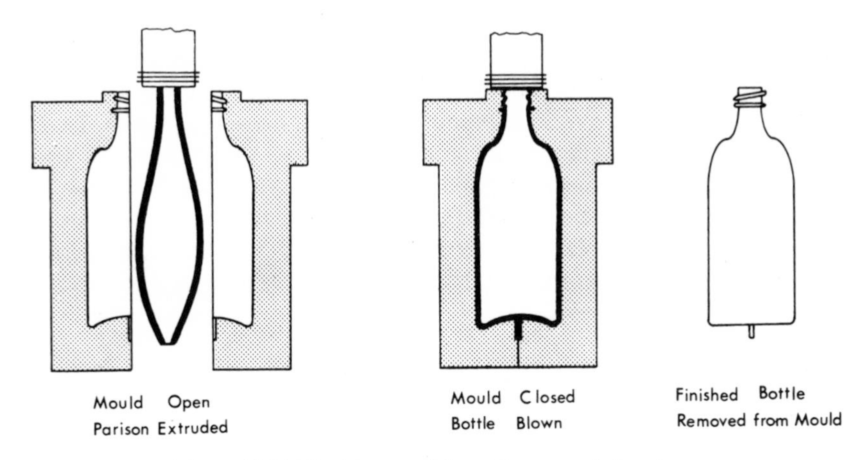

Figure 15.16 *The blow moulding of an extruded parison of thermoplastic to form a narrow-necked container.*

Extrusion

This is the method employed to form thermoplastic materials into continuous sheets, film, tubes, rod, profile shapes, filaments, and to coat wire, cable, and cord. The dry powdered or granulated plastic is first loaded into a hopper and is then fed into a long heating chamber through which it is moved by the action of a continuously rotating screw. When the material reaches the end of this chamber it is forced out through a small orifice or die having the sectional shape of the article to be made (see Fig. 15.17).

In the case of wire and cable coating, the thermoplastic is extruded around a continuous length of wire or cable which, like the plastic, passes through the extrusion die. The coated product is wound on to drums after cooling. In forming wide film or sheeting, the plastic is extruded in the form of a tube;

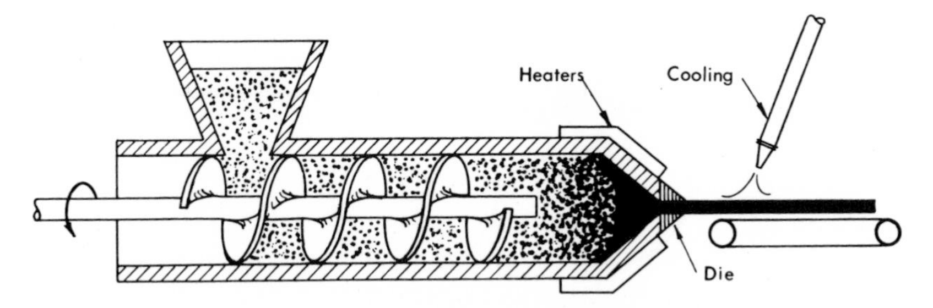

Figure 15.17 *Continuous extrusion: the shape of the die orifice determines the sectional shape of the extrusion and polymerization occurs within the die.*

this is split as it comes from the die, and then stretched and thinned to the required size by rolling.

Calendering

Calendering can be used to form thermoplastics into film and sheeting, and also to coat such materials as cloth and other sheeting. In calendering film and sheeting, the plastic compound is passed between a series of large heated rollers, the thickness being controlled by the setting of the last set of rollers. When coating, the coating compound is passed through the two top rollers while the uncoated material passes through the two bottom rollers where a fine film of the plastic material is affixed to it. The final surface finish of the film depends upon the texture and finish of the final pair of rollers.

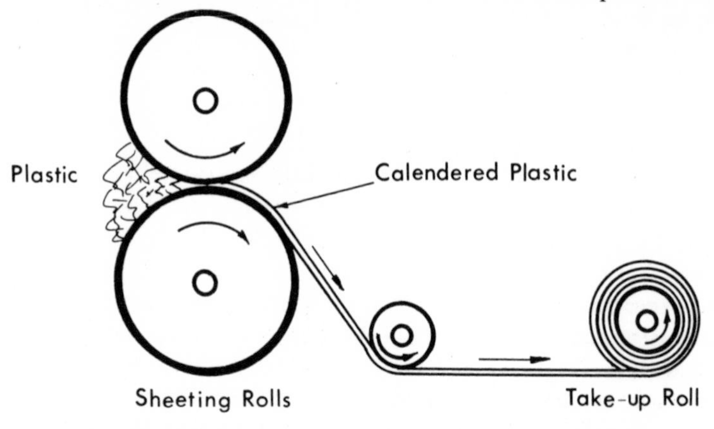

Figure 15.18 *Forming sheet from a thermoplastic by calendering.*

Surface Coating

Thermosetting and thermoplastic materials may both be used as coatings; possible base materials include metal, wood, paper, fabric, leather, glass,

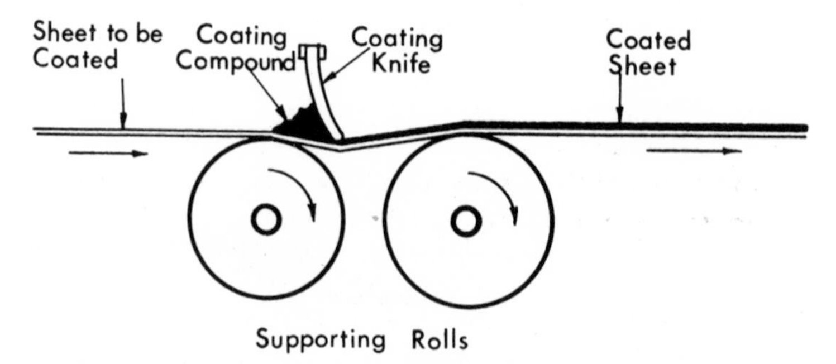

Figure 15.19 *The knife coating of fabric with a polymer.*

concrete, ceramics, and other plastics. Methods of coating vary from knife or spread coating, spraying, roller coating, to dipping or brushing. In spread coating, as illustrated in Figure 15.19, the material to be coated passes over a roller and under a long knife edge. The plastic coating compound, placed on the material just in front of the knife, is thus spread out over the surface of the material by the scraping action of the knife. Coated fabrics are often manufactured in this manner.

NATURAL AND SYNTHETIC RUBBERS

Natural rubber is a polymer of a single material called isoprene, and it exists in certain trees and plants as a colloidal dispersion in the milky liquid known as rubber latex. Crude rubber, often called "crepe rubber", is produced when the latex is coagulated by acetic acid, the resulting spongy mass being passed through rollers which form the material into a sheet. This crude rubber must be further processed by the addition of fillers, plasticisers, and pigments before it is in a form suitable for industry, and even then the processed rubber is usually vulcanised before going into service.

Unvulcanised rubber is susceptible to attack by vegetable and mineral oils, petrol, benzene, carbon tetrachloride, nitric acid, strong sulphuric acid, and other common industrial solvents. Vulcanised rubber has a greater chemical resistance to all these reagents, such resistance generally increasing with the amount of sulphur added.

The extensive use of rubber dates only from 1839 when Charles Goodyear, working in the United States, discovered that crude rubber could be vulcanised by the addition of about 5% sulphur, the reaction taking place when the rubber and the sulphur were mixed together and heated to a suitable temperature. Pigments and fillers are added at the same time, some fillers increasing the tensile strength, abrasion resistance, and resistance to ageing, while others are inert, having influence only on the bulk properties of the rubber. Carbon black is a common filler added to improve the mechanical properties of vulcanised rubber.

The Structure of Rubber

Rubber has a structure consisting of very long-chain molecules which are intertwined and tangled to such an extent that crystallisation is impossible while the material is in an unstressed state. When the rubber is stretched, however, these long-chain molecules begin to straighten and untangle, the net result being that the macromolecules become partly aligned with respect to one another. While in this stressed state the rubber crystallises, and the material stiffens due to increased attractive forces between molecules. When the deforming stress is released, the chains revert to their original tangled state, the material becoming amorphous again. Natural rubber always behaves in this way when stressed, but some synthetic rubbers, such as styrene-butadiene rubber, do not exhibit stress crystallisation; such rubber-like polymers show poor tensile properties unless reinforced with a suitable filler.

Natural rubber is a thermoplastic, softening considerably with increasing temperature. When stressed to any great extent, natural rubber is left with a permanent set of considerable size due to the "sliding" of some chains over others. However, when natural rubber is vulcanised, cross-links of sulphur are formed between adjoining chains. This prevents the sliding of some chains over others, thereby eliminating permanent set after deformation. The carbon-sulphur bonds formed during vulcanisation are, of course, covalent, and thus are strong and permanent. The greater the number of such bonds the less the elasticity of the rubber, until finally ebonite (hard rubber) is formed, having a network structure similar to that of phenol-formaldehyde. Ebonite, unlike other forms of rubber, is not thermoplastic.

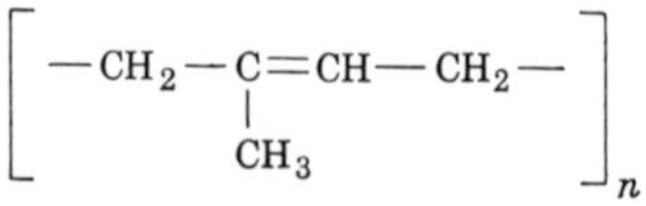

$$\left[-CH_2-\underset{\displaystyle CH_3}{\underset{|}{C}}=CH-CH_2- \right]_n$$

Natural rubber
(isoprene)

```
            CH3
             |     |
   — CH2 — C — CH — CH2 —
             |
             S
             |
   — CH2 — C — CH — CH2 —
             |     |
            CH3    S
            CH3    |
             |     |
   — CH2 — C — CH — CH2 —
             |
```

A

B

Figure 15.20 *Vulcanisation of natural rubber: (A) cross-linking of chain molecules occurs by means of sulphur atoms covalently bonded to carbon atoms of adjoining chains; (B) snarls are connected so that their respective slippages are restricted. (Reproduced with permission from "Nature and Properties of Engineering Materials" by Z. D. Jastrzebski. John Wiley and Sons Inc., N.Y. 1959.)*

Synthetic rubbers may be vulcanised using metallic oxides; for instance, Neoprene uses oxides of zinc and magnesium.

Physical and Mechanical Properties of Natural Rubber

Taking a sample of high-quality natural rubber vulcanised with a small amount of sulphur, we would find that it possesses the following properties.

(1) It will extend to about 8–10 times its original length under the influence of a tensile force, and will carry a load of about 1,000–3,000 psi just prior to breaking. When plotting the stress-strain curve, loads producing certain percentage elongations are determined, plotting being done as for tests on metals. It is immediately apparent that rubber does not obey Hooke's Law. If an unloading curve is plotted, the degree of permanent set can also be ascertained quite readily.

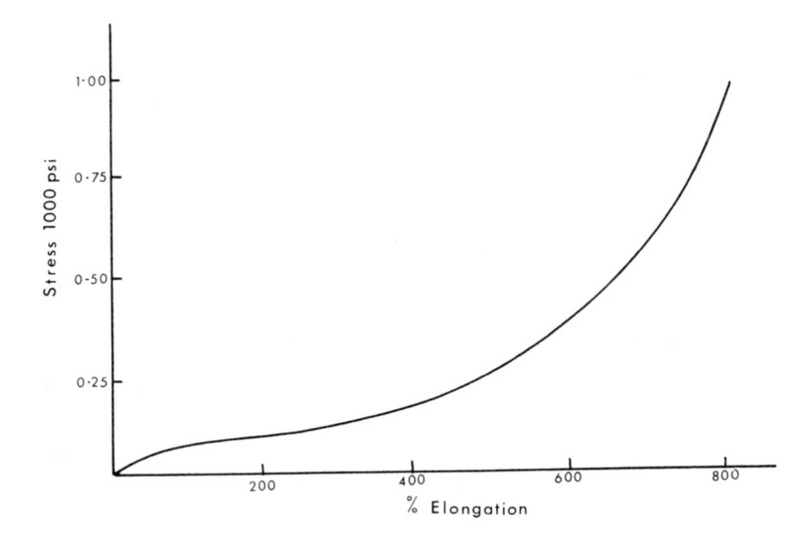

Figure 15.21 *A tensile stress-strain curve for soft rubber (rubber only very lightly vulcanised). Since rubber exhibits a high degree of "elasticity" and yet does not obey Hooke's Law, it is termed an anelastic material and is said to possess anelasticity.*

(2) Vulcanised rubber has extremely good resistance to abrasion; however, this is difficult to assess except on the basis of comparative tests. With certain alterations in composition, however, rubber loses this abrasion resistance (e.g. as in a pencil eraser).

(3) Rubber is a good electrical insulator, but once again composition is important, together with the fact that rubber tends to absorb small amounts of water. Ebonite is an extremely good insulator, being used for all kinds of electrical switchgear.

(4) Resilience, the property which enables rubber to return to its original shape after removal of the deforming load, is important in many applications. It also varies with composition and degree of vulcanisation. Automobile tyres must possess high resilience, otherwise they heat up and the structure of the rubber is destroyed, with unpleasant results.

Synthetic Rubbers

Because rubber must be imported from distant countries, much time and energy has been expended in the development of synthetic rubbers, several of which are of commercial importance today. While most synthetic rubbers are distinctly inferior to natural rubbers, several are intentionally different, possessing certain important properties which make them important for specific applications.

The following list includes some of the more important synthetic rubbers.

(1) *Neoprene*: (polychloroprene), closely related chemically to natural rubber, but possesses superior resistance to oils and greases. It is used for hoses, oil seals and gaskets which are in contact with mineral and vegetable oils or petroleum solvents.

(2) *Styrene Rubber*: (Buna-S or GR-S), produced by the copolymerization of styrene and butadiene, and has similar though somewhat inferior properties to natural rubber. It was developed during World War II and largely replaced natural rubber at that time.

(3) *Polyisoprene*: appears to have the same molecular structure as natural rubber, consequently possesses similar properties, and has similar uses.

(4) *Polyurethanes*: high resistance to many organic solvents, susceptible to acid and alkali attack, and good elastic properties; very suitable for the manufacture of expanded flexible foams.

GLOSSARY OF TERMS

Branched polymer: a polymer in which secondary chains branch from the main molecular chains.

Copolymer: the combination of two polymers into the one chain structure (an analogy may be drawn to some types of metal alloys).

Cross-linking: the bonding together of polymer chains at various points by means of covalent bonds.

Elastomer: a non-crystalline polymer which can be stretched to at least twice its original length by a tensile load and which quickly snaps back to its original length upon removal of the deforming load.

Filler: a relatively inert substance added to a polymer either to merely increase its bulk or to impart special properties to the polymer.
Functionality: the number of reaction sites possessed by a monomer at which polymerization can occur.
Macromolecular: a structure characterised by massive molecules which are either arranged in long chains or in a non-crystalline three-dimensional network.
Mer: the smallest repeating unit in any polymer.
Monomer: a molecule which supplies a single mer to a polymeric structure.
Polymer: a solid composed of long molecular chains having a basic repeating pattern of structure.
Polymerization: the process by which a monomer is transformed into a polymer.
Rubber: a polymeric material that is also an elastomer and which possesses a high elastic yield strain.
Thermoplastic: (thermosoftening polymer): a polymer that softens when heated and rehardens upon subsequent cooling.
Thermosetting polymer: (thermohardening polymer): a polymer which, once set, cannot be resoftened and reworked by heating.
Unsaturated hydrocarbon: a hydrocarbon compound in which some carbon atoms possess multiple covalent bonds.
Vulcanisation: that process which when applied to a rubber causes controlled cross-linking between the molecular chains.

REVIEW QUESTIONS

1. Define the term "polymer" and list the three important characteristics of all polymers.
2. Compare and contrast the processes of addition polymerization and condensation polymerization using examples to illustrate your answer.
3. Define the terms: monomer; mer; unsaturated hydrocarbon; initiator.
4. How does a copolymer differ from a normal polymer, and what similarities, if any, exist between metal alloys and copolymers?
5. What general effects can fillers have in polymers?
6. Sketch the following types of polymer structures and explain how they influence the properties of polymers: (i) network structures; (ii) extensive cross-linking of chains; (iii) branched chains. Use examples to illustrate your answer.
7. What is an elastomer?
8. Distinguish between thermoplastic and thermosetting polymers in terms of their typical properties and molecular structures.

9. List and briefly discuss five important properties of polymers that make them useful engineering materials.

10. Contrast the mechanisms of elastic and plastic deformation in polymers with those mechanisms that operate in metals.

11. Briefly outline the method whereby the following articles can be mass-produced from polymeric materials: (i) flexible narrow-necked containers; (ii) transparent continuous flexible sheeting; (iii) tube; (iv) auto tyres; (v) flexible baby baths. In each case suggest a suitable material and justify your choice.

12. How is natural rubber vulcanised, and what effects does vulcanisation have on the mechanical properties of rubber?

13. Sketch a tensile stress-strain curve for vulcanised rubber and compare it to a tensile stress-strain curve for a ductile metal. Does rubber obey Hooke's Law? If not, how can its elasticity be described?

14. Electric iron handles are to be mass-produced. Suggest a suitable polymer together with a suitable manufacturing process. Explain how the handle would be made, and justify the use of a polymer rather than a metal.

16

Ceramics

CERAMIC materials have already been defined as those containing phases that are compounds of metals and non-metals. Hence, the art and science of ceramics is concerned with the manufacture of solid articles from inorganic non-metallic materials. Thus, ceramics includes such materials as pottery, porcelain, structural clay products, refractories, abrasives, cements, and glass. The wide range of ceramic materials available occurs because of the many chemical combinations of metals and non-metals that can be effected, and in general, ceramics have properties quite unlike those of either metals or organic polymers.

CLASSIFICATION OF CERAMICS

Many different classifications of ceramic materials are in use in industry, with perhaps the most common form describing ceramics in terms of their functions and properties. Table 16.1 is an example of such a classification.

TABLE 16.1

FUNCTIONAL CLASSIFICATION OF SOME CERAMICS

Group	*Examples*
Pure oxide ceramics	MgO, Al_2O_3, SiO_2
Fired-clay products	Bricks, tiles, earthenware pipes, porcelain
Refractories	Silica brick, magnesite
Inorganic glasses	Window glass, lead glass, borosilicate glass
Devitrified glasses	"Pyroceram"—a form of crystallised glass cooking ware
Abrasives	Alumina, carborundum
Cementing materials	Portland cement, lime
Rocks	Granites, sandstone
Minerals	Quartz, calcite

This type of classification is useful to the engineer since the group names are indicative of particular industries and industrial applications; however, few of these groups are mutually exclusive. For instance, alumina (Al_2O_3) is a pure oxide ceramic, an abrasive of great industrial importance and a refractory. A more useful classification in which ceramics are grouped according to structural criteria uses the following four major divisions:

(1) *Crystalline ceramics*—which may be single phase like MgO or multi-phase like of certain refractories from the $MgO–Al_2O_3$ binary system.

(2) *Non-crystalline ceramics*—including all natural and synthetic inorganic glasses, e.g. window glass.

(3) *"Glass-bonded" ceramics*—in which the crystalline phases are held in a glassy (vitreous) matrix, e.g. fired clay products.

(4) *Cements*—which may be crystalline, or which may contain crystalline and non-crystalline phases.

STRUCTURE OF CRYSTALLINE CERAMICS

Crystalline ceramic phases may exhibit purely ionic bonding (as in sodium chloride), purely covalent bonding (as in silicon carbide), or bonds which have both ionic and covalent characteristics (as in SiO_2 and MgO). In general, the crystal structure of any ceramic phase is dependent on the relative sizes of the atoms involved and on their electronic configurations, the latter determining the bonding tendencies of the atoms themselves.

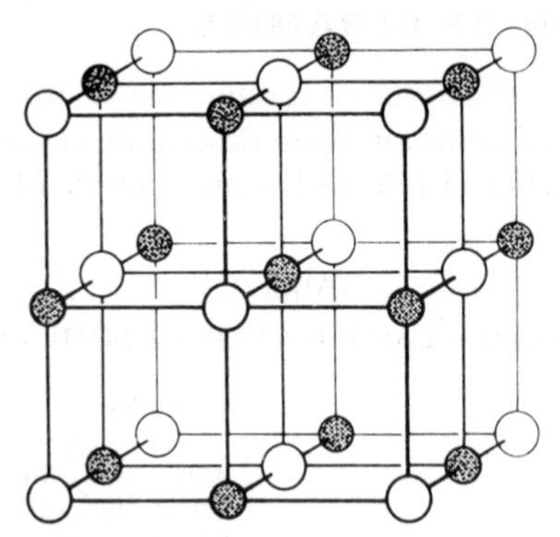

Figure 16.1 *The crystal structure of rock salt (sodium chloride). The shaded spheres represent the centres of the sodium atoms.*

Crystal structures of ceramic phases are therefore much more complex than those of metal crystals; variations of the normal close-packed structures do occur, the anions present forming normal close-packed structures while the cations fill some or all of the interstices present in the structure. The following are the common crystal structures found in crystalline ceramics, particularly those of the oxide type.

Rock Salt Structure: Many halides and oxides crystallise in the cubic structure characteristic of $NaCl$, in which both anions and cations have co-ordination numbers of 6. In order to achieve this structure, the valencies of both anions and cations must be equal. The structure can be thought of as an FCC arrangement of anions with cations filling all available interstitial sites, so that the crystal lattice has four anion and four cation positions per unit cell. Examples: MgO, CaO, BaO, SrO, CdO, MnO, FeO, CoO, NiO, LiF, MnS, AgS.

Wurtzite Structure: The crystal structure of wurtzite (ZnS) and beryllium oxide (BeO) is one in which the larger anions are arranged in an HCP structure, half of the tetrahedral interstices being filled by the smaller cations so that maximum cation separation occurs.

Zincblende Structure: Silicon carbide (βSiC), cadmium sulphide (CdS), aluminium phosphide (AlP), and zincblende (ZnS) form a structure in which the larger anions are arranged in an FCC structure, half of the available tetrahedral interstices being filled with the smaller cations.

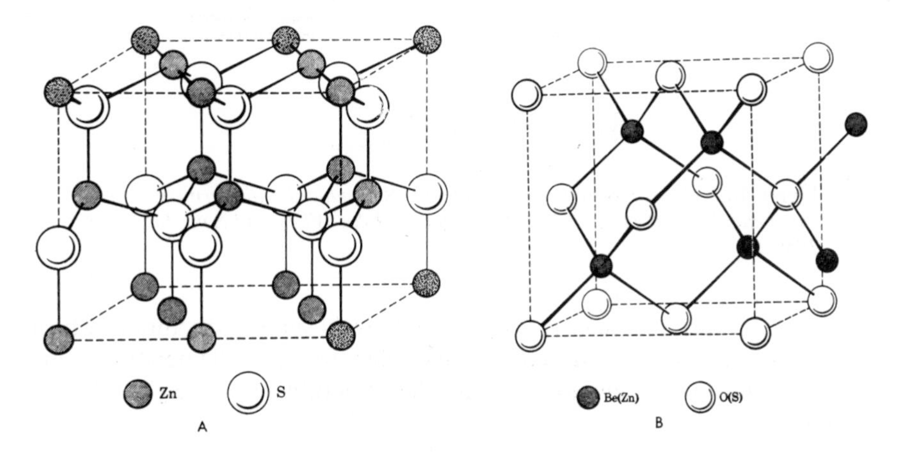

Figure 16.2 *The crystal structures of (A) wurtzite and (B) zincblende. (Reproduced with permission from "Introduction to Ceramics" by W. Kingery. John Wiley and Sons Inc., N.Y. 1960.)*

Spinel Structure: The spinels are an important group of oxides having the general formula AB_2O_4 where A and B are different metals; for example, magnesium aluminate spinel ($MgAl_2O_4$), and zinc ferrospinel ($ZnFe_2O_4$). The crystal structure present in spinels can be viewed as a combination of the rock salt and zincblende types, the features of this structure being, *(a)* the oxygen ions are in an FCC pattern; *(b)* the A^{2+} metal ions take up the tetrahedral sites; and *(c)* the B^{3+} metal ions occupy the octohedral sites.

The unit cell thus consists of thirty-two oxygen ions, sixteen B^{3+} metal ions (in octohedral positions), and eight A^{2+} metal ions (in tetrahedral positions).

Variations of this structure occur for other types of spinels. The magnetic properties of such minerals depend upon cation distributions.

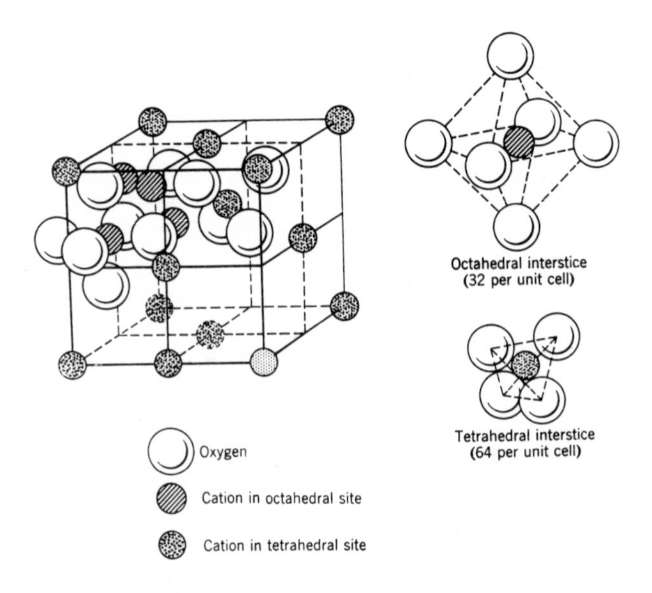

Figure 16.3 *Crystal structure of a spinel. (Reproduced with permission from "Introduction to Ceramics" by W. Kingery. John Wiley and Sons Inc., N.Y. 1960.)*

Cesium Chloride Structure: This is a relatively simple structure in which the chlorine ions are arranged in a simple cubic structure, all interstices being occupied by the cesium ions. Thus, the structure reveals eight-fold co-ordination.

Fluorite Structure: Fluorite (CaF_2) has a structure in which the Ca^{2+} ions are arranged in an FCC pattern while the smaller F^- ions fill some of the available interstitial sites. An interesting feature is the large vacancy in the centre of the unit cell. Uranium dioxide (UO_2) has this structure, the large number of vacant sites making it ideal for a nuclear fuel.

Common Features of Structures

Other oxide structures do exist but the foregoing discussion is sufficient for our purposes. The most striking feature of all of the crystal structures examined is their dependence on primitive cubic or close-packed oxygen arrangements into which cations are fitted in varying interstitial patterns.

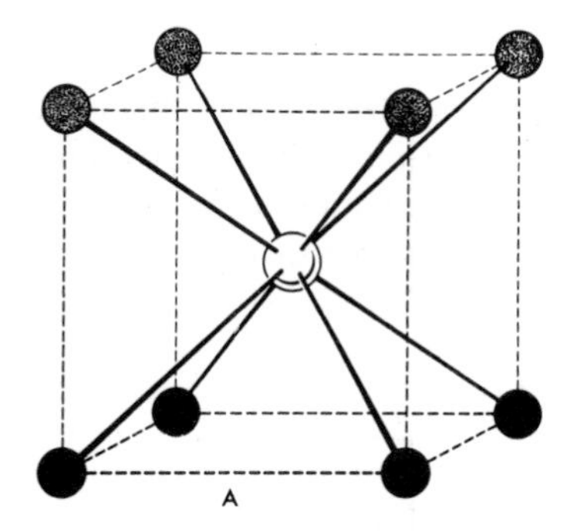

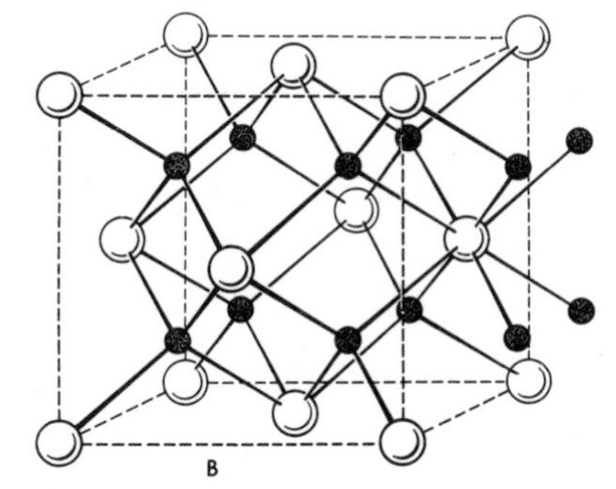

Figure 16.4 *Crystal structures of (A) cesium chloride and (B) fluorite. The shaded spheres represent the fluorine atoms. (Reproduced with permission from "Introduction to Ceramics" by W. Kingery. John Wiley and Sons Inc., N.Y. 1960.)*

This results in crystal structures in which *(a)* the overall symmetry is considerably less than that found in simple metal crystals, and *(b)* the various ions present vary considerably in diameter. For instance, the cation/anion size ratios for simple rock salt structures vary between 0.732 and 0.414, and that of the wurtzite structure is as low as 0.25. These two considerations have considerable importance in the study of the mechanical behaviour of pure oxide and related types of ceramics.

TABLE 16.2

TYPES OF SILICATE STRUCTURES

Silicate Type	*Structural Units*	*Arrangement of Tetrahedra*	*Examples*
Orthosilicates	$(SiO_4)^{4-}$	Independent tetrahedra	Zircon, garnet, the olivine minerals
Pyrosilicates	$(Si_2O_7)^{6-}$	Pairs of adjacent tetrahedra share one corner	Such silicates are rare
Metasilicates	$(SiO_3)_n^{2n-}$	Tetrahedra linked across two corners, hence forming chain or ring structures	Pyroxines and amphiboles
Framework Structures		A three-dimensional framework of tetrahedra	Felspars
Layer Structures	$(Si_2O_5)_n$	The tetrahedra join at their corners forming a layer structure rather than a framework	Clay minerals

Structure of the Silicates

The silicates form a very large and important group of ceramic materials. While most of the silicates are chemically complex, their basic structures possess a simplicity of order not found in many other large groups of materials. The fundamental structural unit of all silicates is the *silicon-oxygen tetrahedron* in which one silicon atom is surrounded by four oxygen atoms such that the oxygens occupy the corners of a tetrahedron which has a silicon atom as its centre. The different silicate structures arise from the various ways in which the silicon-oxygen tetrahedra combine together. Table 16.2 sets out the common groups of silicates and their structural arrangements.

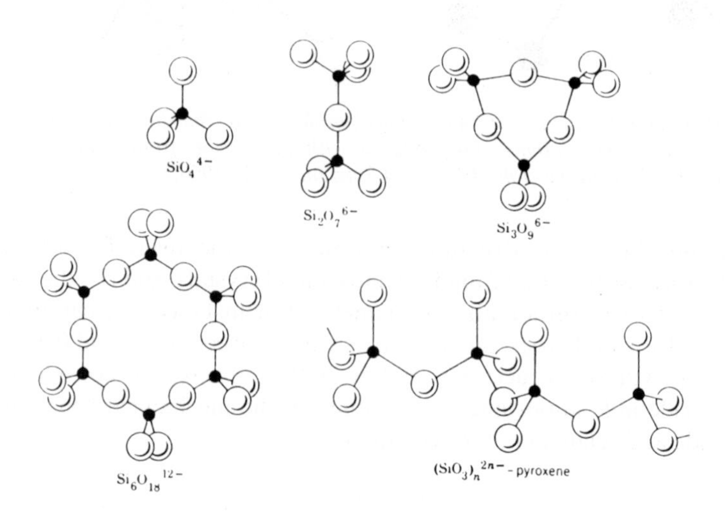

Figure 16.5 *Some common silicate ion and chain structures showing some of the ways in which the* SiO_4^{4-} *cross-link together in silicates. (Reproduced with permission from "Introduction to Ceramics" by W. Kingery. John Wiley and Sons Inc., N.Y. 1960.)*

STRUCTURE AND COMPOSITION OF GLASS

Inorganic glasses are fusion products which have failed to crystallise upon cooling to room temperature. Silica is the best glass-former, and forms the basis of all commercial glasses, but the oxides of boron (B_2O_3), and phosphorus (P_2O_5), also form glasses when suitably heated and cooled. Table 16.3 shows the compositions of four common commercial glasses in terms of the oxide components used in their manufacture.

The oxide components added into a glass batch may be subdivided into groups on the basis of functions that they perform within the glass.

Glass formers or network formers are those oxides which are indispensable to the formation of glass since they form the basis of the random three-dimensional network of the glass.

TABLE 16.3

COMPOSITIONS OF FOUR COMMON GLASSES

Oxide Components	*Soda-Lime Glass*	*Lead Glass*	*Borosilicate Glass*	*High-Silica Glass*
SiO_2	70—75	53—68	73—82	96
Na_2O	12—18	5—10	3—10	—
K_2O	0—1	1—10	0.4—1	—
CaO	5—14	0—6	0—1	—
PbO	—	15—40	0—10	—
B_2O_3	—	—	5—20	3
Al_2O_3	0.5—2.5	0—2	2—3	—
MgO	0—4	—	—	—

Intermediates are those oxides that approach glass forming properties but are not really true glass formers; they can be added in fairly high proportion to the glass batch if required. Intermediates link in with the basic glass network, providing continuity of structure where required.

Modifiers are devoid of glass forming tendencies, but serve to modify the properties of the glass quite considerably.

Table 16.4 lists the common oxides used in glass manufacture under the headings of glass formers, intermediates, and modifiers.

TABLE 16.4

FUNCTIONAL CLASSIFICATION OF OXIDES USED IN GLASS MANUFACTURE

Glass Formers	*Intermediates*	*Modifiers*
SiO_2	Al_2O_3	MgO
B_2O_3	Sb_2O_3	Li_2O
GeO_2	ZrO_2	BaO
P_2O_5	TiO_2	CaO
V_2O_5	PbO	SrO
As_2O_3	BeO	Na_2O
	ZnO	K_2O

It is important to realise that there can be no sharp distinctions drawn between the three classes of oxides listed in Table 16.4, and it is quite common for one oxide to act as an intermediate in one glass composition and as a modifier in another.

Some oxides introduced into a glass batch act as *fluxes* in that they lower the melting point of the batch and render the molten glass workable at reasonable temperatures. Fluxes may reduce the resistance of glass to chemical attack, render it water-soluble, or make it subject to partial or complete devitrification upon cooling. To overcome these problems *stabilisers* are added to the glass batch. Intermediates and modifiers act as fluxes or stabilisers under well-defined conditions. For example, in the common "soda-lime-silicate" glass used for windows, mirrors, and bottles, the silica is the glass former, the soda (Na_2O) and potash (K_2O) act as fluxes, and the stabilisers are lime (CaO), magnesia (MgO), and alumina (Al_2O_3).

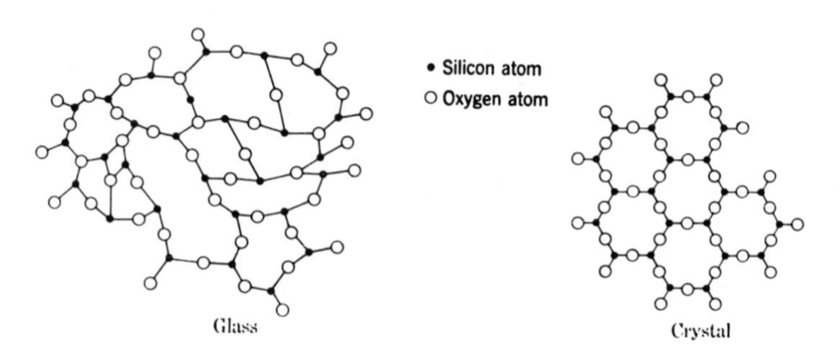

Figure 16.6 *Non-crystalline (glassy) and crystalline forms of silica. (Reproduced with permission from "Textbook of Engineering Materials" by M. Nord. John Wiley and Sons Inc., N.Y. 1952.)*

If an oxide is to form a glass when cooled from a melt, its cation-oxygen bond must be flexible enough to allow the disordered three-dimensional glass network to form, and also strong enough to maintain this network once it has formed. In general, these conditions of cation-oxygen bond strength will only be satisfied when the cation has a relatively high charge combined with a fairly small ionic radius.

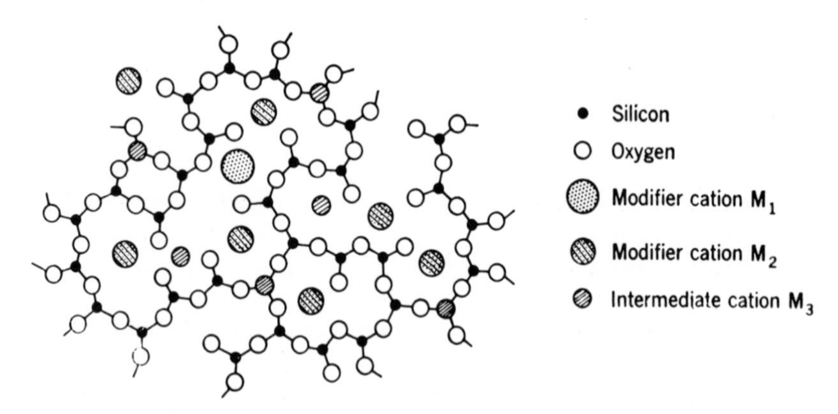

Figure 16.7 *Functions of modifiers and intermediates in a silica glass. (Reproduced with permission from "Glass" in Volume 10 of Kirk-Othmer, Encyclopedia of Chemical Technology, 2nd Edition. John Wiley and Sons Inc., N.Y., 1966.)*

Laboratory (49): *Weigh out the following and grind them together in a mortar and pestle:*

anhydrous sodium carbonate	*2g*
calcium carbonate	*1g*
sand (SiO_2)	*1g*

Take a platinum wire loop, heat in a bunsen flame, and pick up some of the mixture. Heat this mass until it fuses to the wire. Now pick up some more of the mixture and fuse it to the existing bead. Examine the bead and then test its brittleness by hitting it with a hammer. Discuss the functions of the above three substances in the formation of the bead.

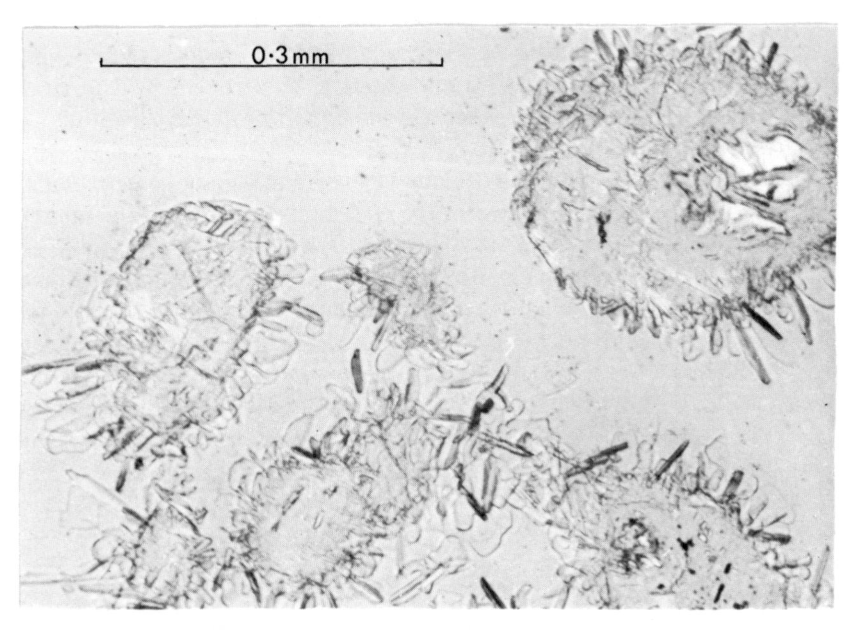

Figure 16.8 *Photomicrograph showing stones in a glass batch. In this case the stones originated from partly melted sand (quartz) grains; the acicular crystals growing from the crystalline inclusions into the glass matrix are high temperature forms of quartz. (Photograph courtesy of Australian Consolidated Industries Ltd.)*

Devitrification of Glass

Under certain conditions any glass melt will become contaminated with crystalline particles. A glass in this condition is termed *devitrified*, which is simply another way of stating that it has partially or completely crystallised. Devitrified glass, unless of the very special variety that is produced deliberately, is undesirable since the crystalline areas are extremely weak and brittle, as well as being translucent. Crystalline segregations in glass are known as *stones*. Figure 16.8 shows a stone in a blown bottle.

Recrystallised Glasses

Recrystallised glass, also known as polycrystalline glass, is commonly produced by adding nucleating agents to the glass batch. The glass can be formed into the desired shape by any of the conventional glass forming processes, and is then heat-treated to promote recrystallisation. Recrystallised glass possesses increased impact strength, hardness, and thermal shock resistance compared with conventional non-crystalline glasses. One commonplace application of a recrystallised glass is in the manufacture of the so-called refrigerator-to-oven cooking dishes.

Glass Fibre

Glass fibre, or FIBREGLASS* as it is commonly known, is glass in fibre form. It is made by one of several processes, each of which involves the drawing out of the filaments from glass in the viscous state. Modern technical developments date from World War I when Germany had to find a substitute for asbestos insulating materials.

The process whereby continuous filament glass fibre is made is of particular interest. The glass batch is melted in the glass tank furnace and the molten glass is moulded into ordinary sized marbles. These marbles are fed into an electric furnace that contains a platinum-alloy bushing containing more than one hundred tiny holes. When the glass marbles reach the holes, the viscous

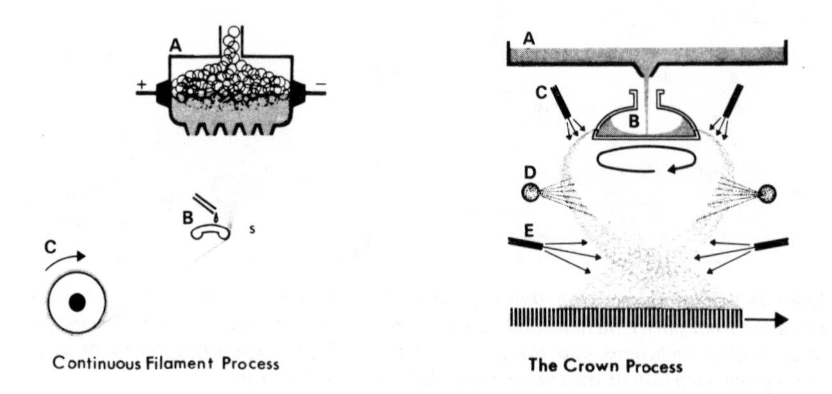

Figure 16.9 *Two methods of manufacturing glass fibre. The diagram of the continuous filament process shows the platinum bushing (A), the collection of individual filaments (B), and the winding of the continuous strand on to a spindle (C). The diagram of the Crown process shows the forehearth (A), the rotating dish (B), the blowing rings (C), the spraying of the polymer binder (D), and the second blowing process that ensures random fibre orientation (E). (Reproduced with permission from "Glass In The Modern World" by F. J. T. Maloney. Aldus Books Ltd., London, 1967.)*

*FIBREGLASS is really a registered trade name.

glass is teased through the holes and drawn down vertically, forming over one hundred individual filaments. These filaments can then be twisted into a single fibre and spun on normal textile spindles as required. About one third of an ounce of glass produces about one hundred miles of filament at the rate of about 6,000 feet per minute.

Glass wool, on the other hand, is produced by an entirely different type of process. The most common technique, known as the Crown process, produces a thick matte of short glass fibres which are held together by a polymer binder. In the Crown method, a thick stream of molten glass is poured into a rapidly rotating steel dish that has hundreds of tiny vents around its periphery. The glass is forced through these vents by centrifugal force, forming relatively short fibres of diameters of about 0.0007 cm. The matte is then passed through curing ovens where the binder sets and is then cut into sizes suitable for insulation.

STRUCTURE AND COMPOSITION OF CLAY-BODY CERAMICS

Ceramics formed from the controlled firing of clays are complex in structure since they always contain both crystalline and non-crystalline (glassy) phases. The properties of clay-body ceramics depend very largely upon the nature of the glassy phase present.

The Nature of Clays

Clays are naturally-occurring inorganic materials that are essentially the weathered remains of various types of rocks. Although great variations in chemical composition and colour occur, all clays possess the following important properties. (*a*) When moist they are plastic; (*b*) they become rigid when dried but regain their plasticity when rewetted; and (*c*) when fired they become mechanically strong, hard, and permanently non-plastic.

Rocks such as granite, pegmatite, syenite, basalt, and dolerite weather down to form clay deposits which contain clay minerals, residual partially altered minerals, and organic matter. Clay minerals are hydrated alumino-silicates, one common theory stating that the clay minerals are derived from the feldspar minerals present in the parent rock. The two most common clay minerals are kaolinite and montmorillonite, which have the following oxide formulae:

$$\text{Kaolinite } Al_2O_3.2SiO_2.2H_2O$$
$$\text{Montmorillonite } Al_2O_3.4SiO_2.nH_2O$$

These conventional oxide formulae of clays do not indicate anything about the actual crystal structures of the clay minerals. All clay mineral crystals are layer or sheet structures in which negative silicate ions are interleaved with positive hydrated metal ions. Kaolinite is typical; the silicate ions are $(Si_2O_5)^{2-}$, and the hydrated metal ions are $[Al_2(OH_4)]^{2+}$. This ionic structure gives rise to the characteristic small "platey" or "flakey" shape of the clay mineral crystals which is largely responsible for the plasticity of moist clay substance. The crystals are themselves very small, with diameters rarely exceeding 10 microns (0.01mm).

The plasticity of clay varies with its water content. In a plastic clay the water molecules are weakly bonded to the clay crystals by secondary bonds; when the clay is deformed, many of these bonds are broken and the water film acts as an internal lubricant, allowing the clay crystals to slide readily over each other. Since many forming processes for clay depend upon plasticity, the control of water content is of the utmost importance.

Clay Bodies

Pure clay mineral is very rarely used for the manufacture of fired ceramics. In most cases, a clay body is made by combining several different clay minerals with a certain proportion of *non-plastics*, such as crushed and ground quartz, feldspar, talc, or "grog", the latter being finely ground previously-fired clay material. In general, non-plastics have one of the following effects: they alter the plasticity of the clay body, thus making it more suitable for various forming processes; they act as fluxes, causing greater degrees of vitrification at various temperatures during firing; and they reduce the drying and firing shrinkages of the clay body.

The Classification of Clay-Body Ceramics

Fired clay-body ceramics are usually classified on the basis of their properties after firing and four different groups are generally recognised. The characteristics of these groups are briefly outlined in Table 16.5.

TABLE 16.5

TYPES AND CHARACTERISTICS OF CLAY-BODY CERAMICS

Body Type	*Apparent Porosity*	*Firing Range*	*Typical Applications*
Earthenware	Usually 6–8% may exceed 15%	800°–950°C, but may exceed 1000°C	Porous drainage pipes, ceramic filters, wall tiles, bricks
Bone China (also termed fine china)	Low—usually less than 1%	1100°–1200°C	Fine tableware
Stoneware	Less than 3%; usually between 1–2%	Greater than 1250°C	Glazed drainage pipes, roofing tiles, tableware, tiles
Porcelain	Less than 1%; often practically zero	1300°–1450°C	Fine tableware, scientific equipment

It is important to realise that stoneware and porcelain, both very low with respect to apparent porosity, do possess significant amounts of total porosity. Thus, when these bodies are used where porosity is a problem, as in sanitary ware and drain pipes, they are always glazed.

Stoneware is distinguished from porcelain in that it is a much coarser body and is opaque, whereas porcelain is characteristically fine and translucent. Bone china, another type of clay ware, is similar in appearance to true porcelain but is fluxed with alkaline fluxes which promote vitrification at temperatures between 1100°–1200°C, the original flux being calcined animal bones (hence the name "bone china"). In practice it is very difficult to distinguish between a "china" and a porcelain.

The Effects of Firing on Clays

When a clay body is dried and fired, its structure alters completely, and a hard, rigid material is formed which is sometimes termed "artificial stone". The main effects of drying and firing upon a typical clay-body ceramic are summarised in Table 16.6.

TABLE 16.6

EFFECTS OF FIRING ON THE CLAY BODY

Stage	*Range*	*Main Alterations in Body*
Drying	Up to 100°C	The "water of plasticity" is driven off the air-dried clay and the clay becomes rigid but brittle.
Dehydration	100° to 700°C	The clay minerals lose their "water of crystallisation", kaolinite losing the bulk of its water of crystallisation between 450°C to 510°C.
Oxidation	550° to 900°C	Iron compounds present in the body are oxidised to Fe_2O_3 and all carbonaceous impurities are burnt out before 800°C. The clay material now becomes meta-kaolin $2Al_2O_3.4SiO_2$, but retains pseudomorphs of its original structure. Sintering begins, and any alkaline fluxes present begin to form a liquid phase, in which the alumina from the body dissolves.
Vitrification	From 950°C upwards	Crystallisation of mullite begins; the mullite ($Al_6Si_2O_{13}$) crystals grow as the temperature is increased, and the glassy phase contracts, causing a severe shrinkage. Mullite appears in two forms—as primary mullite, appearing at temperatures as low as 980°C, and as secondary mullite, appearing as large acicular crystals after the temperature passes 1200°C.

The final structure of the fired and cooled clay body will contain the crystalline phases known as primary and secondary mullite embedded in a glassy matrix, together with altered or unaltered remains of whatever non-plastics were present. Quartz, for example, is a common constituent of fired clays.

Laboratory (50): *Measure out the following and grind them together in a mortar and pestle:*

commercial zircon (300 mesh)	*90g*
air-floated Kaolin clay	*10g*
70% commercial phosphoric acid	*5ml*

Sieve the resulting mixture through a 60 mesh sieve and ram it to shape in a steel die (say to the shape of a small crucible). Fire the rammed mixture to 700°–800°C using a propane torch and allow it to cool. Discuss the changes in structure that have occurred within the ceramic material.

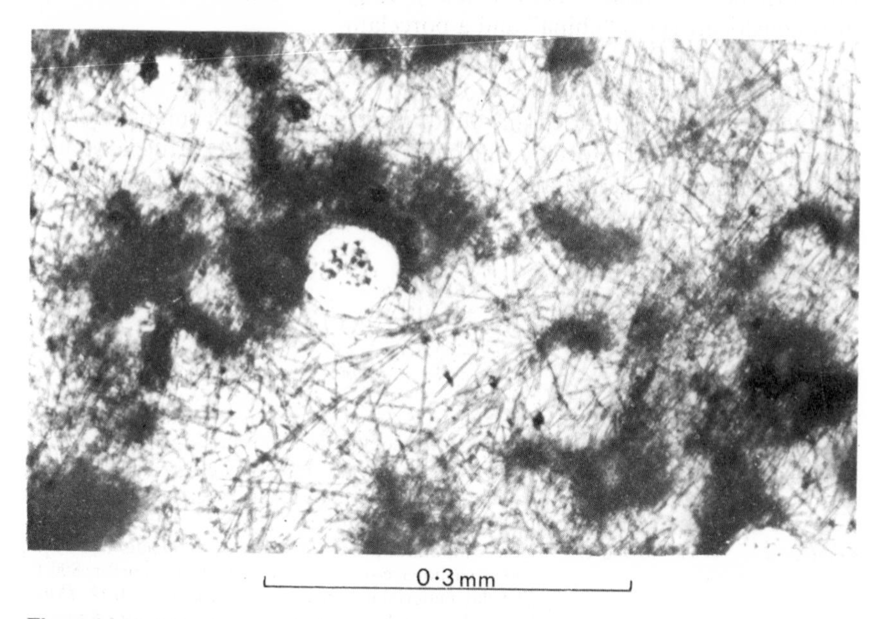

Figure 16.10 *A highly-magnified section of a low alumina refractory brick showing needle-like mullite crystals embedded in a glass matrix. This is an example of a glass-bonded ceramic. (Photograph courtesy of Australian Consolidated Industries Ltd.)*

THE STRUCTURE AND COMPOSITION OF CEMENTS

Inorganic cements are materials that will set and harden when made into paste form with a sufficiency of water, and are classified as either hydraulic or non-hydraulic depending upon the natures of their hardening mechanisms. Hydraulic cements, like Portland cement, will set and harden under water, whereas non-hydraulic cements set in air.

Lime mortar is a non-hydraulic cementing material that has a long history. Lime is made by calcining calcium carbonate at about 1,000°C at which temperature quicklime (CaO) is formed. The quicklime is then slaked with water to form powdered slaked lime ($Ca(OH)_2$) which is mixed with about three parts sand and water to make lime mortar. The lime mortar sets as the excess water dries out, and hardens due to the action of atmospheric carbon dioxide, according to the following reaction.

$$Ca(OH)_2 + CO_2 \longrightarrow CaCO_3 + H_2O$$

Plaster of Paris is formed when gypsum ($CaSO_4.2H_2O$) is heated to about 180°C. Plaster of Paris sets when mixed with water to form a paste.

Keenan's cement, on the other hand, is formed when gypsum is calcined at about 550°C, at which temperature all the water of crystallisation of the gypsum is driven off. Keenan's cement is much stronger than Plaster of Paris and does not reveal any appreciable solubility in water after the initial hardening is complete. It is thus useful in wall construction and in the manufacture of imitation marble.

Portland cement is the most important type of cement used for all forms of construction. Portland cement is made by "burning" together a finely powdered mixture of calcareous and argillaceous materials such as limestone and clay or shale, and grinding the resulting *clinker* to a fine powder. The temperatures reached in the cement kiln almost promote complete fusion of the ingredients, and the cement clinker emerges from the lower end of the rotary kiln as hard round balls. During the grinding process small amounts of gypsum may be added to the cement powder in order to prevent the cement from taking a "flash set" when mixed with water.

Table 16.7 gives an approximate range of oxide compositions for Portland cement.

TABLE 16.7

APPROXIMATE COMPOSITION OF PORTLAND CEMENT (OXIDE ANALYSIS)

Oxide Component	*Percentage Range*
Silica	20—24
Iron oxide	2— 4
Alumina	1—14
Lime	60—65
Magnesia	1— 4
Sulphur trioxide	1— 1.8

While the composition of cement powder can be analysed into oxide components as in Table 16.7, its actual chemical composition is a mixture of four minerals. These are tricalcium silicate ($3CaO.SiO_2$), dicalcium silicate ($2CaO.SiO_2$), tricalcium aluminate ($3CaO.Al_2O_3$) and tetracalcium alumino-ferrite ($4CaO.Al_2O_3.Fe_2O_3$).

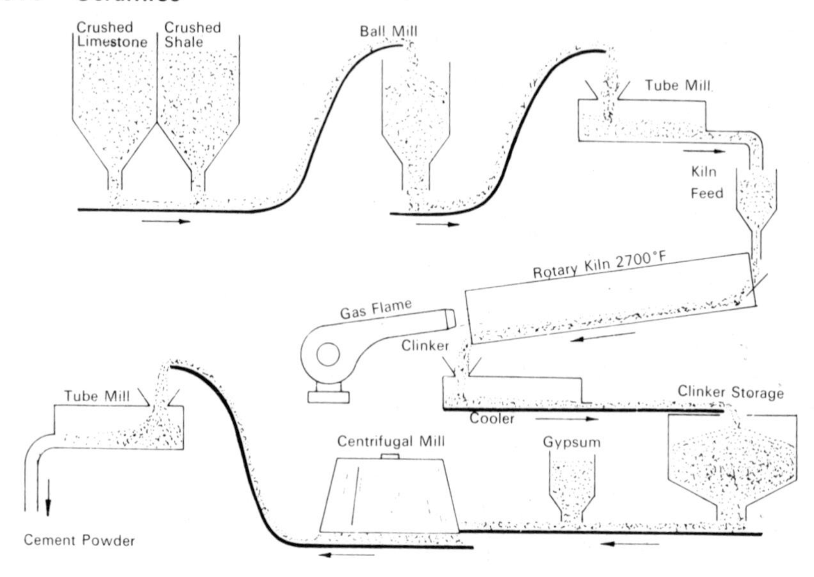

Figure 16.11 *Diagrammatic representation of the manufacture of Portland cement by the dry process.*

Portland cement sets by a series of complex hydration reactions which result in the formation of several hydrates and a relatively complex silicate gel which has been called *tobermorite gel* since it resembles the naturally-occurring silicate rock known as tobermorite. The setting reactions may be represented as follows:

$$2(3CaO.SiO_2)+6H_2O = 3CaO.2SiO_2.3H_2O+3Ca(OH)_2$$
tobermorite gel

$$2(2CaO.SiO_2)+4H_2O = 3CaO.2SiO_2.3H_2O+Ca(OH)_2$$
tobermorite gel

$$4CaO.Al_2O_3.Fe_2O_3+10H_2O+2Ca(OH)_2 = 6CaO.Al_2O_3.Fe_2O_3.12H_2O$$
calcium alumino-ferrite hydrate

$$3CaO.Al_2O_3+12H_2O+Ca(OH)_2 = 3CaO.Al_2O_3.Ca(OH)_2.12H_2O$$
tetracalcium aluminate hydrate

$$3CaO.Al_2O_3+10H_2O+CaSO_4.2H_2O = 3CaO.Al_2O_3.CaSO_4.10H_2O$$
gypsum — calcium monosulpho-aluminate

The tobermorite gel accounts for about 50% of the bulk of the set cement and very largely determines its strength properties.

THE MECHANICAL PROPERTIES OF CERAMIC PHASES

In the following discussion of the mechanical properties of ceramics it will become clear that most ceramics are stronger in compression than in tension. This fact is taken into consideration in the application of ceramics to engineering; for example, ceramics like brick, cement, and glass are always used in compression and not in tension.*

The Mechanical Properties of Crystalline Ceramics

It is clear from Chapter 7 that the plastic deformation of crystalline materials occurs because of the movement of dislocations along well-defined planes in the crystal structure. Crystallographically, the deformation process results in *slip*, in which adjacent crystal elements glide over one another, or in *twinning*, in which homogeneous shearing occurs with a consequent displacement of one part of the crystal with respect to the other. However, since slip is far more likely to occur during plastic deformation, the following discussion is restricted to it alone.

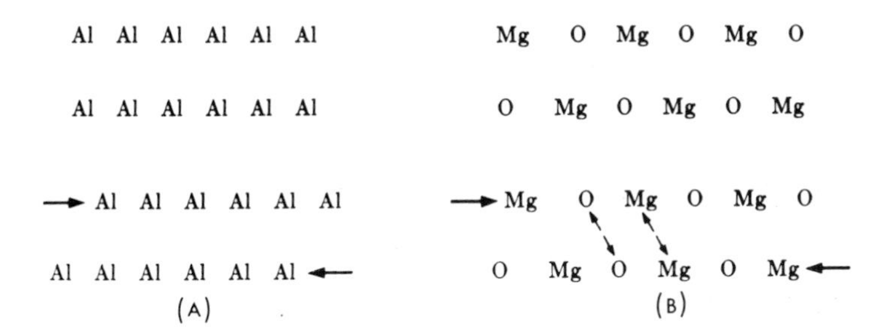

Figure 16.12 *A schematic comparison of slip in (A) a metal and (B) the oxide ceramic MgO. More force is required to displace the atoms in the ceramic since strong repulsive forces between ions have to be overcome. (Reproduced with permission from "Elements of Materials Science", 2nd Edition, by L. H. Van Vlack. Addison-Wesley Publishing Co., Reading. Mass. 1964.)*

Restrictions on the operation of likely slip systems in ceramic crystals result from the following geometric and electrostatic considerations.

(1) The atomic particles in ceramic crystals are significantly different with respect to their atomic or ionic radii. Therefore, slip is mechanically inhibited due to the uneven natures of most of the possible slip planes.

(2) Many ceramic crystals possess significant amounts of ionic bonding, and consequently have crystal structures made up of both positively and

*Reinforced concrete is an exception. See Chapter 17.

negatively charged particles (anions and cations). Slip will be restricted if it tends to bring similarly charged particles closer together.

(3) Ceramic crystals usually possess low symmetry since their structures normally exhibit low co-ordination. This in turn limits the total number of planes that could possibly act as slip planes.

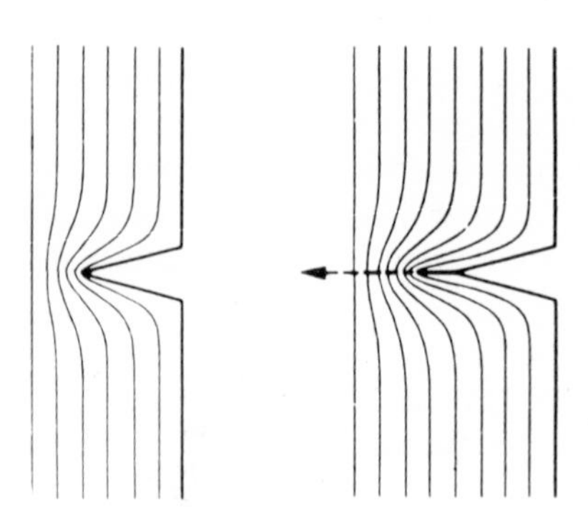

Figure 16.13 *Stress concentrations in a non-ductile material. A non-ductile (brittle) material cannot adjust to these stresses; therefore, while the average stress may be low, the tensile strength may be locally exceeded, causing cracking. Such a crack, once formed, is readily propagated across the section. (Reproduced with permission from "Elements of Materials Science" by L. H. Van Vlack. Addison-Wesley Publishing Co., Reading, Mass. 1959.)*

Thus, while most metallic crystals exhibit good ductility coupled with a distinct tendency to fail in shear, most ceramic crystals are characterised by brittleness and a tendency to fail in tension by cleavage. The restrictions on plastic deformation by slip also result in the high compressive strengths of ceramic crystals, provided no porosity is present. Tensile strengths are theoretically high, but in practice are usually quite low because of the effects of stress concentrations. In ductile materials the effects of stress concentrations such as small surface cracks and flaws are relieved by localised plastic deformation around the defect. In brittle materials such relief is not possible, and crack propagation occurs if the strength of the material is locally exceeded (see Figure 16.13). However, the presence of stress concentrations does not greatly affect compressive strength since loads are transferred across the crack and are not concentrated at the root.

The Mechanical Properties of Glass

Non-crystalline solids do not exhibit the long-range order characteristic of crystals, nor do they possess slip planes along which plastic deformation can proceed. In complete contrast to crystalline materials, which plastically deform by slip, glass plastically deforms by the mechanism of *viscous flow*. The rate at which viscous deformation can occur in glass depends primarily upon the magnitude of the applied stress, but is also somewhat dependent upon composition and structure.

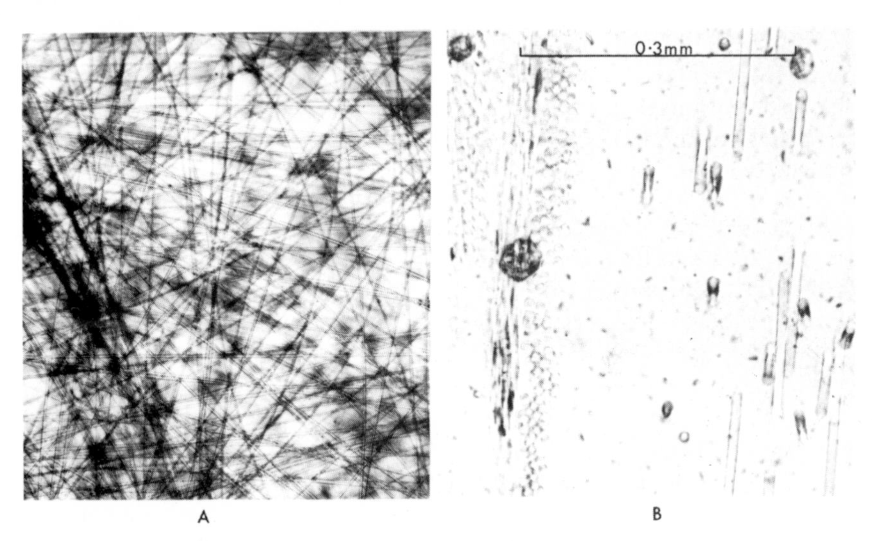

Figure 16.14 *Glass fibres in a polymer matrix: (A) Under low magnification the random orientations of the fibres can be clearly seen; (B) the shapes of the fibres are seen clearly under high magnification (note the circular ends of the fibres along the left-hand side). (Photograph courtesy of Australian Consolidated Industries Ltd.)*

Under the influence of a small applied stress some of the more highly strained bonds within the structure of the glass are broken and the consequent atomic rearrangements produce a small amount of permanent strain. During these internal structural rearrangements more bonds become strained and are consequently broken, and thus viscous flow continues. Stresses required to cause a certain degree of viscous flow are less at high temperatures since many bonds become highly strained as a result of the high temperature.

In actual fact, the rate of viscous flow in glass at ordinary temperatures is extremely low, and if glass is subjected to a suddenly applied stress it is more likely to fail in a brittle manner. However, glass may flow under its own weight over a long period of time; this has been observed in large plate glass windows, which are commonly significantly thicker at the bottom edge.

The Effects of Heat Treatment on Glass

The mechanical properties of glass can be profoundly altered by two different types of heat treatment.

Annealing is used to remove the thermal stresses induced into the glass during manufacturing processes in which cooling rates are characteristically fast. If glass is not annealed after being formed into shape it is typically weak and brittle and cannot stand up to normal service conditions. In practice, annealing is carried out immediately after shaping in a special furnace called a *lehr* in which the glass is reheated to a suitable temperature and is then slowly cooled. At the annealing temperature, viscous flow causes stress relaxation within the glass, leaving it relatively stress-free. This process must be contrasted with the annealing of metals, which recrystallise at the annealing temperature.

Tempering may be applied to glass to make it tougher and more crack resistant on its surface. In common with other brittle materials, glass fails readily by crack propagation under the influence of a tensile load. When glass is tempered, its surface is placed into compression, and surface crack propagation cannot occur as readily while the glass is in this condition. To temper glass, it is necessary to heat it to its annealing temperature and then cool its outside surfaces rapidly by means of blasts of cold air or an oil spray. The outside surfaces thus contract very rapidly and are placed in compression, while the centre cools much more slowly and remains in tension.

The Mechanical Properties of Glass Fibres

Glass fibres, in contrast to normal glass, possess tensile strengths of very high magnitudes, often in excess of 100,000 psi. Two main factors are considered to be responsible for this; firstly, the structure of glass fibres, in common with normal glass structures, does not allow slip to occur. However, unlike normal glass, glass fibres are almost free from surface defects and thus crack propagation under tensile loads is kept to a minimum.

Laboratory (51): *Take samples of window glass, continuous filament glass fibre, and glass fibre matte and examine each one in turn. Hit the piece of window glass and the glass fibre filament with a hammer. List the main properties of window glass and glass fibre.*

THE FORMING AND SHAPING OF CERAMICS

In order to be useful, ceramic materials must be shaped into articles having specific tolerances and properties. The traditional methods are still of major importance, and may be summed up in the following way.

(a) Clay-body ware is either shaped as a suspension in water or as a soft plastic mass which is then dried and fired.

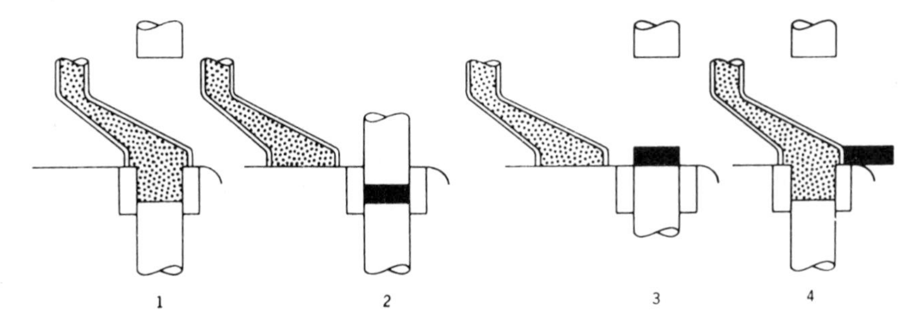

Figure 16.15 *The four basic steps involved in the automatic dry pressing of a ceramic powder. (1) The die cavity is filled from a hopper; (2) the powder is compacted; (3) the powder is ejected; (4) the die cavity is filled again. (Reproduced with permission from "Ceramic Fabrication Processes" edited by W. Kingery. The M.I.T. Press, Boston, 1958.)*

(b) Glass is formed into shape while in the viscous state and is then cooled "in a controlled way" in order to develop specific properties.

(c) Cementing materials are used in a fluid state and allowed to set by hydration or dehydration.

The Forming and Shaping of Powdered or Plastic Ceramics

Forming Powders by Pressing: Many of the more recently-developed ceramics are best prepared as powders and then pressed into shape in a special die-set, after which they are fired or sintered to promote internal cohesion. The ceramic powders may be used either completely dry or slightly damp, and the process is commonly applied to refractories, wall tiles, spark plug insulators, and special electrical and magnetic ceramics. Binders and lubricants are normally added to those ceramic powders that are naturally non-plastic; for example, 1–2% of polyvinyl alcohol may be added to alumina powder when pressing it in carbide dies. Figure 16.12 illustrates the major steps in the dry-pressing of a ceramic powder.

Extrusion: A stiff plastic ceramic mixture can be extruded through a suitably shaped die orifice to form a wide variety of shapes including bricks, pipes, electrical insulators, and hollow tiles. The degree of compaction and hence porosity can be controlled by the die shape and the pressure employed. The most common type of extrusion employs a continuously revolving auger to force the plastic clay through the die, but piston extrusion machines may be used for intermittent extrusion operations.

Throwing Processes: Hand throwing clay on the potter's wheel is one of the oldest known processes applied to plastic clays. In a traditional throwing process, the potter's hands pull, push and guide the soft and plastic clay into shape. This process is now only used by artist potters.

However, mechanised throwing processes have been developed in which moulds and templates take the place of the potter's hands. The most widely used of such processes is termed *jiggering*, and is used to form plates and also

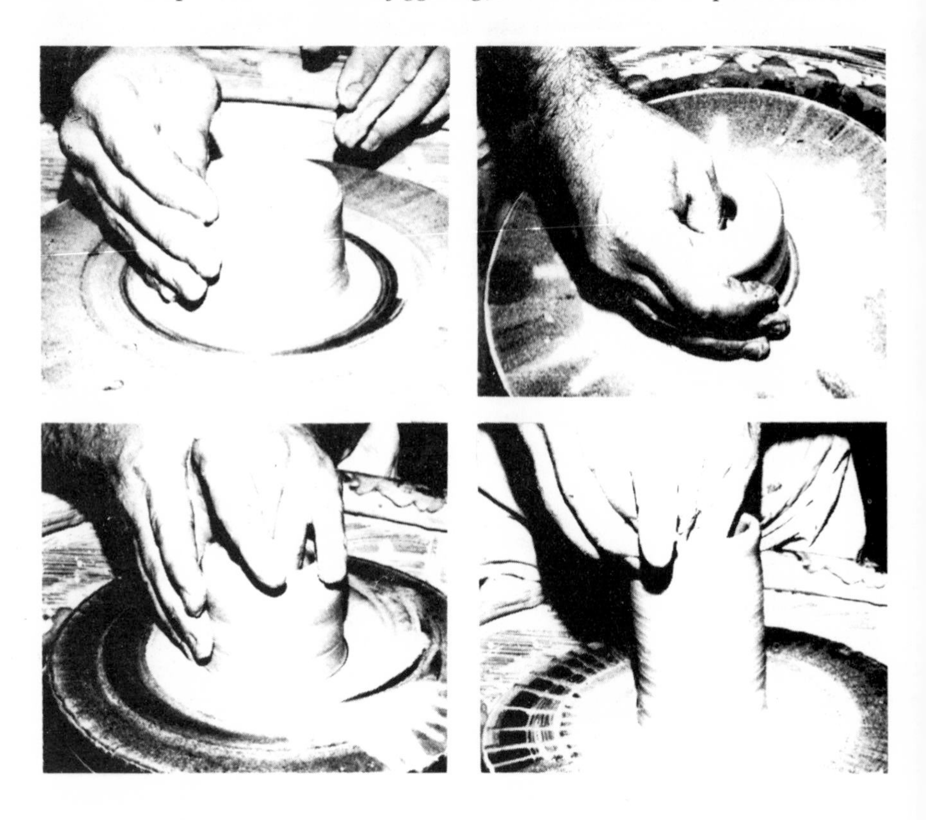

Figure 16.16 *Hand throwing a cylinder on the potter's wheel.*

electrical insulators. In jiggering, a plaster mould is placed on top of the wheel head, a disc of soft plastic clay placed over it, and the wheel rotated at about 400 rpm. A profile tool is then pulled down on to the disc and forms the upper surface of the article while the mould forms the lower surface. (See Fig. 16.17.)

Slip Casting: A thick creamy suspension of clay in water is readily made if a suitable dispersing or deflocculating agent is added to the mix. If the clay slip is then poured into a Plaster of Paris mould, a hard layer of clay will build up inside the mould as the plaster sucks the water from the clay slip. If the process is allowed to continue until the entire contents of the mould solidify,

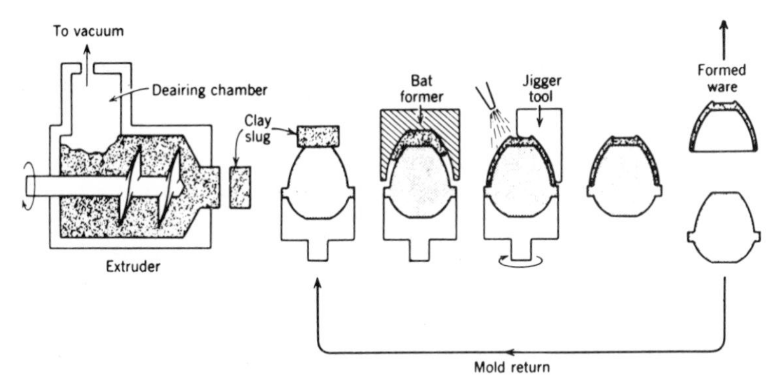

Figure 16.17 *Steps in the manufacture of a vase by jiggering. Note that the slug of plastic clay is extruded and then partly pressed to shape prior to jiggering. (Reproduced with permission from "Ceramics" in Vol. 4 of Kirk-Othmer, "Encyclopedia of Chemical Technology", 2nd Edition. John Wiley and Sons Inc., N.Y., 1964.)*

a solid cast is made; however, if the excess slip is poured out when a suitable wall thickness is developed, *a hollow cast* is formed.

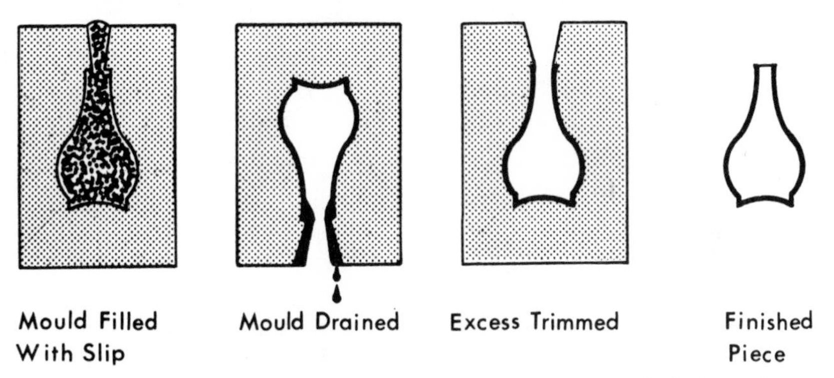

Figure 16.18 *Producing hollow ware by slip casting. Wall thickness is determined by the length of time that the clay slip remains in the plaster mould.*

Unlike either hand or mechanised throwing, slip casting can be used to form relatively complex, non-concentric shapes such as sanitary ware, teapots, and laboratory ware. An added advantage is that a relatively cheap plaster mould can be used to make a large number of identical articles.

Glass Forming Processes

The basic methods of forming glass are extremely simple and depend upon the viscosity of molten glass. Under suitable conditions, glass can be cast into a mould; pressed into shape in a die; blown into shape by air pressure; and rolled, bent, twisted, and stretched as required. As well, at high temperatures, the lower viscosity of glass enables glass-to-glass welds to be effected with ease.

Glass, unlike other materials, is usually manufactured into finished articles by the glass producer; for example, the whole manufacture of a bottle from from furnace to final annealing is always carried out in the one plant.

Almost all glass is manufactured in a *glass tank furnace*, which is a continuous furnace into which the batch materials are introduced at one end and out of which the viscous glass is drawn at the other. A fairly high percentage of cullet (broken scrap glass) is used in the batch, as this assists the melting process; the cullet used must be compatible with the glass batch being melted.

Figure 16.19 *Glass blowing. (Photograph courtesy of Australian Consolidated Industries Ltd.)*

Hand Blowing of Glass: If hand blown ware is to be produced, small batches of glass are usually melted in separate pots. Hand blowing is a very skilled process and long experience is necessary. A gob of glass of the required size is gathered on to the end of the blow pipe and the glass is then blown and rolled to form the general shape of the ware. The external shape may be hand formed using a charred wooden paddle, or alternatively, the semi-finished product may be blown into a mould. The completed article is always fire annealed to remove internal stresses.

Blowing and Moulding: Most glassware of hollow shape is produced by highly automated blowing, moulding, or combination blow-and-mould processes; in each case success depends upon the supply of a gob of glass of the correct mass and viscosity and the rapid execution of the shaping operations. Glass products are always annealed after shaping to remove thermal stresses.

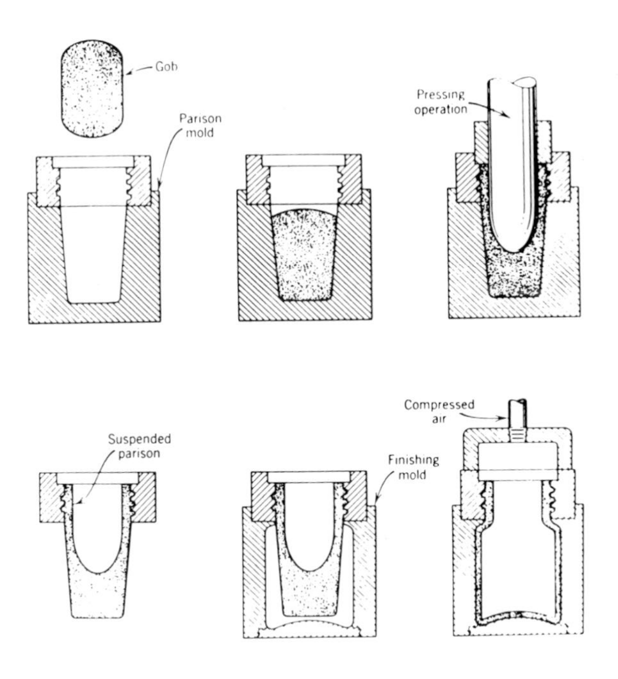

Figure 16.20 *Forming a wide-necked glass container by the press and blow method. (Reproduced by permission from "Introduction to Ceramics" by W. Kingery. John Wiley and Sons Inc., N.Y. 1960.)*

Press moulding is commonly employed only for relatively shallow articles such as dishes and coups. Grey cast iron steel moulds are generally used, the internal shape of the mould forming the external shape of the article and the external shape of the plunger forming the internal shape of the article. To speed the process, the mould is water-cooled to promote faster cooling of the viscous glass after pressing is complete.

The press-and-blow process, normally used to form wide-mouth jars, involves a combination of mechanical pressing plus blowing in a mould. A gob of glass falls under gravity into the parison mould where the neck of the container is pressed into shape. The glass gob, now known as a parison, is then transferred to the finishing mould where the body is blown into shape by a controlled blast of compressed air.

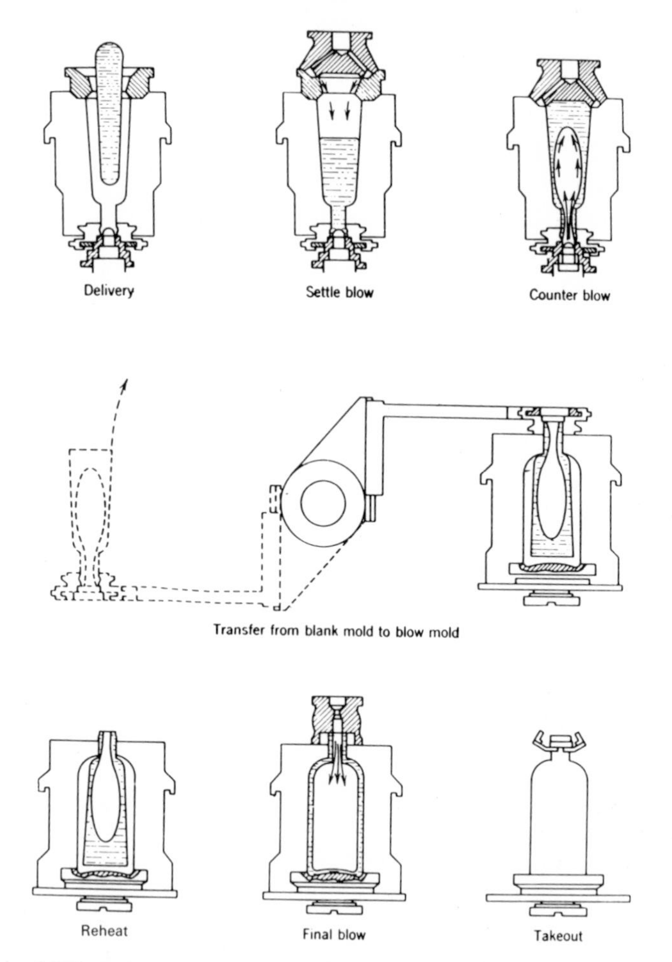

Figure 16.21 *Forming a narrow-necked glass bottle by the blow-and-blow method. (Reproduced with permission from "Handbook of Glass Manufacture", Vol. 1, edited by F. V. Tooley. Ogden Publishing Co., N.Y. 1961.)*

The blow-and-blow process, usually employed for narrow-necked containers, roughly forms the parison by a preliminary blowing operation. This kind of process is usually carried out in an "Individual Section" (I.S.) machine and seven steps, as illustrated in Figure 16.21, are necessary.

(1) *Delivery*: a gob of glass of the required size is fed into the first of the two moulds by means of an air delivery system.

(2) *Settle blow*: the top blowing head comes down over the top of the mould and the gob of glass is blown, by an air blast, into the bottom portion of the mould.

(3) *Counter blow*: the top blowing head is removed and replaced by a baffle and an air blast from beneath the mould partly blows out the gob of glass.

(4) *Transfer*: the blank mould now opens and the partly formed gob (now known as the parison) is transferred to the blow mould.

(5) *Reheat*: the parison is held stationary inside the blow mould for a very short time to allow the cold skin on the glass, formed by contact with the mould surface, to regain heat from the rest of the parison.

(6) *Final blow*: the blowing head is positioned over the neck of the bottle and the parison is blown into its final shape by a controlled air blast.

(7) *Take out*: the blow mould opens and the finished bottle is removed by a pair of tongs and placed on to the conveyor that starts it through the annealing lehr.

One of the most important advantages of the I.S. machine is that the moulds do not move except to open and close as the article being formed moves from one to another. Thus, high-speed forming is possible even when large iron moulds are being used.

Sheet and Plate Glass Production

The four main types of flat glass produced are flat drawn sheet, rolled glass, polished plate, and float glass, and the processes involved are highly automated.

Flat drawn glass is used for windows and other applications where accuracy of thickness and a high surface finish are not vital. In this process, the molten glass is drawn up vertically from the end of a large rectangular tank furnace, the thickness of the drawn sheet being controlled principally by the size of the slot in the steel form held just above the surface of the molten glass. A steel starting dummy is used to commence the drawing action, but this is cracked off once it passes through the pinch rolls that prevent the glass sheet from contracting across its width. Once started, this process is continuous, the cooled and solidified sheet merely being cracked off in suitable lengths once it leaves the pinch rolls. The sheet is annealed in a vertical lehr as it passes through the pinch rolls.

Rolled sheet glass is produced by a process combining continuous casting and rolling. Normally, only rough-cast, figured, and wired safety glass is produced by this technique; however, polished plate usually begins as rolled sheet.

Polished plate glass is expensive to make but does provide plate of exact thickness and high surface finish. The early glassmakers hand polished their plate, but modern processes are fully automated, and produce glass that is highly polished on both sides at the rate of about 1,400 square yards per minute.

The "Float Process" is the most recent advance in the production of plate glass with a high surface finish. In the float process flat, fire-finished, stress-free glass is formed by preventing contact between the glass sheet and anything solid. The glass leaves the glass-tank in a continuous molten strip and is floated immediately on to the surface of a bath of molten tin. The tin bath is surrounded by a non-oxidising atmosphere and while in this atmosphere the glass is heated sufficiently to prevent the development of internal stresses. As the glass sheet leaves the float bath chamber it enters an annealing lehr. Glass produced by the float process is free from distortion and has fire-finished surfaces that are smoother and flatter than those of normal polished plate glass.

GLOSSARY OF TERMS

Ceramic: a material containing phases that are compounds of metals and non-metals.

Clay: hydrated alumino-silicates that result from the weathering of rocks containing feldspars; all clays become plastic when mixed with a sufficiency of water.

Clay body: a mixture of clay mineral and non-plastics.

Devitrification: the reversion of a glass to a more stable crystalline phase.

Earthenware: a fired clay product in which little vitrification occurs and which is consequently very porous.

Glass: an inorganic fusion product that has failed to crystallise upon cooling.

Glass-bonded ceramic: a ceramic material containing crystalline phases embedded in a vitreous matrix.

Hydraulic cement: a cement that can set and harden under water.

Jiggering: a mechanical throwing process in which the potter's hands are replaced by some type of form tool.

Mullite: a crystalline phase of composition $Al_6Si_2O_{13}$ formed when clay is fired to above 950°.

Porcelain: a fired and highly vitrified clay product that has almost no apparent porosity and is translucent in thin sections.

Stoneware: a coarse, highly vitrified clay product that is coarser than porcelain.

Tobermorite gel: the name given to the complex silicate gel that forms when Portland cement hardens.
Viscous flow: a mechanism of deformation that occurs in non-crystalline materials subjected to relatively small applied stresses.
Vitreous: glassy.

REVIEW QUESTIONS

1. Why is the family of ceramic materials exceptionally large?
2. Sketch and describe the crystal structures of MgO and CaF_2. What features have they in common?
3. Sketch and describe the fundamental structural sub-unit of all silicates and explain how complex ions are formed by the linking together of several to many of the sub-units.
4. List three glass-formers and state which is the basis of all commercial glasses.
5. Discuss the functions of glass formers, intermediates, and modifiers in window glass.
6. What is the function of a stabiliser in a glass?
7. Define the terms "devitrification" and "recrystallisation" and point out any important differences in meaning.
8. Describe two methods of glass fibre manufacture and the forms of glass fibre that result from each.
9. What is meant by the term "glass bonded ceramic"?
10. Describe the important characteristics of clay (i) in a state suitable for extrusion; (ii) after firing.
11. How is a clay body manufactured?
12. Distinguish between earthenware, stoneware, and porcelain in terms of their firing temperatures and properties.
13. What is bone china and how does it differ from porcelain?
14. Distinguish between lime mortar, Plaster of Paris, and Keenan's cement, and give some uses of each.
15. Briefly describe the manufacture of Portland cement.
16. Distinguish between hydraulic and non-hydraulic cements.
17. Describe the essential reactions occurring during the setting and hardening of Portland cement.
18. Explain why a typical oxide ceramic such as MgO has little ductility compared with a typical metal crystal such as copper.
19. Briefly explain the differences between crack propagation in brittle and in ductile materials subjected to tensile stresses.
20. What is the effect of (i) annealing and (ii) tempering on the properties of glass?

21. Briefly describe the shaping processes that would be used to manufacture (i) bricks from a plastic clay; (ii) a machine cutting tip from a powdered ceramic; (iii) a dinner plate from plastic clay.
22. What advantages does slip casting offer over jiggering for the production of complex shapes from clay?
23. Describe the manufacture of narrow-necked glass bottles in an I.S. machine.
24. What types of glassware are produced by press moulding?
25. Compare and contrast the manufacture of (i) drawn glass sheet; (ii) plate glass; (iii) float glass.

17

The Modification of Materials

MATERIALS science involves the study of the relationship existing between the structures of materials and their properties. Since the suitability of an engineering material for a specific application is increased if its properties are appropriately varied by controlled structural alterations, the methods used to control the structures of materials are of fundamental importance to the materials scientist. The following techniques are used to vary the structures of materials to obtain desired properties:

(1) the control of phases and microstructure

(2) the development of composite materials

(3) the application of surface coatings.

THE CONTROL OF PHASES AND MICROSTRUCTURE

When studying the relationship between microstructure and properties the following four factors deserve consideration:

The number of phases or microconstituents that are present: Looked at in this way, materials are either single or multi-phase.

The amount of each phase that is present: The proportions of the phases present in a multi-phase material determine the overall properties of the material. Properties in multi-phase materials are either *additive* in the sense that they are weighted averages of the properties of the individual phases (*e.g.* density, specific heat) or they are *interactive* in the sense that they depend upon the interactions of the various phases present (*e.g.* strength, hardness, stiffness).

The natures of the phases present: Both additive and interactive properties depend upon the natures of the individual phases present. For example, the presence of a hard and brittle phase dispersed in a ductile matrix strengthens the material, and fillers like wood flour and carbon are added to polymers to increase strength and stiffness.

TABLE 17.1
THE EFFECTS OF HEAT TREATMENT

Heat Treatment (a) *Metals*	*Purpose*	*Applications*	*Procedure*
Annealing	Removal of work hardening and internal stresses with consequent increase of ductility.	All cold worked metals.	Heat above recrystallisation temperature, soak, and cool under equilibrium conditions.
Normalising	Removal of work hardening effects and internal stresses, but retention of fine grain.	Cold worked or machined low and medium carbon steels.	Heat into austenitic range, soak, and air-cool.
Process Annealing	Removal of work hardening with consequent increase of ductility.	Cold worked low carbon steels (recrystallises ferrite only).	Heat to just below the lower critical temperature and then cool as desired.
Quench Hardening	To achieve maximum hardness.	Applied to medium and high carbon steels, and tool steels.	Heat to austenitic range, soak, and quench cool at a rate that causes a fully martensitic structure to develop.
Tempering	To reduce brittleness in hardened (martensitic) steels by causing a gradual reversion to globular pearlite.	Apply to fully hardened steels.	Heat to below the lower critical temperature and cool slowly. Tempered structures depend upon the tempering temperature and the soaking time at that temperature.
Spheroidising	To soften hardened high carbon or high alloy tool steels.	Applied only to steels that cannot be softened by normal annealing.	Heat to just below the lower critical temperature and soak to spheroidise the carbides present.
Austempering	To harden steels without developing martensite.	Applied to high carbon and tool steels.	Austenitise, quench to just above the temperature at which martensite begins to form, and hold at this temperature until bainite transformation goes to completion.

TABLE 17.1 *continued*

Heat Treatment	*Purpose*	*Applications*	*Procedure*
Martempering	To avoid quench-cracking when hardening.	Applied to high carbon and tool steels.	Austenitise, quench to just above the temperature at which martensite begins to form, hold at this temperature and then quench to room temperature before bainite transformation begins.
Solution Treatment	To form a single phase alloy.	Applied to austenitic steels and *Al-Cu* alloys capable of being precipitation hardened.	Heat the alloy into its single phase area and quench in order to retain this phase.
Precipitation Hardening	To harden non-ferrous metals.	Certain *Al-Cu* and *Al-Si* alloys, and beryllium copper.	Solution treatment followed by reheating to a low temperature in order to begin the precipitation of the second phase.
Malleabilising	To form a more ductile metal.	Applied to white cast iron.	Reheat white cast iron to dissociate carbides and thus form graphite nodules.
(b) Non-Metallic Materials			
Annealing	To relieve residual strain.	Glass.	Heat so that residual stresses can be removed by atomic readjustments.
Tempering	To increase toughness.	Glass.	Heat to annealing temperature and spray surface with oil or cool with air blasts.
Recrystallisation	To improve thermal properties.	Special glass ("Pyroceram")	May occur spontaneously upon cooling, or may require reheating to a certain temperature.

The distribution of phases in relation to one another within a multi-phase material: The relative sizes of the domains occupied by each phase are of great importance when studying structure-property relationships. For example, grainsize and strength are closely related in metals, while the size and shape of graphite particles in cast iron profoundly influences strength properties.

The microstructure of a material may be altered in one of the following ways.

Changes in Composition

By appropriate alloying, solid solutions can be formed between two or more metals, new phases developed, or the proportion of an existing phase altered. For example, the addition of up to 36% zinc to copper causes the formation of a solid solution alloy whose strength increases with increasing zinc content, while the addition of slightly more zinc causes the formation of a new hard and brittle phase which reduces the formability of the metal. In the field of plastics, copolymers can be formed which possess properties superior to those of their constituent mers, and ceramics having particular properties are readily formed by varying the constituents of initial batch materials. For example, a glass having a very high refractive index is formed if a certain proportion of lead oxide is introduced into the batch materials.

Heat Treatment

The hardening and tempering of carbon steels depend upon changes in the nature of the phases present in the steel as well as upon changes in the sizes and distribution of the phases present. However, most non-ferrous metals only undergo alterations to the shapes and sizes of their phases when heat treated, exceptions being those non-ferrous alloys that can be precipitation hardened. Many non-metallic materials are heat treated to alter their properties; for instance, glass and crystalline ceramics are often heat treated in order to improve their mechanical properties. Table 17.1 lists some of the more common effects of heat treatment on various materials.

Forming and Shaping Methods

(a) Hot and Cold Forming: When metals are cold worked, ductile phases are distorted in the direction of rolling and the metal becomes work hardened, while brittle phases may be mechanically broken up and dispersed more evenly throughout the metal. Cold working is the only method of improving the mechanical strength of low carbon steels and most non-ferrous metals (unless further alloying is carried out). Hot worked metals also exhibit directional properties; however, directional effects are slight compared to cold worked metals since hot working causes recrystallisation. Directional properties may be induced in polymers by working procedures that cause molecular orientation. Oriented polymers have improved strength and stiffness in the direction of working. Extruded polymers commonly exhibit molecular orientation.

(b) Sintered Materials: A completely different structure can be developed in a metal or an alloy if it is formed by the technique of *powder metallurgy*. The main feature of sintered metals is their porosity, which can be readily controlled during the forming process, and porous metals find extensive applications as bearings and very fine metal filters. Also, composites not normally possible can be produced by the techniques of powder metallurgy, either directly or by means of a subsequent infiltration process.

COMPOSITE MATERIALS

A composite material is defined as a combination of two or more materials exhibiting properties distinctively different from those of the individual materials used to make the composite. In terms of some specific property like strength, heat resistance, or stiffness, the composite is better than either of the individual component materials or radically different from both of them. Many naturally-occurring materials are composites; wood, for example, consists of long cellulose fibres held together by amorphous lignin, while bone is a composite of the strong but soft protein, collagen, and a hard and brittle mineral consisting essentially of the carbonate and phosphate of lime. Composite materials may be roughly classified as (1) agglomerated materials, (2) laminates, (3) surface-coated materials, or (4) reinforced materials.

Agglomerated Materials

Concrete: This is one of the oldest agglomerated composite materials to be used for engineering construction, and consists of a mixture of aggregate and sand bonded together by the hydrated silicate gel formed when the Portland cement "sets" with water. Each of the materials used to make concrete has a particular function: the aggregate or gravel makes up the bulk of the concrete, the sand fills many of the pores left between the pieces of aggregate, and the Portland cement and water react together to form the silicate binder. The following five factors affect the strength of concrete.

Ratio of Aggregate, Sand and Cement: a very common mix consists of 4 parts aggregate, 2 parts sand and 1 part cement powder (parts by volume). In general the strength of the set concrete depends upon the proportion of cement used; however, the aggregate must be at least as strong as the set cement.

The Water-Cement Ratio: the water added to the concrete is used in the hydration of the cement itself, and any water in excess of the amount required for setting reactions has a weakening effect upon the concrete.

The Nature of the Aggregate and Sand: the bond between the hydrated cement and the aggregate and sand is improved if both the aggregate and sand are sharp-cornered rather than rounded. Strong fine-grained igneous rocks like basalt, dolerite, and quartzite are commonly used for concrete aggregate, the size of which varies with the size of the job. For instance, while 2″ aggregate may be the maximum permissible for a retaining wall, 24″ aggregate may be used in the wall of a dam. Graded aggregate is better than aggregate that is all the same size, since the former allows for closer packing.

Mixing and Laying: under or overmixing gives a poor concrete, and the method of laying is of the utmost importance. Concrete vibrated into place is always stronger than concrete poured and hand screeded.

Curing Time: the hardening of cement occurs over a considerable length of time and it is important to prevent the evaporation of moisture during the initial stages. Concrete is often covered with wet sand or bags for seven days to prevent the evaporation of moisture, and concrete cured under water achieves its maximum strength. A normal concrete mix cured in air would have a compressive strength of about 560 psi after three days, 1450 psi after seven days, 2420 psi after twenty-eight days, and 5600 psi after one year.

Asphalt paving: This is a composite in which rock aggregate is bonded by viscous asphalt; it is used extensively for road surfacing. The material is not as rigid as concrete, this being an advantage for road construction. When asphalt paving is laid over a soil base, the strength characteristics of the soil influence the service behaviour of the paving.

Cermets: These are agglomerates that consist of combinations of metals and ceramics, the metal acting as the binder. Cermets are made using the techniques of powder metallurgy, the sintering temperature usually being above the melting point of the metal powder. Typical cermets include ZrC with iron; TiC, Mo_2C and WC with cobalt; Cr_3C_2, Mo_2B and MgO with nickel, and Al_2O_3 with chromium. The ceramic must be powdered to the correct particle size prior to mixing and pressing. In the finished cermet the ceramic contributes hardness and refractoriness while the metal forms a tough, ductile bond. Cemented carbides, previously mentioned as important high speed, heavy duty cutting tips, are examples of cermets.

Other examples of agglomerated composite materials include abrasive grinding wheels in which the abrasive particles are held together by a vitreous or a resin bond; sintered alloys in which the bond forms by solid diffusion around the particle boundaries; shell moulding sands in which a resin binder is used; and particle board, in which wood chips are held together by a suitable glue.

Laminates

Many different types of laminated materials are made for different applications, the mild steel-stainless steel combination discussed in Chapter 2 being a good example of a modern metal-to-metal laminate. Duralumin and other clad metals can also be considered as laminates; however, since the outer claddings are very thin in comparison with the base metal, it is logical to consider these materials under the heading of surface coatings. Two important non-metallic laminates are plywood and laminated plastic sheeting.

Plywood: This is made by bonding together an odd number of sheets of wood veneer so that the grain directions of alternate sheets are perpendicular to each other. The wood veneer used is normally between 0.002″ and 0.25″ thick, and is usually produced by slicing or peeling a suitable log. Sawn timber possesses directional properties which are dependent upon the fibre orientation; cross-bonding the veneers in plywood overcomes these directional properties, producing a sheet material that has uniform strength in all directions across its surface. The possibility of splitting and uneven shrinkage in plywood is also considerably reduced because of the cross-bonded pattern of veneers. The glues used to bond veneers into plywood are also an important part of this laminate. If a waterproof glue such as a phenolic resin is used the plywood is said to be waterproof and is suitable for external applications; however, plywood made for interior use is bonded with a cheaper non-waterproof glue. The use of plywood also leads to a more economic utilisation of highly prized "cabinet" timbers since only the top veneer need be this type of wood. A small percentage of plywood is metal-faced with thin sheets of aluminium, copper, or bronze. This increases the strength of the plywood and also results in a metal sheet which can be readily bent and formed for panelling and lining, and which looks like a polished metal sheet when in place. Plastic-surfaced plywood is also used for decorative linings.

Laminated plastic sheet: This is usually made from sheets of paper or cloth and a suitable thermosetting resin. The paper or cloth passes through a tank containing the resin solution, between rollers that squeeze out the excess resin, and then through a drying oven in which excess solvents are removed and the resin is partially cured. The impregnated material is then cut into suitable lengths, a number of which are piled together; the pile is then pressed in a suitable hot pressing machine. The heat and pressure soften the resin which flows to fill all interstices within the laminate, and then sets hard as polymerization takes place. The paper or fabric used provides the bulk of the strength, while the resin acts as a semi-rigid binder. Laminated plastic sheet may be machined, drilled, punched, and pressed to shape, and is used in the production of gears, bearings, electrical components, and small cabinets. The resins used include phenolics, polyesters, epoxides, and silicones, the phenolics having limited applications since volatiles given off during the curing operation can cause blistering and the formation of cavities within the sheet.

Reinforced Materials

Reinforced materials form the biggest and most important group of composite materials, the purpose of reinforcement always being the improvement of strength properties. Reinforcement may involve the use of a dispersed phase, or strong fibre, thread, or rod. For example, precipitation hardening depends upon the strengthening effects of a sub-microscopic dispersed phase, and bainite obtains its strength from the fine dispersion of carbide particles throughout a ferrite matrix, while reinforced concrete depends upon the strengthening effects of suitably located steel rods.

Reinforced concrete: This is the most widely used of all construction materials, since it is not only comparatively easy to place into position and finish, but is also maintenance free during its service life. Concrete is extremely weak in tension but stronger in compression; the steel reinforcing placed into reinforced concrete takes all of the tensile load placed upon the structure. The steel usually has a tensile yield of at least 60,000 psi, and must be capable of being bent and welded. Figure 17.1 shows the placement of steel reinforcement in several concrete structures. Reinforced concrete may be prestressed if it is desirable to counteract the initial deflection of a member such as a structural beam. The steel reinforcement is placed under a tensile load and the concrete poured around it using a special steel mould. After the concrete has cured the tensile load and the mould are both removed, whereupon the concrete member tends to undergo an initial deflection. When placed into position the service load overcomes this deflection, thus tending to straighten the member. (See Figure 17.1D.)

Glass-fibre reinforced plastics: These combine the strength of glass fibre with the shock resistance and formability of a plastic. The usual types of reinforcement are the chopped strand mat and the woven fabric, the latter giving increased strength to the composite. Resins used may be of the phenolic type requiring considerable pressure for setting, or of the espoxide or polyester types that polymerize without pressure under the influence of a chemical catalyst. Fibre-reinforced plastics are used for boat hulls, car bodies, truck cabins, aircraft fittings, and other applications where strength, lightness, freedom from corrosion, and ease of manufacture are desirable.

Asbestos-reinforced plastics: These are used in the aircraft industry and offer the advantage of increased stiffness. A phenolic resin is usually employed, and the asbestos is used in the form of a loose matte.

The more recently developed reinforced materials utilize very strong and often brittle whiskers or fibres embedded in a tough but ductile metallic matrix. Ceramic whiskers of alumina have been embedded in a metal such as cobalt to produce a material that retains its strength at high temperatures. Boron and tungsten fibres have been embedded in aluminium and copper to give composites of high stiffness. Silicon carbide and graphite whiskers, of lengths of about 5 microns, are finding increasing applications as similar fibre reinforcing materials.

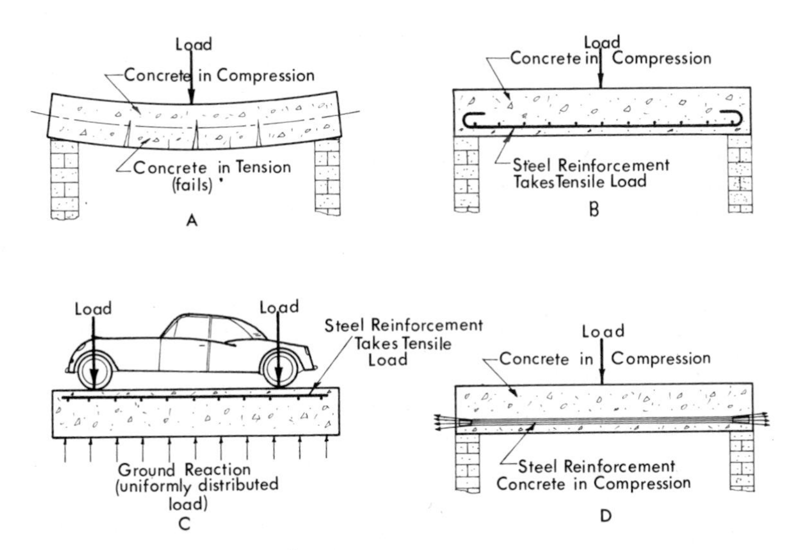

Figure 17.1 *The effects of steel reinforcement in concrete structures. (A) A suspended concrete beam or slab fails in tension when loaded; (B) When the steel reinforcement is placed as shown, it takes the tensile load and failure does not occur. (C) In a concrete slab lying on the ground and loaded as shown, the uniform ground reaction places the top portion of the slab in tension, and thus the reinforcing steel is placed in the top portion of the concrete slab. (D) Pre-tensioned steel reinforcement prevents deflection when a load is applied to a suspended beam or slab. This is one form of prestressed concrete.*

Surface Coatings

The primary function of a surface coating is the protection of the material to which it is applied; however, surface coatings may also perform decorative functions. It is usual to classify surface coatings as metallic coatings, inorganic chemical coatings, and organic chemical coatings.

Metallic coatings: These are usually applied by hot dipping, electroplating, cladding, or spraying techniques and serve to protect the base metal from corrosion. Table 17.4 lists some common metallic coatings used on steel, their method of application, and their specific functions.

TABLE 17.2

METALLIC COATINGS

Coating	*Application*	*Functions*
Zinc	Hot dipping, electrodeposition spraying	Applied to steel to make galvanised iron; protects the steel both mechanically and by acting as a sacrificial anode.
Tin	Hot dipping, electrodeposition	Applied to sheet steel to make tin plate; does not provide galvanic protection, but tinplate containers can be used to hold food since the corrosion products of tin are non-poisonous.
Lead	Electrodeposition, spraying	Applied to sheet steel to form terne plate; resistant to dilute sulphuric acid; does not offer galvanic protection to atmospheric corrosion.
Copper	Electrodeposition, spraying	Usually used as a base for other electrodeposited metals since it is easily polished.
Nickel	Electrodeposition	Usually applied under chromium coatings on steel; offers no galvanic protection.
Chromium	Electrodeposition	Excellent resistance to tarnishing, therefore applied for decorative and protective purposes to steel; also used for hard surfacing bearing surfaces.

Alclad, the composite formed by cladding duralumin with thin sheets of pure aluminium, is a high strength alloy in which the aluminium cladding provides galvanic protection for the more corrosive duralumin.

Inorganic chemical coatings: These are conveniently divided into vitreous coatings, oxide coatings, and phosphate coatings. Vitreous coatings are commonly applied to steel in the form of a powder or frit and are then fused to the steel surface by heat. Such coatings are relatively brittle, but offer absolute protection against corrosion. The glaze applied to a porcelain or stoneware ceramic may also be considered as a protective coating, since the ceramic is protected from moisture absorption. Thus, porcelain insulators are glazed to prevent electrical breakdown due to moisture absorption. Both *oxide* and *phosphate* coatings are obtained by chemical action. If such films are strong and non-porous, they provide protection against corrosion. A familiar example is the phosphating of steel prior to painting.

Organic chemical coatings: These include paints, varnishes, laquers, and enamels and all serve to both protect the base material and to enhance its appearance. It is doubtful whether the application of such coatings constitutes the formation of a true composite material in the modern sense of the term.

CONCLUSION

From the foregoing discussion it is clear that the suitability of a material for its service conditions can be improved by one means or another, and that new materials for particular applications can be developed if the fundamental natures of engineering materials are understood and appreciated. The techniques of alloying, heat treating, laminating, reinforcing, combining, and surface coating are essential if materials are to be adapted to particular conditions.

GLOSSARY OF TERMS

Asbestos: a fibrous silicate material.
Asphalt: a mixture of bitumen, pitch, or coal-tar with sand and aggregate; used for road surfacing.
Cermet: an agglomerate of a metal plus a ceramic in which the metal forms the matrix phase.
Composite: a combination of two or more materials possessing properties different from those of the constituent materials; composite materials may be agglomerated, laminated, or reinforced.
Concrete: an agglomerate consisting of an aggregate plus a hydraulic cement binder.

REVIEW QUESTIONS

1. What three general techniques are used to vary the structures of materials in order to develop in them more desirable properties?
2. Briefly discuss the four factors concerning the microstructure of a material that need consideration when relationships between microstructure and properties are being established.
3. Explain the meaning of the term "composite material".
4. Distinguish between agglomerated materials and laminates, and give some examples of each.
5. What is a cermet and what advantages do cermets have compared to crystalline ceramics?
6. What is meant by the term "reinforced material"? Give examples of (i) dispersion reinforcement; (ii) fibre reinforcement; (iii) rod reinforcement.
7. What is prestressed concrete and what advantages does a prestressed structural beam have over an ordinary reinforced concrete beam?
8. What is the primary function of a surface coating? Give some examples of metallic coatings, inorganic chemical coatings, and organic coatings, and describe the particular purposes of each type that you have listed.

9. Suggest ways in which the deficiencies of the materials listed below can be overcome: (i) the tendency of steel roofing to rust; (ii) the low tensile strength of concrete foundations; (iii) the low tensile strength of a polymer used for a boat hull.

10. List and discuss the advantages of (i) plywood and (ii) flakeboard (particle board) over milled timber.

General Review Questions

1. Outline a classification of crystalline materials based upon the various types of units occupying lattice points within crystal structures. How do patterns of (i) chemical bonding, and (ii) crystal structure, determine the typical properties of each of the above groups of crystalline materials?

2. What particular characteristics are responsible for the tendency of metallic atoms to exist in crystal structures of high co-ordination?

3. Sketch typical tensile stress-strain curves for the following materials: (i) grey cast iron; (ii) annealed mild steel; (iii) concrete. What features of each type of test curve are of significance to the materials scientist?

4. Why is the yield strength of many metals expressed in terms of proof stress?

5. It is necessary to mass produce, by casting, door handles for a medium-sized automobile. Select a suitable material and a suitable manufacturing process, discuss the characteristics of this material, describe the essential features of the process, and justify your selection.

6. A sample of a pure metal is cooled slowly from above its melting point down to room temperature. Describe the process of solidification that occurs in the metal and sketch its final microstructure.

7. An as-cast billet of a single-phase alloy is hot rolled, cold rolled and finally annealed. Describe, using fully-labelled sketches, its structure and properties: (i) in the as-cast condition; (ii) after hot rolling; (iii) after cold rolling; (iv) after annealing.

8. Discuss the differences between the cold drawing of aluminium wire and the direct extrusion of aluminium bar stock.

9. What is the role of crystal dislocations in plastic deformation and in strain hardening?

10. Compare the structure of a bronze bearing made by powder metallurgy with that of an as-cast bearing of similar size. What particular advantages would be offered by the bearing produced by powder metallurgy?

11. Two metals, A and B, exhibit complete liquid solubility and form a continuous series of solid solutions. Sketch the form of their equilibrium diagram and label phase areas. Describe the cooling of a 50/50 alloy of these two metals.

12. Explain the general characteristics desirable in deep-drawing steel; give its approximate carbon content, and sketch and label its microstructure.

13. Hypo-eutectoid steels are generally annealed at temperatures about 30°C above their upper critical temperatures; however hyper-eutectoid steels are annealed at temperatures 30°C above their lower critical temperatures. Give reasons for this difference in heat treating procedures.

14. Explain how the form and dispersion of carbon in cast irons influences their mechanical properties. Use some specific irons to illustrate your answer.

15. What general features must be exhibited by an alloy that is capable of being precipitation-hardened? Use a specific example to illustrate your answer.

16. Waste phenol-formaldehyde (bakelite) has no value, but scrap polythene is collected for re-use. Explain how structural differences between these two materials account for this.

17. Describe the phase changes in a 0.35% carbon steel as it is cooled from its austenitic state to room temperature. Sketch and describe its final microstructure, and list its important properties.

18. Contrast plastic deformation in a pure oxide ceramic with that in a linear polymer.

19. What are cermets and why are they useful materials in modern technology?

20. Compare and contrast the manufacture of cup-shaped articles from (i) plastic clay; (ii) a ductile metal; (iii) glass.

21. Give typical uses of each of the following materials and explain why they have these uses:

(i) tungsten carbide
(ii) alpha brass
(iii) austenitic stainless steel
(iv) poly-methylmethacrylate
(v) stoneware
(vi) nodular iron.

22. What are the particular values of (i) sulphur, and (ii) carbon black, in rubber that is to be vulcanised?

23. Distinguish between the stiffness of a material and its elasticity. Under what service conditions are these properties important in engineering materials?

24. Very often composite materials possess properties that are unique and quite different from those of the constituent materials. Select two composite materials that are of this type, describe their structures, and list their typical properties and uses.

25. It is desired to mass produce dinner plates of high quality. Select a suitable material and manufacturing process; describe the properties of the material, outline the manufacturing process, and justify your answer.

INDEX

Index

PERIODIC TABLE OF THE ELEMEN

3 LITHIUM **Li** 6·94 2.1	4 BERYLLIUM **Be** 9·01 2.2							
11 SODIUM **Na** 22·99 2.8.1	12 MAGNESIUM **Mg** 24·31 2.8.2							
19 POTASSIUM **K** 39·10 2.8.8.1	20 CALCIUM **Ca** 40·08 2.8.8.2	21 SCANDIUM **Sc** 44·96 2.8.9.2	22 TITANIUM **Ti** 47·90 2.8.10.2	23 VANADIUM **V** 50·94 2.8.11.2	24 CHROMIUM **Cr** 52·00 2.8.13.1	25 MANGANESE **Mn** 54·94 2.8.13.2	26 IRON **Fe** 55·85 2.8.14.2	CO C 58 2.8
37 RUBIDIUM **Rb** 85·47 2.8.18.8.1	38 STRONTIUM **Sr** 87·62 2.8.18.8.2	39 YTTRIUM **Y** 88·91 2.8.18.9.2	40 ZIRCONIUM **Zr** 91·22 2.8.18.10.2	41 NIOBIUM **Nb** 92·91 2.8.18.12.1	42 MOLYBDENUM **Mo** 95·94 2.8.18.13.1	43 TECHNETIUM **Tc** [99] 2.8.18.13.2	44 RUTHENIUM **Ru** 101·07 2.8.18.15.1	RHO R 102 2.8.1
55 CAESIUM **Cs** 132·91 2.8.18.18.8.1	56 BARIUM **Ba** 137·34 2.8.18.18.8.2	57-71 LANTHANIDES	72 HAFNIUM **Hf** 178·49 2.8.18.32.10.2	73 TANTALUM **Ta** 180·95 2.8.18.32.11.2	74 TUNGSTEN **W** 183·85 2.8.18.32.12.2	75 RHENIUM **Re** 186·2 2.8.18.32.13.2	76 OSMIUM **Os** 190·2 2.8.18.32.14.2	IRI 19 2.8.1
87 FRANCIUM **Fr** [223] 2.8.18.32.18.8.1	88 RADIUM **Ra** [226] 2.8.18.32.18.8.2	89–103 ACTINIDES						

LANTHANIDES

57 LANTHANUM **La** 138·91 2.8.18.18.9.2	58 CERIUM **Ce** 140·12 2.8.18.20.8.2	59 PRASEODYMIUM **Pr** 140·91 2.8.18.21.8.2	60 NEODYMIUM **Nd** 144·24 2.8.18.22.8.2	61 PROMETHIUM **Pm** [145] 2.8.18.23.8.2	SAM S 15 2.8.1

ACTINIDES

89 ACTINIUM **Ac** [227] 2.8.18.32.18.9.2	90 THORIUM **Th** 232·04 2.8.18.32.18.10.2	91 PROTACTINIUM **Pa** [231] 2.8.18.32.20.9.2	92 URANIUM **U** 238·03 2.8.18.32.21.9.2	93 NEPTUNIUM **Np** [237] 2.8.18.32.22.9.2	PLUT [2 2.8.18.

Key
ATOMIC NUMBER NAME **SYMBOL** ATOMIC WEIGHT ELECTRONIC CONFIGURATION

Valu